Nuclear and Particle Physics

Nuclear and Particle Physics

B. R. Martin

Department of Physics and Astronomy, University College London

John Wiley & Sons, Ltd

Copyright © 2006   John Wiley & Sons Ltd, The Atrium, Southern Gate, Chichester,
West Sussex PO19 8SQ, England

Telephone   (+44) 1243 779777

Email (for orders and customer service enquiries): cs-books@wiley.co.uk
Visit our Home Page on www.wileyeurope.com or www.wiley.com

Other Wiley Editorial Offices

John Wiley & Sons Inc., 111 River Street, Hoboken, NJ 07030, USA

Jossey-Bass, 989 Market Street, San Francisco, CA 94103-1741, USA

Wiley-VCH Verlag GmbH, Boschstr. 12, D-69469 Weinheim, Germany

John Wiley & Sons Australia Ltd, 42 McDougall Street, Milton, Queensland 4064, Australia

John Wiley & Sons (Asia) Pte Ltd, 2 Clementi Loop #02-01, Jin Xing Distripark, Singapore 129809

John Wiley & Sons Canada Ltd, 22 Worcester Road, Etobicoke, Ontario, Canada M9W 1L1

Wiley also publishes its books in a variety of electronic formats. Some content that appears in print may not
be available in electronic books.

Library of Congress Cataloging-in-Publication Data
Martin, B. R. (Brian Robert)
Nuclear and particle physics/B. R. Martin.
    p. cm.
ISBN-13: 978-0-470-01999-3 (HB)
ISBN-10: 0-470-01999-9 (HB)
ISBN-13: 978-0-470-02532-1 (pbk.)
ISBN-10: 0-470-02532-8 (pbk.)
1. Nuclear physics–Textbooks.   2. Particle physics–Textbooks.   I. Title.
QC776.M34 2006
539.7′2–dc22

                                    2005036437

British Library Cataloguing in Publication Data

A catalogue record for this book is available from the British Library

ISBN-13 978-0 470 01999 9 (HB)   ISBN-10 0 470 01999 9 (HB)
      978-0 470 02532 8 (PB)   0 470 02532 8 (PB)

Typeset in 10.5/12.5pt Times by Thomson Press (India) Limited, New Delhi
Printed and bound in Great Britain by Antony Rowe Ltd., Chippenham, Wiltshire
This book is printed on acid-free paper responsibly manufactured from sustainable forestry
in which at least two trees are planted for each one used for paper production.

To Claire

Contents

Preface

It is common practice to teach nuclear physics and particle physics together in an introductory course and it is for such a course that this book has been written. The material is presented so that different selections can be made for a short course of about 25–30 lectures depending on the lecturer's preferences and the students' backgrounds. On the latter, students should have taken a first course in quantum physics, covering the traditional topics in non-relativistic quantum mechanics and atomic physics. A few lectures on relativistic kinematics would also be useful, but this is not essential as the necessary background is given in appendix B and is only used in a few places in the book. I have not tried to be rigorous, or present proofs of all the statements in the text. Rather, I have taken the view that it is more important that students see an overview of the subject which for many – possibly the majority – will be the only time they study nuclear and particle physics. For future specialists, the details will form part of more advanced courses. Nevertheless, space restrictions have still meant that it has been necessary to make a choice of topics covered and doubtless other, equally valid, choices could have been made. This is particularly true in Chapter 8, which deals with applications of nuclear physics, where I have chosen just three major areas to discuss. Nuclear and particle physics have been, and still are, very important parts of the entire subject of physics and its practitioners have won an impressive number of Nobel Prizes. For historical interest, I have noted in the footnotes many of the awards for work related to the field.

Some parts of the book dealing with particle physics owe much to a previous book, *Particle Physics*, written with Graham Shaw of Manchester University, and I am grateful to him and the publisher, John Wiley and Sons, for permission to adapt some of that material for use here. I also thank Colin Wilkin for comments on all the chapters of the book, David Miller and Peter Hobson for comments on Chapter 4 and Bob Speller for comments on the medical physics section of Chapter 8. If errors or misunderstandings still remain (and any such are of course due to me alone) I would be grateful to hear about them. I have set up a website (www.hep.ucl.ac.uk/~brm/npbook.html) where I will post any corrections and comments.

Brian R. Martin
January 2006

Notes

References

References are referred to in the text in the form Ab95, where Ab is the start of the first author's surname and 1995 is the year of publication. A list of references with full publication details is given at the end of the book.

Data

Data for particle physics may be obtained from the biannual publications of the Particle Data Group (PDG) and the 2004 edition of the PDG definitive *Review of Particle Properties* is given in Ei04. The PDG Review is also available at http://pdg.lbl.gov and this site contains links to other sites where compilations of particle data may be found. Nuclear physics data are available from a number of sources. Examples are: the combined Isotopes Project of the Lawrence Berkeley Laboratory, USA, and the Lund University Nuclear Data WWW Service, Sweden (http://ie.lbl.gov/toi.html), the National Nuclear Data Center (NNDC) based at Brookhaven National Laboratory, USA (http://www.nndc.bnl.gov), and the Nuclear Data Centre of the Japan Atomic Energy Research Institute (http://www.nndc.tokai.jaeri.go.jp). All three sites have extensive links to other data compilations. It is important that students have some familiarity with these data compilations.

Problems

Problems are provided for all chapters and appendices except Chapter 9 and Appendices A and D. They are an integral part of the text. The problems are mainly numerical and require values of physical constants that are given in a table following these notes. A few also require input data that may be found in the references given above. Solutions to all the problems are given in Appendix D.

Illustrations

Some illustrations in the text have been adapted from, or are loosely based on, diagrams that have been published elsewhere. In a few cases they have been reproduced exactly as previously published. In all cases this is stated in the captions. I acknowledge, with thanks, permission to use such illustrations from the relevant copyright holders.

Physical Constants and Conversion Factors

Quantity	Symbol	Value
Speed of light in vacuum	c	$2.998 \times 10^8 \, \text{ms}^{-1}$
Planck's constant	h	$4.136 \times 10^{-24} \, \text{GeV s}$
	$\hbar \equiv h/2\pi$	$6.582 \times 10^{-25} \, \text{GeV s}$
	$\hbar c$	$1.973 \times 10^{-16} \, \text{GeV m}$
	$(\hbar c)^2$	$3.894 \times 10^{-31} \, \text{GeV}^2\text{m}^2$
electron charge (magnitude)	e	$1.602 \times 10^{-19} \, \text{C}$
Avogadro's number	N_A	$6.022 \times 10^{26} \, \text{kg-mole}^{-1}$
Boltzmann's constant	k_B	$8.617 \times 10^{-11} \, \text{MeV K}^{-1}$
electron mass	m_e	$0.511 \, \text{MeV}/c^2$
proton mass	m_p	$0.9383 \, \text{GeV}/c^2$
neutron mass	m_n	$0.9396 \, \text{GeV}/c^2$
W boson mass	M_W	$80.43 \, \text{GeV}/c^2$
Z boson mass	M_Z	$91.19 \, \text{GeV}/c^2$
atomic mass unit	$u \equiv (\frac{1}{12}\text{mass}\,^{12}\text{C atom})$	$931.494 \, \text{MeV}/c^2$
Bohr magneton	$\mu_B \equiv e\hbar/2m_e$	$5.788 \times 10^{-11} \, \text{MeV T}^{-1}$
Nuclear magneton	$\mu_N \equiv e\hbar/2m_p$	$3.152 \times 10^{-14} \, \text{MeV T}^{-1}$
gravitational constant	G_N	$6.709 \times 10^{-39} \, \hbar c (\text{GeV}/c^2)^{-2}$
fine structure constant	$\alpha \equiv e^2/4\pi\varepsilon_0\hbar c$	$7.297 \times 10^{-3} = 1/137.04$
Fermi coupling constant	$G_F/(\hbar c)^3$	$1.166 \times 10^{-5} \, \text{GeV}^{-2}$
strong coupling constant	$\alpha_s(M_Z c^2)$	0.119

$1 \, \text{eV} = 1.602 \times 10^{-19} \, \text{J}$
$1 \, \text{fermi} = 1 \, \text{fm} \equiv 10^{-15} \, \text{m}$
$1 \, \text{Tesla} = 1\text{T} = 0.561 \times 10^{30} \, \text{MeV}/c^2\text{C}^{-1}\text{s}^{-1}$

$1 \, \text{eV}/c^2 = 1.783 \times 10^{-36} \, \text{kg}$
$1 \, \text{barn} = 1 \, \text{b} \equiv 10^{-28} \, \text{m}^2$
$1 \, \text{year} = 3.1536 \times 10^7 \, \text{s}$

1

Basic Concepts

1.1 History

Although this book will not follow a strictly historical development, to 'set the scene' this first chapter will start with a brief review of the most important discoveries that led to the separation of nuclear physics from atomic physics as a subject in its own right and later work that in its turn led to the emergence of particle physics from nuclear physics.[1]

1.1.1 The origins of nuclear physics

Nuclear physics as a subject distinct from atomic physics could be said to date from 1896, the year that Henri Becquerel observed that photographic plates were being fogged by an unknown radiation emanating from uranium ores. He had accidentally discovered *radioactivity*: the fact that some nuclei are unstable and spontaneously decay. In the years that followed, the phenomenon was extensively investigated, notably by the husband and wife team of Pierre and Marie Curie and by Ernest Rutherford and his collaborators,[2] and it was established that there were three distinct types of radiation involved: these were named (by Rutherford) α-, β- and γ-rays. We know now that α-rays are bound states of two protons and two neutrons (we will see later that they are the nuclei of helium atoms), β-rays are electrons and γ-rays are photons, the quanta of electromagnetic radiation, but the historical names are still commonly used.

---

[1] An interesting account of the early period, with descriptions of the personalities involved, is given in Se80. An overview of the later period is given in Chapter 1 of Gr87.
[2] The 1903 Nobel Prize in Physics was awarded jointly to Becquerel for his discovery and to Pierre and Marie Curie for their subsequent research into radioactivity. Rutherford had to wait until 1908, when he was awarded the Nobel Prize in Chemistry for his 'investigations into the disintegration of the elements and the chemistry of radioactive substances'.

Nuclear and Particle Physics   B. R. Martin
© 2006 John Wiley & Sons, Ltd

At about the same time as Becquerel's discovery, J. J. Thomson was extending the work of Perrin and others on the radiation that had been observed to occur when an electric field was established between electrodes in an evacuated glass tube, and in 1897 he was the first to definitively establish the nature of these 'cathode rays'. We now know the emanation consists of free *electrons*, (the name 'electron' had been coined in 1894 by Stoney) denoted e^- (the superscript denotes the electric charge) and Thomson measured their mass and charge.[3] The view of the atom at that time was that it consisted of two components, with positive and negative electric charges, the latter now being the electrons. Thomson suggested a model where the electrons were embedded and free to move in a region of positive charge filling the entire volume of the atom – the so-called 'plum pudding model'.

This model could account for the stability of atoms, but could not account for the discrete wavelengths observed in the spectra of light emitted from excited atoms. Neither could it explain the results of a classic series of experiments performed in 1911 at the suggestion of Rutherford by his collaborators, Geiger and Marsden. These consisted of scattering α-particles by very thin gold foils. In the Thomson model, most of the α-particles would pass through the foil, with only a few suffering deflections through small angles. Rutherford suggested they should look for large-angle scattering and to their surprise they found that some particles were indeed scattered through very large angles, even greater than $90°$. Rutherford showed that this behaviour was not due to multiple small-angle deflections, but could only be the result of the α-particles encountering a very small positively charged central *nucleus*. (The reason for these two different behaviours is discussed in Appendix C.)

To explain the results of these experiments Rutherford formulated a 'planetary' model, where the atom was likened to a planetary system, with the electrons (the 'planets') occupying discrete orbits about a central positively charged nucleus (the 'Sun'). Because photons of a definite energy would be emitted when electrons moved from one orbit to another, this model could explain the discrete nature of the observed electromagnetic spectra when excited atoms decayed. In the simplest case of hydrogen, the nucleus is a single *proton* (p) with electric charge $+e$, where e is the magnitude of the charge on the electron[4], orbited by a single electron. Heavier atoms were considered to have nuclei consisting of several protons. This view persisted for a long time and was supported by the fact that the masses of many naturally occurring elements are integer multiples of a unit that is about 1 per cent smaller than the mass of the hydrogen atom. Examples are carbon and nitrogen, with masses of 12.0 and 14.0 in these units. However, it could not explain why not all atoms obeyed this rule. For example, chlorine has a mass of 35.5 in these

---

[3]J. J. Thomson received the 1906 Nobel Prize in Physics for his discovery. A year earlier, Philipp von Lenard had received the Physics Prize for his work on cathode rays.

[4]Why the charge on the proton should have exactly the same magnitude as that on the electron is a very long-standing puzzle, the solution to which is suggested by some as yet unproven, but widely believed, theories of particle physics that will be discussed briefly in Chapter 9.

units. At about the same time, the concept of *isotopism* (a name coined by Soddy) was conceived. *Isotopes* are atoms whose nuclei have different masses, but the same charge. Naturally occurring elements were postulated to consist of a mixture of different isotopes, giving rise to the observed masses.[5]

The explanation of isotopes had to wait 20 years until a classic discovery by Chadwick in 1932. His work followed earlier experiments by Irène Curie (the daughter of Pierre and Marie Curie) and her husband Frédéric Joliot.[6] They had observed that neutral radiation was emitted when α-particles bombarded beryllium and later work had studied the energy of protons emitted when paraffin was exposed to this neutral radiation. Chadwick refined and extended these experiments and demonstrated that they implied the existence of an electrically neutral particle of approximately the same mass as the proton. He had discovered the *neutron* (*n*) and in so doing had produced almost the final ingredient for understanding nuclei.[7]

There remained the problem of reconciling the planetary model with the observation of stable atoms. In classical physics, the electrons in the planetary model would be constantly accelerating and would therefore lose energy by radiation, leading to the collapse of the atom. This problem was solved by Bohr in 1913. He applied the newly emerging quantum theory and the result was the now well-known Bohr model of the atom. Refined modern versions of this model, including relativistic effects described by the Dirac equation (the relativistic analogue of the Schrödinger equation that applies to electrons), are capable of explaining the phenomena of atomic physics. Later workers, including Heisenberg, another of the founders of quantum theory,[8] applied quantum mechanics to the nucleus, now viewed as a collection of neutrons and protons, collectively called *nucleons*. In this case, however, the force binding the nucleus is not the electromagnetic force that holds electrons in their orbits, but is a short-range[9] force whose magnitude is independent of the type of nucleon, proton or neutron (i.e. charge-independent). This binding interaction is called the *strong nuclear force*.

These ideas still form the essential framework of our understanding of the nucleus today, where nuclei are bound states of nucleons held together by a strong charge-independent short-range force. Nevertheless, there is still no single theory that is capable of explaining all the data of nuclear physics and we shall see that different models are used to interpret different classes of phenomena.

---

[5]Frederick Soddy was awarded the 1921 Nobel Prize in Chemistry for his work on isotopes.
[6]Irène Curie and Frédéric Joliot received the 1935 Nobel Prize in Chemistry for 'synthesizing new radioactive elements'.
[7]James Chadwick received the 1935 Nobel Prize in Physics for his discovery of the neutron.
[8]Werner Heisenberg received the 1932 Nobel Prize in Physics for his contributions to the creation of quantum mechanics and the idea of *isospin symmetry*, which we will discuss in Chapter 3.
[9]The concept of range will be discussed in more detail in Section 1.5.1, but for the present it may be taken as the effective distance beyond which the force is insignificant.

1.1.2   The emergence of particle physics: the standard model and hadrons

By the early 1930s, the 19th century view of atoms as indivisible *elementary particles* had been replaced and a larger group of physically smaller entities now enjoyed this status: electrons, protons and neutrons. To these we must add two electrically neutral particles: the *photon* (γ) and the *neutrino* (ν). The photon was postulated by Planck in 1900 to explain black-body radiation, where the classical description of electromagnetic radiation led to results incompatible with experiments.[10] The neutrino was postulated by Fermi in 1930 to explain the apparent non-conservation of energy observed in the decay products of some unstable nuclei where β-rays are emitted, the so-called *β-decays*. Prior to Fermi's suggestion, β-decay had been viewed as a parent nucleus decaying to a daughter nucleus and an electron. As this would be a two-body decay, it would imply that the electron would have a unique momentum, whereas experiments showed that the electron actually had a momentum *spectrum*. Fermi's hypothesis of a third particle (the neutrino) in the final state solved this problem, as well as a problem with angular momentum conservation, which was apparently also violated if the decay was two-body. The β-decay data implied that the neutrino mass was very small and was compatible with the neutrino being massless.[11] It took more than 25 years before Fermi's hypothesis was confirmed by Reines and Cowan in a classic experiment in 1956 that detected free neutrinos from β-decay.[12]

The 1950s also saw technological developments that enabled high-energy beams of particles to be produced in laboratories. As a consequence, a wide range of controlled scattering experiments could be performed and the greater use of computers meant that sophisticated analysis techniques could be developed to handle the huge quantities of data that were being produced. By the 1960s this had resulted in the discovery of a very large number of unstable particles with very short lifetimes and there was an urgent need for a theory that could make sense of all these states. This emerged in the mid 1960s in the form of the so-called *quark model*, first suggested by Murray Gell-Mann and independently and simultaneously by George Zweig, who postulated that the new particles were bound states of three families of more fundamental physical particles.

---

[10]X-rays had already been observed by Röntgen in 1895 (for which he received the first Nobel Prize in Physics in 1901) and γ-rays were seen by Villard in 1900, but it was Planck who first made the startling suggestion that electromagnetic energy was quantized. For this he was awarded the 1918 Nobel Prize in Physics. Many years later, he said that his hypothesis was an 'act of desperation' as he had exhausted all other possibilities.

[11]However, in Section 3.1.4 we will discuss recent evidence that neutrinos have very small, but non-zero, masses.

[12]A description of this experiment is given in Chapter 12 of Tr75. Frederick Reines shared the 1995 Nobel Prize in Physics for his work in neutrino physics and particularly for the detection of the electron neutrino.

Gell-Mann called these *quarks* (q).[13] Because no free quarks were detected experimentally, there was initially considerable scepticism for this view. We now know that there is a fundamental reason why quarks cannot be observed as free particles (it will be discussed in Chapter 5), but at the time many physicists looked upon quarks as a convenient mathematical description, rather than physical particles.[14] However, evidence for the existence of quarks as real particles came in the 1960s from a series of experiments analogous to those of Rutherford and his co-workers, where high-energy beams of electrons and neutrinos were scattered from nucleons. (These experiments will also be discussed in Chapter 5.) Analysis of the angular distributions of the scattered particles showed that the nucleons were themselves bound states of three point-like charged entities, with properties consistent with those hypothesized in the quark model. One of these properties was unexpected and unusual: quarks have fractional electric charges, in practice $-\frac{1}{3}e$ and $+\frac{2}{3}e$. This is essentially the picture today, where elementary particles are now considered to be a small number of physical entities, including quarks, the electron, neutrinos, the photon and a few others we shall meet, but no longer nucleons.

The best theory of elementary particles we have at present is called, rather prosaically, the *standard model*. This aims to explain all the phenomena of particle physics, except those due to gravity, in terms of the properties and interactions of a small number of *elementary* (or *fundamental*) *particles*, which are now defined as being point-like, without internal structure or excited states. Particle physics thus differs from nuclear physics in having a single theory to interpret its data.

An elementary particle is characterized by, amongst other things, its mass, its electric charge and its *spin*. The latter is a permanent angular momentum possessed by all particles in quantum theory, even when they are at rest. Spin has no classical analogue and is not to be confused with the use of the same word in classical physics, where it usually refers to the (orbital) angular momentum of extended objects. The maximum value of the spin angular momentum about any axis is $s\hbar(\hbar \equiv h/2\pi)$, where h is Planck's constant and s is the *spin quantum number*, or *spin* for short. It has a fixed value for particles of any given type (for example $s = \frac{1}{2}$ for electrons) and general quantum mechanical principles restrict the possible values of s to be $0, \frac{1}{2}, 1, \frac{3}{2}, \ldots$. Particles with half-integer spin are called *fermions* and those with integer spin are called *bosons*. There are three families of elementary particles in the standard model: two spin-$\frac{1}{2}$ families of fermions called *leptons* and *quarks*; and one family of spin-1 *bosons*. In addition,

---

[13]Gell-Mann received the 1969 Nobel Prize in Physics for 'contributions and discoveries concerning the classification of elementary particles and their interactions'. For the origin of the word 'quark', he cited the now famous quotation 'Three quarks for Muster Mark' from James Joyce's book *Finnegans Wake*. Zweig had suggested the name 'aces', which with hindsight might have been more appropriate, as later experiments revealed that there were four and not three families of quarks.

[14]This was history repeating itself. In the early days of the atomic model many very distinguished scientists were reluctant to accept that atoms existed, because they could not be 'seen' in a conventional sense.

at least one other spin-0 particle, called the *Higgs boson*, is postulated to explain the origin of mass within the theory.[15]

The most familiar elementary particle is the electron, which we know is bound in atoms by the *electromagnetic interaction*, one of the four forces of nature.[16] One test of the elementarity of the electron is the size of its magnetic moment. A charged particle with spin necessarily has an intrinsic magnetic moment $\boldsymbol{\mu}$. It can be shown from the Dirac equation that a point-like spin-$\frac{1}{2}$ particle of charge q and mass m has a magnetic moment $\boldsymbol{\mu} = (q/m)\mathbf{S}$, where $\mathbf{S}$ is its spin vector, and hence $\boldsymbol{\mu}$ has magnitude $\mu = q\hbar/2m$. The magnetic moment of the electron very accurately obeys this relation, confirming that electrons are elementary.

The electron is a member of the family of leptons. Another is the neutrino, which was mentioned earlier as a decay product in β-decays. Strictly speaking, this particle should be called the *electron neutrino*, written ν_e, because it is always produced in association with an electron (the reason for this is discussed in Section 3.1.1). The force responsible for β-decay is an example of a second fundamental force, the *weak interaction*. Finally, there is the third force, the (fundamental) *strong interaction*, which, for example, binds quarks in nucleons. The strong nuclear force mentioned in Section 1.1.1 is not the same as this fundamental strong interaction, but is a consequence of it. The relation between the two will be discussed in more detail later.

The standard model also specifies the origin of these three forces. In classical physics the electromagnetic interaction is propagated by electromagnetic waves, which are continuously emitted and absorbed. While this is an adequate description at long distances, at short distances the quantum nature of the interaction must be taken into account. In quantum theory, the interaction is transmitted discontinuously by the exchange of photons, which are members of the family of fundamental spin-1 bosons of the standard model. Photons are referred to as the *gauge bosons*, or 'force carriers', of the electromagnetic interaction. The use of the word 'gauge' refers to the fact that the electromagnetic interaction possesses a fundamental symmetry called *gauge invariance*. For example, Maxwell's equations of classical electromagnetism are invariant under a specific phase transformation of the electromagnetic fields – the gauge transformation.[17] This property is common to all the three interactions of nature we will be discussing and has profound consequences, but we will not need its details in this book. The weak and strong interactions are also mediated by the exchange of spin-1 gauge bosons. For the weak interaction these are the W^+, W^- and Z^0 bosons (again the superscripts denote the electric charges) with masses about 80–90 times the mass of the proton.

---

[15]In the theory without the Higgs boson, all elementary particles are predicted to have zero mass, in obvious contradiction with experiment. A solution to this problem involving the Higgs boson will be discussed briefly in Chapter 9.

[16]Gravity is so weak that it can be neglected in nuclear and particle physics at presently accessible energies. Because of this, we will often refer in practice to the *three* forces of nature.

[17]See, for example, Appendix C.2 of Ma97.

For the strong interaction, the force carriers are called *gluons*. There are eight gluons, all of which have zero mass and are electrically neutral.[18]

In addition to the elementary particles of the standard model, there are other important particles we will be studying. These are the *hadrons*, the bound states of quarks. Nucleons are examples of hadrons,[19] but there are several hundred more, not including nuclei, most of which are unstable and decay by one of the three interactions. It was the abundance of these states that drove the search for a simplifying theory that would give an explanation for their existence and led to the quark model in the 1960s. The most common unstable example of a hadron is the *pion*, which exists in three electrical charge states, written (π^+, π^0, π^-). Hadrons are important because free quarks are unobservable in nature and so to deduce their properties we are forced to study hadrons. An analogy would be if we had to deduce the properties of nucleons by exclusively studying the properties of nuclei.

Since nucleons are bound states of quarks and nuclei are bound states of nucleons, the properties of nuclei should, in principle, be deducible from the properties of quarks and their interactions, i.e. from the standard model. In practice, however, this is far beyond present calculational techniques and sometimes nuclear and particle physics are treated as two almost separate subjects. However, there are many connections between them and in introductory treatments it is still useful to present both subjects together.

The remaining sections of this chapter are devoted to introducing some of the basic theoretical tools needed to describe the phenomena of both nuclear and particle physics, starting with a key concept: antiparticles.

1.2 Relativity and Antiparticles

Elementary particle physics is also called high-energy physics. One reason for this is that if we wish to produce new particles in a collision between two other particles, then because of the relativistic mass–energy relation $E = mc^2$, energies are needed at least as great as the rest masses of the particles produced. The second reason is that to explore the structure of a particle requires a probe whose wavelength λ is smaller than the structure to be explored. By the de Broglie relation $\lambda = h/p$, this implies that the momentum p of the probing particle, and hence its energy, must be large. For example, to explore the internal structure of the proton using electrons requires wavelengths that are much smaller than the

---

[18]Note that the word 'electric' has been used when talking about charge. This is because the weak and strong interactions also have associated 'charges' which determine the strengths of the interactions, just as the electric charge determines the strength of the electromagnetic interaction. This will be discussed in more detail in later chapters.

[19]The magnetic moments of the proton and neutron do not obey the prediction of the Dirac equation and this is evidence that nucleons have structure and are not elementary. The proton magnetic moment was first measured by Otto Stern using a molecular beam method that he developed and for this he received the 1943 Nobel Prize in Physics.

classical radius of the proton, which is roughly 10^{-15} m. This in turn requires electron energies that are greater than 10^3 times the rest energy of the electron, implying electron velocities very close to the speed of light. Hence any explanation of the phenomena of elementary particle physics must take account of the requirements of the theory of special relativity, in addition to those of quantum theory. There are very few places in particle physics where a non-relativistic treatment is adequate, whereas the need for a relativistic treatment is far less in nuclear physics.

Constructing a quantum theory that is consistent with special relativity leads to the conclusion that for every particle of nature, there must exist an associated particle, called an *antiparticle*, with the same mass as the corresponding particle. This important theoretical prediction was first made by Dirac and follows from the solutions of the equation he first wrote down to describe relativistic electrons.[20] The *Dirac equation* is of the form

$$i\hbar \frac{\partial \mathbf{\Psi}(\mathbf{x},t)}{\partial t} = H(\mathbf{x},\hat{\mathbf{p}})\mathbf{\Psi}(\mathbf{x},t), \tag{1.1}$$

where $\hat{\mathbf{p}} = -i\hbar\nabla$ is the usual quantum mechanical momentum operator and the Hamiltonian was postulated by Dirac to be

$$H = c\boldsymbol{\alpha}\cdot\hat{\mathbf{p}} + \beta mc^2. \tag{1.2}$$

The coefficients $\boldsymbol{\alpha}$ and β are determined by the requirement that the solutions of Equation (1.1) are also solutions of the *Klein–Gordon equation* [21]

$$-\hbar^2\frac{\partial^2\mathbf{\Psi}(\mathbf{x},t)}{\partial t^2} = -\hbar^2 c^2\nabla^2\mathbf{\Psi}(\mathbf{x},t) + m^2 c^4\mathbf{\Psi}(\mathbf{x},t). \tag{1.3}$$

This leads to the conclusion that $\boldsymbol{\alpha}$ and β cannot be simple numbers; their simplest forms are 4×4 matrices. Thus the solutions of the Dirac equation are four-component wavefunctions (called *spinors*) with the form[22]

$$\mathbf{\Psi}(\mathbf{x},t) = \begin{pmatrix} \mathbf{\Psi}_1(\mathbf{x},t) \\ \mathbf{\Psi}_2(\mathbf{x},t) \\ \mathbf{\Psi}_3(\mathbf{x},t) \\ \mathbf{\Psi}_4(\mathbf{x},t) \end{pmatrix}. \tag{1.4}$$

---

[20]Paul Dirac shared the 1933 Nobel Prize in Physics with Erwin Schrödinger. The somewhat cryptic citation stated 'for the discovery of new productive forms of atomic theory'.

[21]This is a relativistic equation, which is 'derived' by starting from the relativistic mass–energy relation $E^2 = p^2 c^2 + m^2 c^4$ and using the usual quantum mechanical operator substitutions, $\hat{\mathbf{p}} = -i\hbar\nabla$ and $E = i\hbar\partial/\partial t$.

[22]The details may be found in most quantum mechanics books, for example, pp. 475–477 of Sc68.

The interpretation of Equation (1.4) is that the four components describe the two spin states of a negatively charged electron with positive energy and the two spin states of a corresponding particle having the same mass but with negative energy. Two spin states arise because in quantum mechanics the projection in any direction of the spin vector of a spin-$\frac{1}{2}$ particle can only result in one of the two values $\pm\frac{1}{2}$, called 'spin up' and 'spin down', respectively. The two energy solutions arise from the two solutions of the relativistic mass–energy relation $E = \pm\sqrt{p^2c^2 + m^2c^4}$. The latter states can be shown to behave in all respects as *positively* charged electrons (called *positrons*), but with positive energy. The positron is referred to as the antiparticle of the electron. The discovery of the positron by Anderson in 1933, with all the predicted properties, was a spectacular verification of the Dirac prediction.

Although Dirac originally made his prediction for electrons, the result is general and is true whether the particle is an elementary particle or a hadron. If we denote a particle by P, then the antiparticle is in general written with a bar over it, i.e. $\bar{P}$. For example, the antiparticle of the proton is the antiproton $\bar{p}$,[23] with negative electric charge; and associated with every quark, q, is an antiquark, $\bar{q}$. However, for some very common particles the bar is usually omitted. Thus, for example, in the case of the positron e^+, the superscript denoting the charge makes explicit the fact that the antiparticle has the opposite electric charge to that of its associated particle. Electric charge is just one example of a *quantum number* (spin is another) that characterizes a particle, whether it is elementary or composite (i.e. a hadron).

Many quantum numbers differ in sign for particle and antiparticle, and electric charge is an example of this. We will meet others later. When brought together, particle–antiparticle pairs, each of mass m, can annihilate, releasing their combined rest energy $2mc^2$ as photons or other particles. Finally, we note that there is symmetry between particles and antiparticles, and it is a convention to call the electron the particle and the positron its antiparticle. This reflects the fact that the normal matter contains electrons rather than positrons.

1.3 Symmetries and Conservation Laws

Symmetries and the invariance properties of the underlying interactions play an important role in physics. Some lead to conservation laws that are universal. Familiar examples are translational invariance, leading to the conservation of linear momentum; and rotational invariance, leading to conservation of angular momentum. The latter plays an important role in nuclear and particle physics as it leads to a scheme for the classification of states based, among other quantum

---

[23]Carl Anderson shared the 1936 Nobel Prize in Physics for the discovery of the positron. The 1958 Prize was awarded to Emilio Segrè and Owen Chamberlain for their discovery of the antiproton.

numbers, on their spins.[24] Another very important invariance that we have briefly mentioned is gauge invariance. This fundamental property of all three interactions restricts the forms of the interactions in a profound way that initially is contradicted by experiment. This is the prediction of zero masses for all elementary particles, mentioned earlier. There are theoretical solutions to this problem whose experimental verification (or otherwise) is probably the most eagerly awaited result in particle physics today.

In nuclear and particle physics we need to consider additional symmetries of the Hamiltonian and the conservation laws that follow and in the remainder of this section we discuss two of the most important of these that we will need later – *parity* and *charge conjugation*.

1.3.1   Parity

Parity was first introduced in the context of atomic physics by Eugene Wigner in 1927.[25] It refers to the behaviour of a state under a spatial reflection, i.e. $\mathbf{x} \rightarrow -\mathbf{x}$. If we consider a single-particle state, represented for simplicity by a non-relativistic wavefunction $\psi(\mathbf{x}, t)$, then under the parity operator, $\hat{P}$,

$$\hat{P}\psi(\mathbf{x}, t) \equiv P\psi(-\mathbf{x}, t). \tag{1.5}$$

Applying the operator again, gives

$$\hat{P}^2\psi(\mathbf{x}, t) = P\hat{P}\psi(-\mathbf{x}, t) = P^2\psi(\mathbf{x}, t), \tag{1.6}$$

implying $P = \pm 1$. If the particle is an eigenfunction of linear momentum $\mathbf{p}$, i.e.

$$\psi(\mathbf{x}, t) \equiv \psi_p(\mathbf{x}, t) = \exp[i(\mathbf{p} \cdot \mathbf{x} - Et)], \tag{1.7}$$

then

$$\hat{P}\psi_p(\mathbf{x}, t) = P\psi_p(-\mathbf{x}, t) = P\psi_{-p}(\mathbf{x}, t) \tag{1.8}$$

and so a particle at rest, with $\mathbf{p} = \mathbf{0}$, is an eigenstate of parity. The eigenvalue $P = \pm 1$ is called the *intrinsic parity*, or just the *parity*, of the state. By considering a multiparticle state with a wavefunction that is the product of single-particle wavefunctions, it is clear that parity is a multiplicative quantum number.

The strong and electromagnetic interactions, but not the weak interactions, are invariant under parity, i.e. the Hamiltonian of the system remains unchanged under

---

[24]These points are explored in more detail in, for example, Chapter 4 of Ma97.
[25]Eugene Wigner shared the 1963 Nobel Prize in Physics, principally for his work on symmetries.

a parity transformation on the position vectors of all particles in the system. Parity is therefore conserved, by which we mean that the total parity quantum number remains unchanged in the interaction. Compelling evidence for parity conservation in the strong and electromagnetic interactions comes from the absence of transitions between nuclear states and atomic states, respectively, that would violate parity conservation. The evidence for non-conservation of parity in the weak interaction will be discussed in detail in Chapter 6.

There is also a contribution to the total parity if the particle has an orbital angular momentum l. In this case its wave function is a product of a radial part $R_{n\ell}$ and an angular part $Y_l^m(\theta, \phi)$:

$$\psi_{lmn}(\mathbf{x}) = R_{nl} Y_l^m(\theta, \phi), \tag{1.9}$$

where n and m are the principal and magnetic quantum numbers and $Y_l^m(\theta, \phi)$ is a spherical harmonic. It is straightforward to show from the relations between Cartesian (x, y, z) and spherical polar co-ordinates (r, θ, ϕ), i.e.

$$x = r \sin\theta \cos\phi, \quad y = r \sin\theta \sin\phi, \quad z = r \cos\theta, \tag{1.10}$$

that the parity transformation $\mathbf{x} \rightarrow -\mathbf{x}$ implies

$$r \rightarrow r, \quad \theta \rightarrow \pi - \theta, \quad \phi \rightarrow \pi + \phi, \tag{1.11}$$

and from this it can be shown that

$$Y_l^m(\theta, \phi) \rightarrow Y_l^m(\pi - \theta, \pi + \phi) = (-)^l Y_l^m(\theta, \phi). \tag{1.12}$$

Equation (1.12) may easily be verified directly for specific cases; for example, for the first three spherical harmonics,

$$Y_0^0 = \left(\frac{1}{4\pi}\right)^{\frac{1}{2}}, \quad Y_1^0 = \left(\frac{3}{4\pi}\right)^{\frac{1}{2}} \cos\theta, \quad Y_1^{\pm 1} = \left(\frac{3}{8\pi}\right)^{\frac{1}{2}} \sin\theta \, e^{\pm i\phi}. \tag{1.13}$$

Hence

$$\hat{P}\psi_{lmn}(\mathbf{x}) = P\psi_{lmn}(-\mathbf{x}) = P(-)^l \psi_{lmn}(\mathbf{x}), \tag{1.14}$$

i.e. $\psi_{lmn}(\mathbf{x})$ is an eigenstate of parity with eigenvalue $P(-1)^l$.

An analysis of the Dirac Equation (1.1) for relativistic electrons, shows that it is invariant under a parity transformation only if $P(e^+ e^-) = -1$. This is a general result for all fermion–antifermion pairs, so it is a convention to assign $P = +1$ to all leptons and $P = -1$ to their antiparticles. We will see in Chapter 3 that in strong interactions quarks can only be created as part of a quark–antiquark pair, so the intrinsic parity of a single quark cannot be measured. For this reason, it is also a convention to assign $P = +1$ to quarks. Since quarks are fermions, it follows from the Dirac result that $P = -1$ for antiquarks. The intrinsic parities of hadrons then follow from their structure in terms of quarks and the orbital angular momentum between the constituent quarks, using Equation (1.14). This will be explored in Chapter 3 as part of the discussion of the quark model.

1.3.2   Charge conjugation

Charge conjugation is the operation of changing a particle into its antiparticle. Like parity, it gives rise to a multiplicative quantum number that is conserved in strong and electromagnetic interactions, but violated in the weak interaction. In discussing charge conjugation, we will need to distinguish between states such as the photon γ and the neutral pion π^0 that do not have distinct antiparticles and those such as the π^+ and the neutron, which do. Particles in the former class we will collectively denote by a and those of the latter type will be denoted by b. It is also convenient at this point to extend our notation for states. Thus we will represent a state of type a having a wavefunction ψ_a by $|a, \psi_a\rangle$ and similarly for a state of type b. Then under the charge conjugation operator, $\hat{C}$,

$$\hat{C}|a, \psi_a\rangle = C_a|a, \psi_a\rangle, \tag{1.15a}$$

and

$$\hat{C}|b, \psi_b\rangle = |\bar{b}, \psi_{\bar{b}}\rangle, \tag{1.15b}$$

where C_a is a phase factor analogous to the phase factor in Equation (1.5).[26] Applying the operator twice, in the same way as for parity, leads to $C_a = \pm 1$. From Equation (1.15a), we see that states of type a are eigenstates of $\hat{C}$ with eigenvalues ± 1, called their *C-parities*. States with distinct antiparticles can only form eigenstates of $\hat{C}$ as linear combinations.

As an example of the latter, consider a $\pi^+\pi^-$ pair with orbital angular momentum L between them. In this case

$$\hat{C}|\pi^+\pi^-; L\rangle = (-1)^L|\pi^+\pi^-; L\rangle, \tag{1.16}$$

because interchanging the pions reverses their relative positions in the spatial wavefunction. The same factor occurs for spin-$\frac{1}{2}$ fermion pairs $f\bar{f}$, but in addition there are two other factors. The first is $(-1)^{S+1}$, where S is the total spin of the pair.

---

[26]A phase factor could have been inserted in Equation (1.15b), but it is straightforward to show that the relative phase of the two states b and $\bar{b}$ cannot be measured and so a phase introduced in this way would have no physical consequences.

This follows directly from the structure of the spin wavefunctions:

$$\left.\begin{array}{cc} \uparrow_1\uparrow_2 & S_z = 1 \\ \frac{1}{\sqrt{2}}(\uparrow_1\downarrow_2 + \downarrow_1\uparrow_2) & S_z = 0 \\ \downarrow_1\downarrow_2 & S_z = -1 \end{array}\right\} S = 1 \tag{1.17a}$$

and

$$\frac{1}{\sqrt{2}}(\uparrow_1\downarrow_2 - \downarrow_1\uparrow_2) \quad S_z = 0 \quad S = 0, \tag{1.17b}$$

where $\uparrow_i$ ($\downarrow_i$) represents particle i having spin 'up' ('down') in the z-direction. A second factor (-1) arises whenever fermions and antifermions are interchanged. This has its origins in quantum field theory.[27] Combining these factors, finally we have

$$\hat{C}|f\bar{f}; J, L, S\rangle = (-1)^{L+S}|f\bar{f}; J, L, S\rangle, \tag{1.18}$$

for fermion–antifermion pairs having total, orbital and spin angular momentum quantum numbers J, L and S, respectively.

1.4 Interactions and Feynman Diagrams

We now turn to a discussion of particle interactions and how they can be described by the very useful pictorial methods of Feynman diagrams.

1.4.1 Interactions

Interactions involving elementary particles and/or hadrons are conveniently summarized by 'equations' by analogy with chemical reactions, in which the different particles are represented by symbols which usually – but not always – have a superscript to denote their electric charge. In the interaction

$$\nu_e + n \rightarrow e^- + p, \tag{1.19}$$

for example, an electron neutrino ν_e collides with a neutron n to produce an electron e^- and a proton p, whereas the equation

$$e^- + p \rightarrow e^- + p \tag{1.20}$$

---

[27]See, for example, pp. 249–250 of Go86.

represents an electron and proton interacting to give the same particles in the final state, but in general travelling in different directions. In such equations, conserved quantum numbers must have the same total values in initial and final states.

Particles may be transferred from initial to final states and vice versa, when they become antiparticles. Thus the process

$$\pi^- + p \rightarrow \pi^- + p, \tag{1.21a}$$

also implies the reaction

$$p + \bar{p} \rightarrow \pi^+ + \pi^-, \tag{1.22}$$

which is obtained by taking the proton from the final state to an antiproton in the initial state and the negatively charged pion in the initial state to a positively charged pion in the final state.

The interactions in Equations (1.20) and (1.21a), in which the particles remain unchanged, are examples of *elastic scattering*, in contrast to the reactions in Equations (1.19) and (1.22), where the final-state particles differ from those in the initial state. Collisions between a given pair of initial particles do not always lead to the same final state, but can lead to different final states with different probabilities. For example, the collision of a negatively charged pion and a proton can give rise to elastic scattering (Equation (1.21a)) and a variety of other reactions, such as

$$\pi^- + p \rightarrow n + \pi^0 \quad \text{and} \quad \pi^- + p \rightarrow p + \pi^- + \pi^- + \pi^+, \tag{1.21b}$$

depending on the initial energy. In particle physics it is common to refer (rather imprecisely) to such interactions as 'inelastic' scattering.

Similar considerations apply to nuclear physics, but the term *inelastic scattering* is reserved for the case where the final state is an excited state of the parent nucleus A, that subsequently decays, for example via photon emission, i.e.

$$a + A \rightarrow a + A^*; \qquad A^* \rightarrow A + \gamma, \tag{1.23}$$

where a is a projectile and A^* is an excited state of A. A useful shorthand notation used in nuclear physics for the general reaction $a + A \rightarrow b + B$ is $A(a, b)B$. It is usual in nuclear physics to further subdivide types of interactions according to the underlying mechanism that produced them. We will return to this in Section 2.9, as part of a more general discussion of nuclear reactions.

Finally, many particles are unstable and spontaneously decay to other, lighter (i.e. having less mass) particles. An example of this is the free neutron (i.e. one not bound in a nucleus), which decays by the β-decay reaction

$$n \rightarrow p + e^- + \bar{\nu}_e, \tag{1.24}$$

with a mean lifetime of about 900 s.[28] The same notation can also be used in nuclear physics. For example, many nuclei decay via the β-decay reaction. Thus, denoting a nucleus with Z protons and N nucleons as (Z, N), we have

$$(Z, N) \rightarrow (Z - 1, N) + e^+ + \nu_e. \tag{1.25}$$

This is also a weak interaction. This reaction is effectively the decay of a proton bound in a nucleus. Although a *free* proton cannot decay by the β-decay $p \rightarrow n + e^+ + \nu_e$ because it violates energy conservation (the final-state particles have greater total mass than the proton), a proton bound in a nucleus can decay because of its binding energy. This will be explained in Chapter 2.

1.4.2   Feynman diagrams

The forces producing all the above interactions are due to the exchange of particles and a convenient way of illustrating this is to use *Feynman diagrams*. There are mathematical rules and techniques associated with these that enable them to be used to calculate the quantum mechanical probabilities for given reactions to occur, but in this book Feynman diagrams will only be used as a convenient very useful pictorial description of reaction mechanisms.

We first illustrate them at the level of elementary particles for the case of electromagnetic interactions, which arise from the emission and/or absorption of photons. For example, the dominant interaction between two electrons is due to the exchange of a single photon, which is emitted by one electron and absorbed by the other. This mechanism, which gives rise to the familiar Coulomb interaction at large distances, is illustrated in the Feynman diagram Figure 1.1(a).

In such diagrams, we will use the convention that particles in the initial state are shown on the left and particles in the final state are shown on the right. (Some authors take time to run along the y-axis.) Spin-$\frac{1}{2}$ fermions (such as the electron)

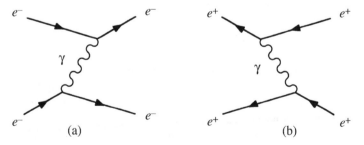

Figure 1.1   One-photon exchange in (a) $e^- + e^- \rightarrow e^- + e^-$ and (b) $e^+ + e^+ \rightarrow e^+ + e^+$

---

[28]The reason that this decay involves an antineutrino rather than a neutrino will become clear in Chapter 3.

are drawn as solid lines and photons are drawn as wiggly lines. Arrowheads pointing to the right indicate that the solid lines represent electrons. In the case of photon exchange between two positrons, which is shown in Figure 1.1(b), the arrowheads on the antiparticle (positron) lines are conventionally shown as pointing to the left. In interpreting these diagrams, it is important to remember: (1) that the direction of the arrows on fermion lines does *not* indicate the particle's direction of motion, but merely whether the fermions are particles or antiparticles, and (2) that particles in the initial state are always to the left and particles in the final state are always to the right.

A feature of the above diagrams is that they are constructed from combinations of simple three-line vertices. This is characteristic of electromagnetic processes. Each vertex has a line corresponding to a single photon being emitted or absorbed, while one fermion line has the arrow pointing toward the vertex and the other away from the vertex, guaranteeing charge conservation at the vertex, which is one of the rules of Feynman diagrams.[29] For example, a vertex like Figure 1.2 would correspond to a process in which an electron emitted a photon and turned into a positron. This would violate charge conservation and is therefore forbidden.

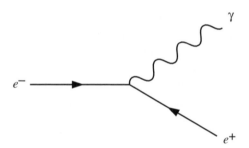

Figure 1.2   The forbidden vertex $e^- \rightarrow e^+ + \gamma$

Feynman diagrams can also be used to describe the fundamental weak and strong interactions. This is illustrated by Figure 1.3(a), which shows contributions to the elastic weak scattering reaction $e^- + \nu_e \rightarrow e^- + \nu_e$ due to the exchange of a Z^0, and by Figure 1.3(b), which shows the exchange of a gluon g (represented by a coiled line) between two quarks, which is a strong interaction.

Feynman diagrams can also be drawn at the level of hadrons. As an illustration, Figure 1.4 shows the exchange of a charged pion (shown as a dashed line) between a proton and a neutron. We shall see later that this mechanism is a major contribution to the strong nuclear force between a proton and a neutron.

We turn now to consider in more detail the relation between exchanged particles and forces.

---

[29]Compare Kirchhoff's laws in electromagnetism.

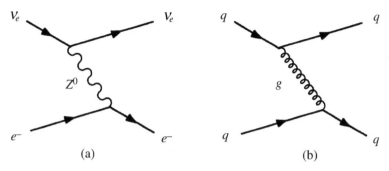

Figure 1.3 (a) Contributions of (a) Z^0 exchange to the elastic weak scattering reaction $e^- + \nu_e \rightarrow e^- + \nu_e$, and (b) gluon exchange contribution to the strong interaction $q + q \rightarrow q + q$

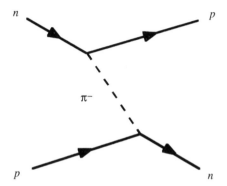

Figure 1.4 Single-pion exchange in the reaction $p + n \rightarrow n + p$

1.5 Particle Exchange: Forces and Potentials

This section starts with a discussion of the important relationship between forces and particle exchanges and then relates this to potentials. Although the idea of a potential has its greatest use in non-relativistic physics, nevertheless it is useful to illustrate concepts and is used in later sections as an intermediate step in relating theoretical Feynman diagrams to measurable quantities. The results can easily be extended to more general situations.

1.5.1 Range of forces

At each vertex of a Feynman diagram, charge is conserved by construction. We will see later that, depending on the nature of the interaction (strong, weak or electromagnetic), other quantum numbers are also conserved. However, it is easy to show that energy and momentum *cannot* be conserved simultaneously.

Consider the general case of a reaction $A + B \rightarrow A + B$ mediated by the exchange of a particle X, as shown in Figure 1.5. In the rest frame of the incident

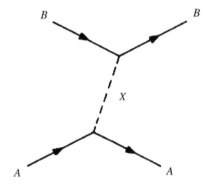

Figure 1.5 Exchange of a particle X in the reaction $A + B \rightarrow A + B$

particle A, the lower vertex represents the *virtual* process ('virtual' because X does not appear as a real particle in the final state),

$$A(M_A c^2, \mathbf{0}) \rightarrow A(E_A, \mathbf{p}_A c) + X(E_X, -\mathbf{p}_A c), \tag{1.26}$$

where E_A is the *total* energy of particle A and $\mathbf{p}_A$ is its three-momentum.[30] Thus, if we denote by P_A the four-vector for particle A,

$$P_A = (E_A/c, \mathbf{p}_A) \tag{1.27}$$

and

$$P_A^2 = E_A^2/c^2 - \mathbf{p}_A^2 = M_A^2 c^2. \tag{1.28}$$

Applying this to the diagram and imposing momentum conservation, gives

$$E_A = (p^2 c^2 + M_A^2 c^4)^{1/2}, \quad E_X = (p^2 c^2 + M_X^2 c^4)^{1/2}, \tag{1.29}$$

where $p = |\mathbf{p}|$. The energy difference between the final and initial states is given by

$$\begin{aligned} \Delta E = E_X + E_A - M_A c^2 &\rightarrow 2pc, \quad p \rightarrow \infty \\ &\rightarrow M_X c^2, \quad p \rightarrow 0 \end{aligned} \tag{1.30}$$

and thus $\Delta E \geq M_X c^2$ for all p, i.e. energy is not conserved. However, by the Heisenberg uncertainty principle, such an energy violation is allowed, but only for a time $\tau \leq \hbar/\Delta E$, so we immediately obtain

$$r \leq R \equiv \hbar/M_X c \tag{1.31}$$

---

[30]A resumé of relativistic kinematics is given in Appendix B.

as the maximum distance over which X can propagate before being absorbed by particle B. This maximum distance is called the *range* of the interaction and this was the sense of the word used in Section 1.1.1.

The electromagnetic interaction has an infinite range, because the exchanged particle is a massless photon. In contrast, the weak interaction is associated with the exchange of very heavy particles – the W and Z bosons. These lead to ranges that from Equation (1.31) are of the order of $R_{W,Z} \approx 2 \times 10^{-18}$ m. The fundamental strong interaction has infinite range because, like the photon, gluons have zero mass. On the other hand, the strong nuclear force, as exemplified by Figure 1.4 for example, has a much shorter range of approximately $(1-2) \times 10^{-15}$ m. We will comment briefly on the relation between these two different manifestations of the strong interaction in Section 7.1.

1.5.2 The Yukawa potential

In the limit that M_A becomes large, we can regard B as being scattered by a static potential of which A is the source. This potential will in general be spin dependent, but its main features can be obtained by neglecting spin and considering X to be a spin-0 boson, in which case it will obey the Klein–Gordon equation

$$-\hbar^2 \frac{\partial^2 \phi(\mathbf{x},t)}{\partial t^2} = -\hbar^2 c^2 \nabla^2 \phi(\mathbf{x},t) + M_X^2 c^4 \phi(\mathbf{x},t). \tag{1.32}$$

The static solution of this equation satisfies

$$\nabla^2 \phi(\mathbf{x}) = \frac{M_X^2 c^4}{\hbar^2} \phi(\mathbf{x}), \tag{1.33}$$

where $\phi(\mathbf{x})$ is interpreted as a static potential. For $M_X = 0$ this equation is the same as that obeyed by the electrostatic potential, and for a point charge $-e$ interacting with a point charge $+e$ at the origin, the appropriate solution is the Coulomb potential

$$V(r) = -e\phi(r) = -\frac{e^2}{4\pi\varepsilon_0} \frac{1}{r}, \tag{1.34}$$

where $r = |\mathbf{x}|$ and ε_0 is the dielectric constant. The corresponding solution in the case where $M_X^2 \neq 0$ is easily verified by substitution to be

$$V(r) = -\frac{g^2}{4\pi} \frac{e^{-r/R}}{r}, \tag{1.35}$$

where R is the range defined earlier and g, the so-called *coupling constant*, is a parameter associated with each vertex of a Feynman diagram and represents the basic strength of the interaction.[31] For simplicity, we have assumed equal strengths for the couplings of particle X to the particles A and B.

The form of $V(r)$ in Equation (1.35) is called a *Yukawa potential*, after the physicist who first introduced the idea of forces due to massive particle exchange in 1935.[32] As $M_X \to 0$, $R \to \infty$ and the Coulomb potential is recovered from the Yukawa potential, while for very large masses the interaction is approximately point-like (zero range). It is conventional to introduce a dimensionless parameter α_X by

$$\alpha_X = \frac{g^2}{4\pi\hbar c}, \tag{1.36}$$

that characterizes the strength of the interaction at short distances $r \leq R$. For the electromagnetic interaction this is the *fine structure constant*

$$\alpha \equiv e^2/4\pi\varepsilon_0\hbar c \approx 1/137 \tag{1.37}$$

that governs the splittings of atom energy levels.[33]

The forces between hadrons are also generated by the exchange of particles. Thus, in addition to the electromagnetic interaction between charged hadrons, all hadrons, whether charged or neutral, experience a strong *short-range* interaction, which in the case of two nucleons, for example, has a range of about 10^{-15} m, corresponding to the exchange of a particle with an effective mass of about $\frac{1}{7}$th the mass of the proton. The dominant contribution to this force is the exchange of a single pion, as shown in Figure 1.4. This nuclear strong interaction is a complicated effect that has its origins in the fundamental strong interactions between the quark distributions within the two hadrons. Two neutral atoms also experience an electromagnetic interaction (the van der Waals force), which has its origins in the fundamental Coulomb forces, but is of much shorter range. Although an analogous mechanism is not in fact responsible for the nuclear strong interaction, it is a useful reminder that the force between two *distributions* of particles can be much more complicated than the forces between their components. We will return to this point when we discuss the nature of the nuclear potential in more detail in Section 7.1.

1.6   Observable Quantities: Cross Sections and Decay Rates

We have mentioned earlier that Feynman diagrams can be turned into probabilities for a process by a set of mathematical rules (the *Feynman Rules*) that can be

---

[31]Although we call g a (point) coupling *constant*, in general it will have a dependence on the momentum carried by the exchanged particle. We ignore this in what follows.

[32]For this insight, Hideki Yukawa received the 1949 Nobel Prize in Physics.

[33]Like g, the coupling α_X will in general have a dependence on the momentum carried by particle X. In the case of the electromagnetic interaction, this dependence is relatively weak.

derived from the quantum theory of the underlying interaction.[34] We will not pursue this in detail in this book, but rather will show in principle their relation to *observables*, i.e. things that can be measured, concentrating on the cases of two-body scattering reactions and decays of unstable states.

1.6.1 Amplitudes

The intermediate step is the *amplitude f*, the modulus squared of which is directly related to the probability of the process occurring. It is also called the *invariant amplitude* because, as we shall show, it is directly related to observable quantities and these have to be the same in all inertial frames of reference. To get some qualitative idea of the structure of *f*, we will use non-relativistic quantum mechanics and assume that the coupling constant g^2 is small compared with $4\pi\hbar c$, so that the interaction is a small perturbation on the free particle solution, which will be taken as plane waves.

If we expand the amplitude in a power series in g^2 and keep just the first term (i.e. lowest-order perturbation theory), then the amplitude for a particle in an initial state with momentum $\mathbf{q}_i$ to be scattered to a final state with momentum $\mathbf{q}_f$ by a potential $V(\mathbf{x})$ is proportional to

$$f(\mathbf{q}^2) = \int d^3\mathbf{x}\, V(\mathbf{x}) \exp[i\mathbf{q} \cdot \mathbf{x}/\hbar], \tag{1.38}$$

i.e. the Fourier transform of the potential, where $\mathbf{q} \equiv \mathbf{q}_i - \mathbf{q}_f$ is the momentum transfer.[35]

The integration may be done using polar co-ordinates. Taking $\mathbf{q}$ in the x-direction, gives

$$\mathbf{q} \cdot \mathbf{x} = |\mathbf{q}| r \cos\theta \tag{1.39}$$

and

$$d^3\mathbf{x} = r^2 \sin\theta\, d\theta\, dr\, d\phi, \tag{1.40}$$

where $r = |\mathbf{x}|$. For the Yukawa potential, the integral in Equation (1.38) gives

$$f(\mathbf{q}^2) = \frac{-g^2\hbar^2}{\mathbf{q}^2 + M_X^2 c^2}. \tag{1.41}$$

---

[34]In the case of the electromagnetic interaction, the theory is called Quantum Electrodynamics (QED) and is spectacularly successful in explaining experimental results. Richard Feynman shared the 1965 Nobel Prize in Physics with Sin-Itiro Tomonoga and Julian Schwinger for their work on formulating quantum electrodynamics. The Feynman Rules are discussed in an accessible way in Gr87.
[35]See, for example, Chapter 11 of Ma92.

This amplitude corresponds to the exchange of a single particle, as shown for example in Figures 1.3 and 1.4. The structure of the amplitude, which is quite general, is a numerator proportional to the product of the couplings at the two vertices (or equivalently α_X in this case), and a denominator that depends on the mass of the exchanged particle and its momentum transfer squared. The denominator is called the *propagator* for particle X. In a relativistic calculation, the term $\mathbf{q}^2$ becomes q^2, where q is the *four-momentum* transfer.

Returning to the zero-range approximation, one area where it is used extensively is in weak interactions, particularly applied to nuclear decay processes. In these situations, $M_X = M_{W,Z}$ and $f \rightarrow -G_F$, where G_F, the so-called *Fermi coupling constant*, is given from Equation (1.36) by

$$\frac{G_F}{(\hbar c)^3} = \frac{4\pi\alpha_W}{(M_W c^2)^2} = 1.166 \times 10^{-5} \text{GeV}^{-2}. \tag{1.42}$$

The numerical value is obtained from analyses of decay processes, including that of the neutron and a heavier version of the electron called the *muon*, whose properties will be discussed in Chapter 3.

All the above is for the exchange of a single particle. It is also possible to draw more complicated Feynman diagrams that correspond to the exchange of more than one particle. An example of such a diagram for elastic $e^- e^-$ scattering, where two photons are exchanged, is shown in Figure 1.6.

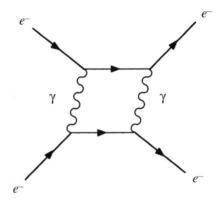

Figure 1.6 Two-photon exchange in the reaction $e^- + e^- \rightarrow e^- + e^-$

The number of vertices in any diagram is called the *order n*, and when the amplitude associated with any given Feynman diagram is calculated, it always contains a factor of $(\sqrt{\alpha})^n$. Since the probability is proportional to the square of the modulus of the amplitude, the former will contain a factor α^n. The probability associated with the single-photon exchange diagrams of Figure 1.1 thus contain a factor of α^2 and the contribution from two-photon exchange is of the order of α^4.

As $\alpha \sim 1/137$, the latter is usually very small compared with the contribution from single-photon exchange.

This is a general feature of electromagnetic interactions: because the fine structure constant is very small, in most cases only the lowest-order diagrams that contribute to a given process need be taken into account, and more complicated higher-order diagrams with more vertices can (to a good approximation) be ignored in many applications.

1.6.2 Cross-sections

The next step is to relate the amplitude to measurables. For scattering reactions, the appropriate observable is the *cross-section*. In a typical scattering experiment, a beam of particles is allowed to hit a target and the rates of production of various particles in the final state are counted.[36] It is clear that the rates will be proportional to: (a) the number N of particles in the target illuminated by the beam, and (b) the rate per unit area at which beam particles cross a small surface placed in the beam at rest with respect to the target and perpendicular to the beam direction. The latter is called the *flux* and is given by

$$J = n_b v_i, \tag{1.43}$$

where n_b is the number density of particles in the beam and v_i their velocity in the rest frame of the target. Hence the rate W_r at which a specific reaction r occurs in a particular experiment can be written in the form

$$W_r = JN\sigma_r, \tag{1.44a}$$

where σ_r, the constant of proportionality, is called the *cross-section* for reaction r. If the beam has a cross-sectional area S, its intensity is $I = JS$ and so an alternative expression for the rate is

$$W_r = N\sigma_r I/S = I\sigma_r n_t t, \tag{1.44b}$$

where n_t is the number of target particles per unit volume and t is the thickness of the target. If the target consists of an isotopic species of atomic mass M_A (in atomic mass units-defined in Section 1.7 below), then $n_t = \rho N_A/M_A$, where ρ is the density of the target and N_A is Avogadro's constant. Thus, Equation (1.44b) may be written

$$W_r = I\sigma_r(\rho t)N_A/M_A, \tag{1.44c}$$

---

[36]The practical aspects of experiments are discussed in Chapter 4.

where (ρt) is a measure of the amount of material in the target, expressed in units of mass per unit area. The form of Equation (1.44c) is particularly useful for the case of thin targets, commonly used in experiments. In the above, the product JN is called the *luminosity L*, i.e.

$$L \equiv JN \tag{1.45}$$

and contains all the dependencies on the densities and geometries of the beam and target. The cross-section is independent of these factors.

It can be seen from the above equations that the cross-section has the dimensions of an area; the rate per target particle $J\sigma_r$ at which the reaction occurs is equal to the rate at which beam particles would hit a surface of area σ_r, placed in the beam at rest with respect to the target and perpendicular to the beam direction. Since the area of such a surface is unchanged by a Lorentz transformation in the beam direction, the cross-section is the same in all inertial frames of reference, i.e. it is a Lorentz invariant.

The quantity σ_r is better named the *partial cross-section*, because it is the cross-section for a particular reaction r. The *total cross-section* σ is defined by

$$\sigma \equiv \sum_r \sigma_r. \tag{1.46}$$

Another useful quantity is the *differential cross-section*, $d\sigma_r(\theta, \phi)/d\Omega$, which is defined by

$$dW_r \equiv JN \frac{d\sigma_r(\theta, \phi)}{d\Omega} d\Omega, \tag{1.47}$$

where dW_r is the measured rate for the particles to be emitted into an element of solid angle $d\Omega = d\cos\theta\, d\phi$ in the direction (θ, ϕ), as shown in Figure 1.7.

The total cross-section is obtained by integrating the partial cross-section over all angles, i.e.

$$\sigma_r = \int_0^{2\pi} d\phi \int_{-1}^{1} d\cos\theta \frac{d\sigma_r(\theta, \phi)}{d\Omega}. \tag{1.48}$$

The final step is to write these formulae in terms of the scattering amplitude $f(\mathbf{q}^2)$ appropriate for describing the scattering of a non-relativistic spinless particle from a potential.

To do this it is convenient to consider a single beam particle interacting with a single target particle and to confine the whole system in an arbitrary volume V (which cancels in the final result). The incident flux is then given by

$$J = n_b v_i = v_i/V \tag{1.49}$$

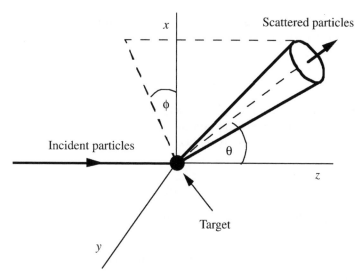

Figure 1.7 Geometry of the differential cross-section: a beam of particles is incident along the z-axis and collides with a stationary target at the origin; the differential cross-section is proportional to the rate for particles to be scattered into a small solid angle $d\Omega$ in the direction (θ, ϕ)

and since the number of target particles is $N = 1$, the differential rate is

$$dW_r = \frac{v_i}{V}\frac{d\sigma_r(\theta, \phi)}{d\Omega}d\Omega. \tag{1.50}$$

In quantum mechanics, provided the interaction is not too strong, the transition rate for any process is given in perturbation theory by the Born approximation[37]

$$dW_r = \frac{2\pi}{\hbar}\left|\int d^3x \psi_f^* V(\mathbf{x})\psi_i\right|^2 \rho(E_f). \tag{1.51}$$

The term $\rho(E_f)$ is the *density-of-states factor* (see below) and we take the initial and final state wavefunctions to be plane waves:

$$\psi_i = \frac{1}{\sqrt{V}}\exp[i\mathbf{q}_i \cdot \mathbf{x}/\hbar], \quad \psi_f = \frac{1}{\sqrt{V}}\exp[i\mathbf{q}_f \cdot \mathbf{x}/\hbar], \tag{1.52}$$

---

[37]This equation is a form of the *Second Golden Rule*. It is discussed in Appendix A.

where the final momentum $\mathbf{q}_f$ lies within a small solid angle $d\Omega$ located in the direction (θ, ϕ) (see Figure 1.7.). Then, by direct integration,

$$dW_r = \frac{2\pi}{\hbar V^2} |f(\mathbf{q}^2)|^2 \rho(E_f), \tag{1.53}$$

where $f(\mathbf{q}^2)$ is the scattering amplitude defined in Equation (1.38).

The density of states $\rho(E_f)$ that appears in Equation (1.51) is the number of possible final states with energy lying between E_f and $E_f + dE_f$ and is given by[38]

$$\rho(E_f) = \frac{V}{(2\pi\hbar)^3} \frac{q_f^2}{v_f} d\Omega. \tag{1.54}$$

If we use this and Equation (1.53) in Equation (1.50), we have

$$\frac{d\sigma}{d\Omega} = \frac{1}{4\pi^2\hbar^4} \frac{q_f^2}{v_i v_f} |f(\mathbf{q}^2)|^2. \tag{1.55}$$

Although this result has been derived in the laboratory system, because we have taken a massive target it is also valid in the centre-of-mass system. For a finite mass target it would be necessary to make a Lorentz transformation on Equation (1.55). The expression is also true for the general two-body relativistic scattering process $a + b \rightarrow c + d$.

All the above is for spinless particles, so finally we have to generalize Equation (1.55) to include the effects of spin. Suppose the initial-state particles a and b, have spins s_a and s_b and the final-state particles c and d have spins s_c and s_d. The total numbers of spin substates available to the initial and final states are g_i and g_f, respectively, given by

$$g_i = (2s_a + 1)(2s_b + 1) \quad \text{and} \quad g_f = (2s_c + 1)(2s_d + 1). \tag{1.56}$$

If the initial particles are unpolarized (which is the most common case in practice), then we must average over all possible initial spin configurations (because each is equally likely) and sum over the final configurations. Thus, Equation (1.55) becomes

$$\frac{d\sigma}{d\Omega} = \frac{g_f}{4\pi^2\hbar^4} \frac{q_f^2}{v_i v_f} |\mathcal{M}_{fi}|^2, \tag{1.57}$$

where

$$|\mathcal{M}_{fi}|^2 \equiv \overline{|f(\mathbf{q}^2)|^2} \tag{1.58}$$

and the bar over the amplitude denotes a spin-average of the squared matrix element.

---

[38]The derivation is given in detail in Appendix A.

1.6.3   Unstable states

In the case of an unstable state, the observable of interest is its *lifetime at rest* τ, or equivalently its *natural decay width*, given by $\Gamma = \hbar/\tau$, which is a measure of the rate of the decay reaction. In general, an initial unstable state will decay to several final states and in this case we define Γ_f as the *partial width* for channel f and

$$\Gamma = \sum_f \Gamma_f \tag{1.59}$$

as the *total decay width*, while

$$B_f \equiv \Gamma_f/\Gamma \tag{1.60}$$

is defined as the *branching ratio* for decay to channel f.

The energy distribution of an unstable state to a final state f has the *Breit–Wigner* form

$$N_f(W) \propto \frac{\Gamma_f}{(W - M)^2 c^4 + \Gamma^2/4}, \tag{1.61}$$

where M is the mass of the decaying state and W is the invariant mass of the decay products.[39] The Breit–Wigner formula is shown in Figure 1.8 and is the same formula that describes the widths of atomic and nuclear spectral lines. (The overall factor depends on the spins of the particles involved.) It is a symmetrical bell-shaped curve with a maximum at $W = M$ and a full width Γ at half the maximum height of the curve. It is proportional to the number of events with invariant mass W.

If an unstable state is produced in a scattering reaction, then the cross section for that reaction will show an enhancement described by the same Breit–Wigner formula. In this case we say we have produced a *resonance state*. In the vicinity of a resonance of mass M, and width Γ, the cross-section for the reaction $i \rightarrow f$ has the form

$$\sigma_{fi} \propto \frac{\Gamma_i \Gamma_f}{(E - Mc^2)^2 + \Gamma^2/4}, \tag{1.62}$$

where E is the total energy of the system. Again, the form of the overall constant will depend on the spins of the particles involved. Thus, for example, if the resonance particle has spin j and the spins of the initial particles are s_1 and s_2, then

$$\sigma_{fi} = \frac{\pi\hbar^2}{q_i^2} \frac{2j + 1}{(2s_1 + 1)(2s_2 + 1)} \frac{\Gamma_i \Gamma_f}{(E - Mc^2)^2 + \Gamma^2/4}. \tag{1.63}$$

---

[39]This form arises from a state that decays exponentially with time, although a proper proof of this is quite lengthy. See, for example, Appendix B of Ma97.

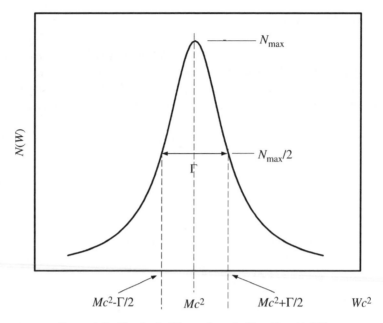

Figure 1.8   The Breit--Wigner formula (Equation (1.61))

In practice there will also be kinematical and angular momentum effects that will distort this formula from its perfectly symmetric shape.

An example of resonance formation in $\pi^- p$ interactions is given in Figure 1.9, which shows the $\pi^- p$ total cross-section in the centre-of-mass energy range

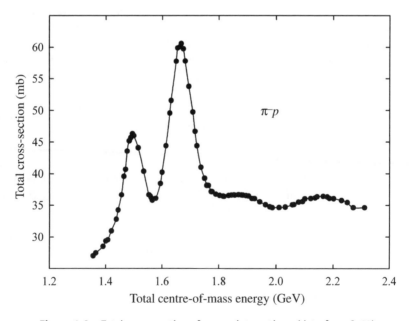

Figure 1.9   Total cross-sections for $\pi^- p$ interactions (data from Ca68)

1.2–2.4 GeV. (The units used in the plots will become clear after the next section.) Two enhancements can be seen that are of the approximate Breit–Wigner resonance form and there are two other maxima at higher energies. In principle, the mass and width of a resonance may be obtained by using a Breit–Wigner formula and varying M and Γ to fit the cross-section in the region of the enhancement. In practice, more sophisticated methods are used that fit a wide range of data, including differential cross-sections, simultaneously and also take account of non-resonant contributions to the scattering. The widths obtained from such analyses are of the order of 100 MeV, with corresponding interaction times of order 10^{-23} s, which is consistent with the time taken for a relativistic pion to transit the dimension of a proton. Resonances are also a feature of interactions in nuclear physics and we will return to this in Section 2.9 when we discuss nuclear reaction mechanisms.

1.7 Units: Length, Mass and Energy

Most branches of science introduce special units that are convenient for their own purposes. Nuclear and particle physics are no exceptions. Distances tend to be measured in femtometres (fm) or, equivalently *fermis*, with 1 fm $\equiv 10^{-15}$ m. In these units, the radius of the proton is about 0.8 fm. The range of the strong nuclear force between protons and neutrons is of the order of 1–2 fm, while the range of the weak force is of the order of 10^{-3} fm. For comparison, the radii of atoms are of the order of 10^5 fm. A common unit for area is the *barn* (b) defined by 1 b $= 10^{-28}$ m^2. For example, the total cross-section for pp scattering (a strong interaction) is a few tens of millibarns (mb) (compare also the $\pi^- p$ total cross-section in Figure 1.9), whereas the same quantity for νp scattering (a weak interaction) is a few tens of femtobarns (fb), depending on the energies involved. Nuclear cross-sections are very much larger and increase approximately like $A^{2/3}$, where A is the total number of nucleons in the nucleus.

Energies are invariably specified in terms of the electron volt (eV) defined as the energy required to raise the electric potential of an electron or proton by 1 V. In SI units, $1\,\text{eV} = 1.6 \times 10^{-19}$ J. The units $1\,\text{keV} = 10^3\,\text{eV}$, $1\,\text{MeV} = 10^6\,\text{eV}$, $1\,\text{GeV} = 10^9\,\text{eV}$ and $1\,\text{TeV} = 10^{12}\,\text{eV}$ are also in general use. In terms of these units, atomic ionization energies are typically a few eV, the energies needed to bind nucleons in heavy nuclei are typically 7–8 MeV per particle, and the highest particle energies produced in the laboratory are of the order of 1 TeV for protons. Momenta are specified in eV/c, MeV/c, etc..

In order to create a new particle of mass M, an energy at least as great as its rest energy Mc^2 must be supplied. The rest energies of the electron and proton are 0.51 MeV and 0.94 GeV respectively, whereas the W and Z^0 bosons have rest energies of 80 GeV and 91 GeV, respectively. Correspondingly, their masses are conveniently measured in MeV/c^2 or GeV/c^2, so that, for example,

$$M_e = 0.51\,\text{MeV}/c^2, \quad M_p = 0.94\,\text{GeV}/c^2,$$
$$M_W = 80.3\,\text{GeV}/c^2, \quad M_Z = 91.2\,\text{GeV}/c^2 \tag{1.64}$$

In SI units, $1\,\text{MeV}/c^2 = 1.78 \times 10^{-30}\,\text{kg}$. In nuclear physics it is also common to express masses in *atomic mass units* (*u*), where $1\,\text{u} = 1.661 \times 10^{-27}\,\text{kg} = 931.5\,\text{MeV}/c^2$.

Although practical calculations are expressed in the above units, it is usual in particle physics to make theoretical calculations in units chosen such that $\hbar \equiv h/2\pi = 1$ and $c = 1$ (called *natural units*) and many books do this. However, as this book is about both nuclear and particle physics, only practical units will be used.

A table giving numerical values of fundamental and derived constants, together with some useful conversion factors is given on page XV.

Problems

1.1 'Derive' the Klein–Gordon equation using the information in Footnote 21 and verify that Equation (1.35) is a static solution of the equation.

1.2 Verify that the spherical harmonic $Y_1^1 = \sqrt{\frac{3}{8}}\sin\theta e^{i\phi}$ is an eigenfunction of parity with eigenvalue $P = -1$.

1.3 A proton and antiproton at rest in an S-state annihilate to produce $\pi^0\pi^0$ pairs. Show that this reaction cannot be a strong interaction.

1.4 Suppose that an intrinsic *C*-parity factor is introduced into Equation (1.15b), which then becomes $\hat{C}|b, \psi_b\rangle = C_b|\bar{b}, \psi_{\bar{b}}\rangle$. Show that the eigenvalue corresponding to any eigenstate of $\hat{C}$ is independent of C_b, so that C_b cannot be measured.

1.5 Consider the reaction $\pi^- d \to nn$, where *d* is a spin-1 S-wave bound state of a proton and a neutron called the deuteron and the initial pion is at rest. Deduce the intrinsic parity of the negative pion.

1.6 Write down equations in symbol form that describe the following interactions:

(a) elastic scattering of an electron antineutrino and a positron;

(b) inelastic production of a pair of neutral pions in proton–proton interactions;

(c) the annihilation of an antiproton with a neutron to produce three pions.

1.7 Draw a lowest-order Feynman diagram for the following processes:

(a) $\nu_e\nu_\mu$ elastic scattering;

(b) $n \to p + e^- + \bar{\nu}_e$;

(c) $e^+e^- \rightarrow e^+e^-$;

(d) a fourth-order diagram for the reaction $\gamma + \gamma \rightarrow e^+ + e^-$.

1.8 Calculate the energy–momentum transfer between two particles equivalent to a distance of approach of (a) 1fm and (b) 10^{-3} fm. Assuming that the intrinsic strengths of the fundamental weak and electromagnetic interactions are approximately equal, compare the relative sizes of the invariant (scattering) amplitudes for weak and electromagnetic processes at these two energy–momentum transfers.

1.9 Verify by explicit integration that $f(\mathbf{q}) = -g^2\hbar^2[\mathbf{q}^2 + m^2c^2]^{-1}$ is the amplitude corresponding to the Yukawa potential

$$V(r) = -\frac{g^2}{4\pi}\frac{e^{-r/R}}{r},$$

where $R = \hbar/mc$ is the range and $r = |\mathbf{x}|$.

1.10 Two beams of particles consisting of n bunches with $N_i(i = 1, 2)$ particles in each, traverse circular paths and collide 'head-on'. Show that in this case the general expression Equation (1.45) for the luminosity reduces to $L = nN_1N_2f/A$, where A is the cross-sectional area of the beam and f is the frequency, i.e. $f = 1/T$, where T is the time taken for the particles to make one traversal of the ring.

1.11 A thin (density $1\,\text{mg cm}^{-2}$) target of ^{24}Mg ($M_A = 24.3$ atomic mass units) is bombarded with a 10 nA beam of alpha particles. A detector subtending a solid angle of 2×10^{-3} sr, records 20 protons/s. If the scattering is isotropic, determine the cross-section for the $^{24}\text{Mg}(\alpha, p)$ reaction.

2

Nuclear Phenomenology

In this chapter we start to examine some of the things that can be learned from experiments, beginning with basic facts about nuclei, including what can be deduced about their shapes and sizes. Then we discuss the important topic of nuclear stability and the phenomenology of the various ways that unstable nuclei decay to stable states. Finally, we briefly review the classification of reactions in nuclear physics. Before that we need to introduce some notation.

2.1 Mass Spectroscopy and Binding Energies

Nuclei are specified by:

Z – atomic number = the number of protons,

N – neutron number = the number of neutrons,

A – mass number = the number of nucleons, so that $A = Z + N$.

We will also refer to A as the *nucleon number*. The charge on the nucleus is $+Ze$, where e is the absolute value of the electric charge on the electron. Nuclei with combinations of these three numbers are also called *nuclides* and are written ^{A}Y or $^{A}_{Z}Y$, where Y is the chemical symbol for the element. Some other common nomenclature is:

nuclides with the same mass number are called *isobars*,

nuclides with the same atomic number are called *isotopes*,

nuclides with the same neutron number are called *isotones*.

Nuclear and Particle Physics   B. R. Martin
© 2006 John Wiley & Sons, Ltd

The concept of isotopes was introduced in Chapter 1. For example, stable isotopes of carbon are ^{12}C and ^{13}C, and the unstable isotope used in dating ancient objects (see later in this chapter) is ^{14}C; all three have $Z = 6$.

Just as in the case of electrons in atoms, the forces that bind the nucleons in nuclei contribute to the total mass of an atom $M(Z, A)$ and in terms of the masses of the proton M_p and neutron M_n

$$M(Z,A) < Z\,(M_p + m_e) + N\,M_n. \tag{2.1}$$

The *mass deficit* is defined as

$$\Delta M(Z,A) \equiv M(Z,A) - Z\,(M_p + m_e) - N\,M_n \tag{2.2}$$

and $-\Delta Mc^2$ is called the *binding energy B*. Binding energies may be calculated if masses are measured accurately. One way of doing this is by using the techniques of *mass spectroscopy*. The principle of the method is shown in Figure 2.1.

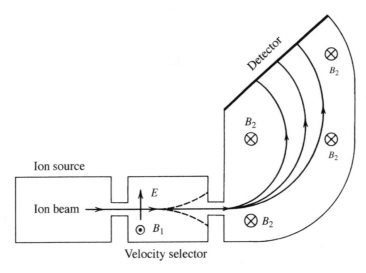

Figure 2.1   Schematic diagram of a mass spectrometer (adapted from Kr88 Copyright John Wiley & Sons, Inc.)

A source of ions of charge q, containing various isotopes passes through a region where there are uniform electric (E) and magnetic (B_1) fields at right angles. The electric field will exert a force qE in one direction and the magnetic field will exert a force qvB_1 in the opposite direction, where v is the speed of the ions. By balancing these forces, ions of a specific speed $v = E/B_1$ can be selected and allowed to pass through a collimating slit. Ions with other velocities (shown as dashed lines) are deflected. The beam is then allowed to continue through a second uniform magnetic field B_2 where it will be bent into a circular path of radius ρ, given by

$$mv = qB_2\rho \tag{2.3}$$

and since q, B_2 and v are fixed, particles with a fixed ratio q/m will bend in a path with a unique radius. Hence isotopes may be separated and focused onto a detector (e.g. a photographic plate). In the common case where $B_1 = B_2 = B$,

$$\frac{q}{m} = \frac{E}{B^2\rho}. \tag{2.4}$$

In practice, to achieve high accuracy, the device is used to measure mass differences rather than absolute values of mass.[1]

Conventional mass spectroscopy cannot be used to find the masses of very short-lived nuclei and in these cases the masses are determined from kinematic analysis of nuclear reactions as follows. Consider the inelastic reaction $A(a, a)A^*$, where A^* is the short-lived nucleus whose mass is to be determined. The kinematics of this are:

$$a(E_i, \mathbf{p}_i) + A(m_A c^2, \mathbf{0}) \rightarrow a(E_f, \mathbf{p}_f) + A^*(\tilde{E}, \tilde{\mathbf{p}}), \tag{2.5}$$

where we use tilded quantities to denote the energy, mass, etc. of A^*. Equating the total energy before the collision

$$E_{tot}(\text{initial}) = E_i + m_a c^2 + m_A c^2 \tag{2.6a}$$

to the total energy after the collision

$$E_{tot}(\text{final}) = E_f + \tilde{E} + m_a c^2 + \tilde{m} c^2 \tag{2.6b}$$

gives the following expression for the change in energy of the nucleus:

$$\Delta E \equiv (\tilde{m} - m_A)c^2 = E_i - E_f - \tilde{E} = \frac{p_i^2}{2m_a} - \frac{p_f^2}{2m_a} - \frac{\tilde{p}^2}{2\tilde{m}}, \tag{2.7}$$

where we have assumed non-relativistic kinematics. If the initial momentum of the projectile is along the x-direction and the scattering angle is θ, then from momentum conservation,

$$(\tilde{p})_x = p_i - p_f \cos\theta, \qquad (\tilde{p})_y = p_f \sin\theta \tag{2.8}$$

---

[1]Practical details of mass spectroscopy may be found in, for example, Chapter 3 of Kr88.

and using these in Equation (2.7) gives

$$\Delta E = E_i\left(1 - \frac{m_a}{\tilde{m}}\right) - E_f\left(1 + \frac{m_a}{\tilde{m}}\right) + \frac{2m_a}{\tilde{m}}\left(E_iE_f\right)^{1/2}\cos\theta. \qquad (2.9)$$

This formula can be used iteratively to deduce ΔE and hence the mass of the excited nucleus A^*, from measurements of the initial and final energy of the projectile by initially setting $\tilde{m} = m_A$ on the right-hand side because ΔE is small in comparison with m_A. One final point is that the energies in Equation (2.9) are measured in the laboratory system, whereas the final energies (masses) will be needed in the centre-of-mass system.[2] The necessary transformation is easily found to be

$$E_{CM} = E_{lab}(1 + m_a/m_A)^{-1}. \qquad (2.10)$$

A similar formula to Equation (2.9) may be derived for the general reaction $A(a,b)B$:

$$\Delta E = E_i\left(1 - \frac{m_a}{m_B}\right) - E_f\left(1 + \frac{m_b}{m_B}\right) + \frac{2}{m_B}\left(m_a m_b E_i E_f\right)^{1/2}\cos\theta + Q, \qquad (2.11)$$

where Q is the kinetic energy released in the reaction.

A commonly used quantity of interest is the *binding energy per nucleon B/A*. This is shown schematically in Figure 2.2 for nuclei that are stable or long-lived.

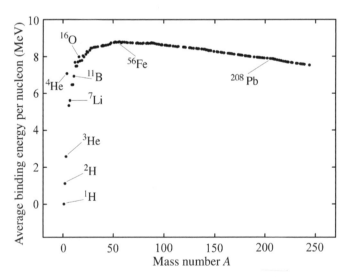

Figure 2.2   Binding energy per nucleon as a function of mass number A for stable and long-lived nuclei

---

[2]A discussion of these two systems is given in Appendix C.

This shows that B/A peaks at a value of 8.7 MeV for a mass number of about 56 (close to iron) and thereafter falls very slowly. Excluding very light nuclei, the binding energy per nucleon is between 7 and 9 MeV over a wide range of the periodic table. In Section 2.5 we will discuss a model that provides an explanation for the shape of this curve.

2.2   Nuclear Shapes and Sizes

The shape and size of a nucleus may be found from scattering experiments; i.e. a projectile is scattered from the nucleus and the angular distribution of the scattered particles examined, as was done by Rutherford and his collaborators when they deduced the existence of the nucleus. The interpretation is simplest in those cases where the projectile itself has no internal structure, i.e. it is an elementary particle, and electrons are often used. In this case the relevant force is electromagnetic and we learn about the *charge distribution* in the nucleus. The first experiments of this type were performed by Hofstader and his collaborators in the late 1950s.[3] If instead of an electron a hadron is used as the projectile, the force is dominantly the nuclear strong interaction and we find information about the *matter density*. Neutrons are commonly used so that Coulomb effects are absent. We discuss these two cases in turn.

2.2.1   Charge distribution

To find the amplitude for electron–nucleus scattering, we should in principle solve the Schrödinger (or Dirac) equation using a Hamiltonian that includes the full electromagnetic interaction and use nuclear wavefunctions. This can only be done numerically. However, in Appendix C we derive a simple formula that describes the electromagnetic scattering of a charged particle in the so-called *Born approximation*, which assumes $Z\alpha \ll 1$ and uses plane waves for the initial and final states. This leads to the *Rutherford cross-section*, which in its relativistic form may be written

$$\left(\frac{d\sigma}{d\Omega}\right)_{\text{Rutherford}} = \frac{Z^2\alpha^2(\hbar c)^2}{4E^2\sin^4(\theta/2)}, \tag{2.12}$$

where E is the total initial energy of the projectile and θ is the angle through which it is scattered. Note that Equation (2.12) is of order α^2 because it corresponds to the exchange of a single photon. Although Equation (2.12) has a limited range of applicability, it is useful to discuss the general features of electron scattering.

Equation (2.12) actually describes the scattering of a spin-0 point-like projectile of unit charge from a fixed point-like target with electric charge Ze, i.e. the charge

---

[3]Robert Hofstader shared the 1961 Nobel Prize in Physics for his pioneering electron scattering experiments.

distribution of the target is neglected. It therefore needs to be modified in a number of ways before it can be used in practice. We will state the modifications without proof.

Firstly, taking account of the electron spin leads to the so-called *Mott cross-section*

$$\left(\frac{d\sigma}{d\Omega}\right)_{\text{Mott}} = \left(\frac{d\sigma}{d\Omega}\right)_{\text{Rutherford}} [1 - \beta^2 \sin^2(\theta/2)], \qquad (2.13)$$

where $\beta = v/c$ and v is the velocity of the initial electron. At higher energies, the recoil of the target needs to be taken into account and this introduces a factor E'/E on the right-hand side of Equation (2.13), where E' is the final energy of the electron. At higher energies we also need to take account of the interaction with the magnetic moment of the target in addition to its charge. The final form for the differential cross-section is

$$\left(\frac{d\sigma}{d\Omega}\right)_{\text{spin} \frac{1}{2}} = \left(\frac{d\sigma}{d\Omega}\right)_{\text{Mott}} \frac{E'}{E} \left[1 + 2\tau \tan^2 \frac{\theta}{2}\right], \qquad (2.14)$$

where

$$\tau = \frac{-q^2}{4M^2 c^2} \qquad (2.15)$$

and M is the target mass. Because the energy loss of the electron to the recoiling nucleus is no longer negligible, $\mathbf{q}$, the previous momentum transfer, has been replaced by the four-momentum transfer q, whose square is

$$q^2 = (p - p')^2 = 2m_e^2 c^2 - 2(EE'/c^2 - |\mathbf{p}||\mathbf{p}'| \cos\theta) \approx -\frac{4EE'}{c^2} \sin^2(\theta/2), \quad (2.16)$$

where $p(p')$ is the four-momentum of the initial (final) electron. (Because $q^2 \leq 0$, it is common practice to replace it with $Q^2 = -q^2$, so as to work with positive quantities.[4]) For the rest of this discussion it will be sufficient to ignore the magnetic interaction, although we will use a variant of the full form (2.16) in Chapter 6.

The final modification is due to the spatial extension of the nucleus. If the spatial charge distribution within the nucleus is written $f(\mathbf{x})$ then we define the *form factor* $F(\mathbf{q}^2)$ by

$$F(\mathbf{q}^2) \equiv \frac{1}{Ze} \int e^{i\mathbf{q}\cdot\mathbf{x}/\hbar} f(\mathbf{x}) \, d^3\mathbf{x} \quad \text{with} \quad Ze = \int f(\mathbf{x}) \, d^3\mathbf{x}, \qquad (2.17)$$

---

[4]To remove any confusion, in the non-relativistic case, which we use in the rest of this chapter, q is interpreted to be $q = |\mathbf{q}| \geq 0$ where $\mathbf{q} \equiv \mathbf{p} - \mathbf{p}'$, as was used in Section 1.6.1. We will need the four-momentum definition of q in Chapter 6.

i.e. the Fourier transform of the charge distribution.[5] In the case of a spherically symmetric charge distribution, the angular integrations in Equations (2.17) may be done using spherical polar coordinates to give

$$F(\mathbf{q}^2) = \frac{4\pi\hbar}{Zeq} \int_0^\infty r\rho(r)\sin\left(\frac{qr}{\hbar}\right)dr, \qquad (2.18)$$

where $q = |\mathbf{q}|$ and $\rho(r)$ is the radial charge distribution. The final form of the experimental cross-section in this approximation is given by[6]

$$\left(\frac{d\sigma}{d\Omega}\right)_{expt} = \left(\frac{d\sigma}{d\Omega}\right)_{Mott} |F(\mathbf{q}^2)|^2. \qquad (2.19)$$

Two examples of measured cross-sections are shown in Figure 2.3. Striking features are the presence of a number of well-defined minima superimposed on a

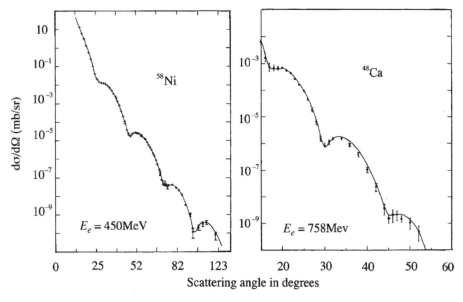

Figure 2.3 Elastic differential cross-sections as a function of the scattering angle for 450 MeV electrons from ^{58}Ni and 758 MeV electrons from ^{48}Ca; the solid lines are fits as described in the text (adapted from Si75 (^{58}Ni data) and Be67 (^{48}Ca data), Copyright American Physical Society)

---

[5]Strictly this formula assumes that the recoil of the target nucleus is negligible and the interaction is relatively weak, so that perturbation theory may be used.
[6]If the magnetic interaction were included, another form factor would be necessary, as is the case in high-energy electron scattering discussed in Chapter 6.

rapid decrease in the cross-section with angle. These features are common to all elastic data, although not all nuclei show so many minima as those shown.

The minima are due to the form factor and we can make this plausible by taking the simple case where the nuclear charge distribution is represented by a hard sphere such that

$$\rho(r) = \text{constant}, \quad r \leq a$$
$$= 0 \qquad\qquad r > a \tag{2.20}$$

where a is a constant. In this case, evaluation of Equation (2.18) gives

$$F(\mathbf{q}^2) = 3[\sin(b) - b\cos(b)]b^{-3}, \tag{2.21}$$

where $b \equiv qa/\hbar$. Thus $F(\mathbf{q}^2)$ will be zero at values of b for which $b = \tan(b)$. In practice, as we will see below, $\rho(r)$ is not a hard sphere, and although it is approximately constant for much of the nuclear volume, it falls smoothly to zero at the surface. Smoothing the edges of the radial charge distribution (2.20) modifies the positions of the zeros, but does not alter the argument that the minima in the cross-sections are due to the spatial distribution of the nucleus. Their actual positions and depths result from a combination of the form factor and the form of the point-like amplitude. We shall see below that the minima can tell us about the size of the nucleus.

If one measures the cross-section for a fixed energy at various angles (and hence various q^2), the form factor can in principle be extracted using Equation (2.19) and one might attempt to find the charge distribution from the inverse Fourier transform

$$f(\mathbf{x}) = \frac{Ze}{(2\pi)^3} \int F(\mathbf{q}^2) e^{-i\mathbf{q}\cdot\mathbf{x}/\hbar} d^3\mathbf{q}. \tag{2.22}$$

However, $\mathbf{q}^2$ only has a finite range for a fixed initial electron energy and even within this range the rapid fall in the cross-section means that in practice measurements cannot be made over a sufficiently wide range of angles for the integral in Equation (2.22) to be evaluated accurately. Thus, even within the approximations used, reliable charge distributions cannot be found from Equation (2.22). Therefore different strategies must be used to deduce the charge distribution. In one approach, plausible – but very general – parameterized forms (for example a sum of Gaussians) are chosen for the charge distribution and are used to modify the point-like electromagnetic interaction. The resulting Schrödinger (or Dirac) equation is solved numerically to produce an amplitude, and hence a cross-section, for electron–nucleus scattering. The parameters of the charge distribution are then varied to give a good fit of the experimental data. The solid curves in Figure 2.3 are obtained in this way.

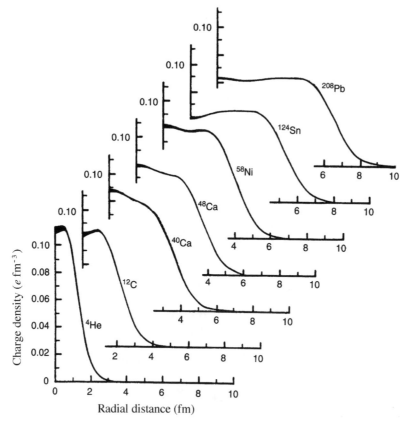

Figure 2.4 Radial charge distributions ρ_{ch} of various nuclei, in units of $e\ \text{fm}^{-3}$; the thickness of the curves near $r = 0$ is a measure of the uncertaintity in ρ_{ch} (adapted from Fr83)

Some radial charge distributions for various nuclei obtained by these methods are shown in Figure 2.4. They are well represented by the form

$$\rho_{ch}(r) = \frac{\rho_{ch}^0}{1 + e^{(r-a)/b}}, \tag{2.23}$$

where a and b for medium and heavy nuclei are found to be

$$a \approx 1.07 A^{1/3}\ \text{fm} \quad \text{and} \quad b \approx 0.54\ \text{fm.} \tag{2.24}$$

From this we can deduce that the charge density is approximately constant in the nuclear interior and falls fairly rapidly to zero at the nuclear surface, as anticipated above. The value of ρ_{ch}^0 is in the range 0.06–0.08 for medium to heavy nuclei and decreases slowly with increasing mass number.

A useful quantity is the *mean square charge radius*,

$$\langle r^2 \rangle \equiv \int_0^\infty r^2 \rho_{ch}(r) \, dr. \tag{2.25}$$

This can be found from the form factor as follows. Expanding Equation (2.17) for $F(\mathbf{q}^2)$ gives

$$F(\mathbf{q}^2) = \frac{1}{Ze} \int f(\mathbf{x}) \sum_{n=0}^\infty \frac{1}{n!} \left(\frac{i|\mathbf{q}|r\cos\theta}{\hbar} \right)^n d^3\mathbf{x} \tag{2.26}$$

and after carrying out the angular integrations this becomes

$$F(\mathbf{q}^2) = \frac{4\pi}{Ze} \int_0^\infty f(r) \, r^2 dr - \frac{4\pi \mathbf{q}^2}{6Ze\hbar^2} \int_0^\infty f(r) \, r^4 dr + \cdots . \tag{2.27}$$

From the normalization of $f(\mathbf{x})$, we finally have

$$F(\mathbf{q}^2) = 1 - \frac{\mathbf{q}^2}{6\hbar^2} \langle r^2 \rangle + \cdots \tag{2.28}$$

and thus the mean square charge radius can be found from

$$\langle r^2 \rangle = -6\hbar^2 \frac{dF(\mathbf{q}^2)}{d\mathbf{q}^2} \bigg|_{\mathbf{q}^2=0}, \tag{2.29}$$

provided the form factor can be measured at very small values of $\mathbf{q}^2$. For medium and heavy nuclei $\langle r^2 \rangle^{1/2}$ is given approximately by[7]

$$\langle r^2 \rangle^{1/2} = 0.94 A^{1/3} \, \text{fm.} \tag{2.30}$$

The nucleus is often approximated by a homogeneous charged sphere. The radius R of this sphere is then quoted as the nuclear radius. The relation of this to the mean square radius is $R^2 = \frac{5}{3} \langle r^2 \rangle$, so that

$$R_{charge} = 1.21 \, A^{1/3} \, \text{fm.} \tag{2.31}$$

2.2.2 Matter distribution

Electrons cannot be used to obtain the distributions of neutrons in the nucleus. We could, however, take the presence of neutrons into account by multiplying $\rho_{ch}(r)$

---

[7]The constant comes from a fit to a range of data, e.g. the compilation for $55 \leq A \leq 209$ given in Ba77.

by A/Z. Then we find an almost identical nuclear density in the nuclear interior for all nuclei, i.e. the decrease in ρ_{ch}^0 with increasing A is compensated by the increase in A/Z with increasing A. The interior nuclear density is given by

$$\rho_{nucl} \approx 0.17 \text{ nucleons/fm}^3. \tag{2.32}$$

Likewise, the effective nuclear matter radius for medium and heavy nuclei is

$$R_{nuclear} \approx 1.2 A^{1/3} \text{ fm}. \tag{2.33}$$

These are important results that will be used extensively later in this chapter and elsewhere in this book.

To probe the nuclear (i.e. matter) density of nuclei experimentally, a strongly interacting particle, i.e. a hadron, has to be used as the projectile. At high energies, where elastic scattering is only a small part of the total interaction, the nucleus behaves more like an absorbing sphere. In this case, the incident particle of momentum p will have an associated quantum mechanical wave of wavelength $\lambda = h/p$ and will suffer diffraction-like effects, as in optics. To the extent that we are dealing at high energies purely with the nuclear strong interaction (i.e. neglecting the Coulomb interaction), the nucleus can be represented by a black disk of radius R and the differential cross-section will have a Fraunhofer-like diffraction form, i.e.

$$\frac{d\sigma}{d\Omega} \propto \left[\frac{J_1(qR)}{qR} \right]^2, \tag{2.34}$$

where $qR \approx pR\theta$ for small θ and J_1 is a first-order Bessel function. For large qR,

$$[J_1(qR)]^2 \approx \left(\frac{2}{\pi qR} \right) \sin^2 \left(qR - \frac{\pi}{4} \right), \tag{2.35}$$

which has zeros at intervals $\Delta\theta = \pi/pR$. The plausibility of this interpretation is borne out by experiment, an example of which is shown in Figure 2.5. The data show a succession of roughly equally spaced minima as suggested by Equation (2.35).

To go further requires solving the equations of motion, but this is far more problematical than in the electron case because the hadrons are more likely to be absorbed as they pass through the nucleus and the effective potential is far less well known. However, the analogy with optics can be pursued further in the so-called *optical model*. The essential idea in this model is that a hadron incident on a nucleus may be elastically scattered, or it may cause a variety of different reactions. As in the discussion above, if the incident particle is represented by a wave, then in classical language it may be scattered or it may be absorbed. In optics this is analogous to the refraction and absorption of a light wave by a medium of complex refractive index, and just as the imaginary part of the refractive index takes account of the absorption of the light wave, so in the

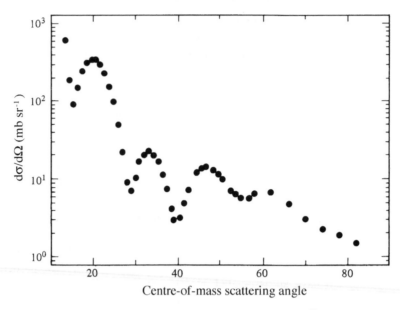

Figure 2.5   Elastic differential cross-sections for 52 MeV deuterons on ^{54}Fe (adapted from Hi68, copyright Elsevier, with permission)

nuclear case the imaginary part of a complex potential describing the interaction takes account of all the inelastic reactions. It is an essential feature of the model that the properties of nuclei are mainly determined by their size, as this implies that the same potential can account for the interaction of particles of different energies with different nuclei. Apart from the theoretical basis provided by analogy with classical optics, the model is essentially phenomenological, in that the values of the parameters of the optical potentials are found by optimizing the fit to the experimental data. This type of semi-phenomenological approach is common in both nuclear and particle physics.

In practice, the Schrödinger equation is solved using a parameterized complex potential where the real part is a sum of the Coulomb potential (for charged projectiles), an attractive nuclear potential and a spin-orbit potential, and the imaginary part is assumed to cause the incoming wave of the projectile to be attenuated within the nucleus, thereby allowing for inelastic effects. Originally, mathematical forms like Equation (2.23) were used to parameterize the real and imaginary parts of the potential, but subsequent work indicated substantial differences between the form factors of the real and imaginary parts of the potential and so different forms are now used for the imaginary part. The free parameters of the total potential are adjusted to fit the data.

The optical model has achieved its greatest success in the scattering of nucleons, but analyses using data obtained from light nuclei targets are also possible. A wide range of scattering data can be accounted for to a high degree of precision by

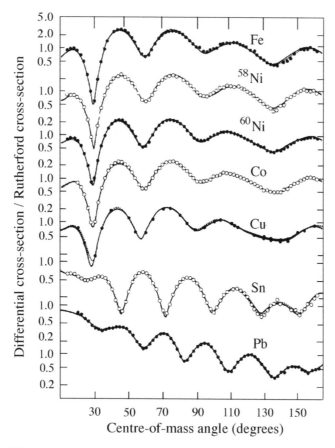

Figure 2.6 Differential cross-sections (normalized to the Rutherford cross-section) for the elastic scattering of 30.3 MeV protons, for a range of nuclei compared with optical model calculations; the solid and dashed lines represent the results using two different potentials (adapted from Sa67, copyright Elsevier, with permission)

the model and examples of this are shown in Figure 2.6. The corresponding wavefunctions are extensively used to extract information on nuclear structure. The conclusions are in accord with those above deduced indirectly from electron data.

2.3 Nuclear Instability

Stable nuclei only occur in a very narrow band in the $Z-N$ plane close to the line $Z = N$ (see Figure 2.7). All other nuclei are unstable and decay spontaneously in various ways. Isobars with a large surplus of neutrons gain energy by converting a

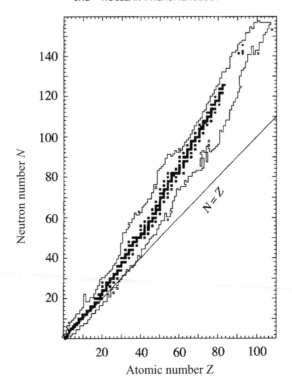

Figure 2.7   The distribution of stable nuclei: the squares are the stable and long-lived nuclei occurring in nature; other known nuclei lie within the jagged lines and are unstable. [adapted from Ch97.)

neutron into a proton; conversely, a nucleus with a large surplus of protons converts protons to neutrons. These are examples of β-decays, already mentioned. A related process is where an atomic electron is captured by the nucleus and a proton is thereby converted to a neutron within the nucleus. This is *electron capture* and like β-decay is a weak interaction. The electron is usually captured from the innermost shell and the process competes with β-decay in heavy nuclei because the radius of this shell (the K-shell) is close to the nuclei radius. The presence of a third particle in the decay process, the neutrino (as first suggested by Fermi), means that the emitted electrons (or positrons) have a continuous energy spectrum. The derivation and analysis of the electron momentum spectrum will be considered in Chapter 7 when we discuss the theory of β-decay.

The maximum of the curve of binding energy per nucleon is at approximately the position of iron (Fe) and nickel (Ni), which are therefore the most stable nuclides. In heavier nuclei, the binding energy is smaller because of the larger Coulomb repulsion. For still heavier nuclear masses, nuclei can decay spontaneously into two or more lighter nuclei, provided the mass of the parent nucleus is larger than the sum of the masses of the daughter nuclei.

Most such nuclei decay via two-body decays and the commonest case is when one of the daughter nuclei is a ^{4}He nucleus (i.e. an α-particle: ^{4}He $\equiv 2p2n$, with $A = 4$, $Z = N = 2$). The α-particle is favoured in such decays because it is a very stable, tightly bound structure. Because this is a two-body decay, the α-particle has a unique energy and the total energy released, the so-called *Q-value*, is given by:

$$Q_\alpha = (M_P - M_D - M_\alpha)c^2 = E_D + E_\alpha, \qquad (2.36)$$

where the subscripts refer to parent and daughter nuclei and the α-particle, and E is a kinetic energy.

The term *fission* is used to describe the rare cases where the two daughters have similar masses. If the decay occurs without external action, it is called *spontaneous fission* to distinguish it from *induced fission*, where some external stimulus is required to initiate the decay. Spontaneous fission only occurs with a probability greater than that for α-emission for nuclei with $Z \geq 110$. The reason for this is discussed in Section 2.7.

Finally, nuclei may decay by the emission of photons, with energies in the γ-ray part of the electromagnetic spectrum (*gamma emission*). This occurs when an excited nuclear state decays to a lower state and is a common way whereby excited states lose energy. The lower energy state is often the ground state. A competing process is *internal conversion*, where the nucleus de-excites by ejecting an electron from a low-lying atomic orbit. Both are electromagnetic processes. Electromagnetic decays will be discussed in more detail in Chapter 7.

2.4 Radioactive Decay

Before looking in more detail at different classes of instability, we will consider the general formalism describing the rate of radioactive decay. The probability per unit time that a given nucleus will decay is called its *decay constant* λ and is related to the *activity* $\mathscr{A}$ by

$$\mathscr{A} = -dN/dt = \lambda N, \qquad (2.37)$$

where $N(t)$ is the number of radioactive nuclei in the sample at time t. The activity is measured in becquerels (Bq), which is one decay per second.[8] The probability here refers to the total probability, because λ could be the sum of decay probabilities for a number of distinct final states in the same way that the total decay width of an unstable particle is the sum of its partial widths. Integrating Equation (2.37) gives

$$\mathscr{A}(t) = \lambda N_0 \exp(-\lambda t), \qquad (2.38)$$

where N_0 is the initial number of nuclei, i.e. the number at $t = 0$.

---

[8]An older unit, the curie (1 Ci $= 3.7 \times 10^{10}$ Bq) is also still in common use. A typical laboratory radioactive source has an activity of a few tens of kBq, i.e. μCi.

The *mean lifetime* τ of an unstable state, such as a radioactive nucleus or a hadron, follows from the general definition of a mean $\bar{x}$ of a distribution $f(x)$:

$$\bar{x} \equiv \left[\int xf(x)\,dx\right] \bigg/ \left[\int f(x)\,dx\right]. \tag{2.39}$$

Thus

$$\tau \equiv \frac{\int t\,dN(t)}{\int dN(t)} = \frac{\int\limits_0^\infty t\,\exp[-\lambda t]\,dt}{\int\limits_0^\infty \exp[-\lambda t]\,dt} = \frac{1}{\lambda}. \tag{2.40}$$

This is the quantity we simply called 'the lifetime' in Chapter 1. The mean lifetime is always used in particle physics, but another measure more commonly used in nuclear physics is the *half-life* $t_{\frac{1}{2}}$, defined as the time for the number of nuclei to fall by one half. Thus $t_{\frac{1}{2}} = \ln 2/\lambda = \tau \ln 2$. In this book, the term *lifetime* will be used for the mean lifetime, both for radioactive nuclei and unstable hadrons, unless explicitly stated otherwise.

A well-known use of the radioactive decay law is in dating ancient specimens using the known properties of radioactive nuclei. For organic specimens, carbon is usually used. Carbon-14 is a radioactive isotope of carbon that is produced by the action of cosmic rays on nitrogen in the atmosphere.[9] If the flux of cosmic rays remains roughly constant over time, then the ratio of ^{14}C to the stable most abundant isotope ^{12}C reaches an equilibrium value of about $1:10^{12}$. Both isotopes will be taken up by living organisms in this ratio, but when the organism dies there is no further interaction with the environment and the ratio slowly changes with time as the ^{14}C nuclei decay by β-decay to ^{14}N with a lifetime of 8.27×10^3 years. Thus, if the ratio of ^{14}C to ^{12}C is measured, the age of the specimen may be estimated.[10] The actual measurements can be made very accurately because modern mass spectrometers can directly measure very small differences in the concentrations of ^{14}C and ^{12}C using only milligrams of material. Nevertheless, in practice, corrections are made to agree with independent calibrations if possible, using, for example, tree-ring growth data, because cosmic ray activity is not strictly constant with time.

In many cases the products of radioactive decay are themselves radioactive and so a decay chain results. Consider a decay chain $A \rightarrow B \rightarrow C \rightarrow \cdots$, with decay constants λ_A, λ_B, λ_C etc.. The variation of species A with time is given by Equation (2.38), i.e.

$$N_A(t) = N_A(0)\exp(-\lambda_A t), \tag{2.41}$$

---

[9]Cosmic rays are high-energy particles, mainly protons, that impinge on the Earth's atmosphere from space. The products of the secondary reactions they produce may be detected at the Earth's surface. Victor Hess shared the 1936 Nobel Prize in Physics for the discovery of cosmic radiation.

[10]This method of using radioactive carbon to date ancient objects was devised by Willard Libby, for which he received the 1960 Nobel Prize in Chemistry.

but the differential equation for $N_B(t)$ will have an extra term in it to take account of the production of species B from the decay of species A:

$$dN_B(t)/dt = -\lambda_B N_B + \lambda_A N_A. \tag{2.42}$$

The solution of this equation may be verified by substitution to be

$$N_B(t) = \frac{\lambda_A}{\lambda_B - \lambda_A} N_A(0)[\exp(-\lambda_A t) - \exp(-\lambda_B t)]. \tag{2.43}$$

Similar equations may be found for decay sequences with more than two stages. Thus, for a three-stage sequence

$$N_C(t) = \lambda_A \lambda_B N_A(0) \left[\frac{\exp(-\lambda_A t)}{(\lambda_B - \lambda_A)(\lambda_C - \lambda_A)} \right.$$
$$\left. + \frac{\exp(-\lambda_B t)}{(\lambda_A - \lambda_B)(\lambda_C - \lambda_B)} + \frac{\exp(-\lambda_C t)}{(\lambda_A - \lambda_C)(\lambda_B - \lambda_C)} \right] \tag{2.44}$$

As an example, the variation of the components as a function of time is shown in Figure 2.8 for the specific case:

$$^{79}_{38}\text{Sr} \rightarrow {}^{79}_{37}\text{Rb} + e^+ + \nu_e \qquad (2.25\,\text{min})$$
$$\hookrightarrow {}^{79}_{36}\text{Kr} + e^+ + \nu_e \qquad (22.9\,\text{min}) \tag{2.45}$$
$$\hookrightarrow {}^{79}_{35}\text{Br} + e^+ + \nu_e \qquad (35.04\,\text{hours})$$

where the final nucleus is stable.

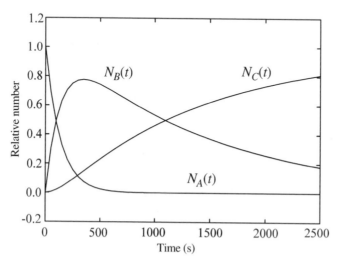

Figure 2.8 Time variation of the relative numbers of nuclei in the decay chain (2.45)

This illustrates the general features that whereas $N_A(t)$ for the initial species falls monotonically with time and $N_C(t)$ for the final stable species rises monotonically, $N_B(t)$ for an intermediate species rises to a maximum before falling. Note that at any time the sum of the components is a constant, as expected.

In the following sections we consider the phenomenology of the various types of radioactivity in more detail and in Chapter 7 we will return to discuss various models and theories that provide an understanding of these phenomena.

2.5   Semi-Empirical Mass Formula: The Liquid Drop Model

Apart from the lightest elements and a few special isolated very stable nuclei, the binding energy data of Figure 2.2 can be fitted by a simple formula containing just a few free parameters. This is the *semi-empirical mass formula* (SEMF), first written down in 1935 by Weizsäcker. It is a *semi*-empirical formula, because although it contains a number of constants that have to be found by fitting experimental data, the formula does have a theoretical basis. This arises from the two properties common to all nuclei (except those with very small A values) that we have seen earlier: (1) the interior mass densities are approximately equal, and (2) their total binding energies are approximately proportional to their masses. There is an analogy here with a classical model of a liquid drop, where for drops of various sizes: (1) interior densities are the same, and (2) latent heats of vaporization are proportional to their masses.[11] However, the analogy of a nucleus as an incompressible liquid droplet, with the nucleons playing the role of individual molecules within the droplet, cannot be taken too far because nucleons of course obey the laws of quantum, not classical, physics.

The semi-empirical mass formula will be taken to apply to *atomic* masses, as these are the masses actually observed in experiment. The atomic mass $M(Z, A)$ may then be written as the sum of six terms $f_i(Z,A)$:

$$M(Z, A) = \sum_{i=0}^{5} f_i(Z, A). \tag{2.46}$$

The first of these is the *mass of the constituent nucleons and electrons,*

$$f_0(Z, A) = Z(M_p + m_e) + (A - Z)M_n. \tag{2.47}$$

The remaining terms are various corrections, which we will write in the form a_i multiplied by a function of Z and A with $a_i > 0$.

The most important correction is the *volume* term,

$$f_1(Z, A) = -a_1 A. \tag{2.48}$$

---

[11]Latent heat is the average energy required to disperse the liquid drop into a gas and so is analogous to the binding energy per nucleon.

This arises from the fact that the strong nuclear force is short-range and each nucleon therefore feels the effect of only the nucleons immediately surrounding it (the force is said to be *saturated*), independent of the size of the nucleus. Recalling the important result deduced in Section 2.2 that the nuclear radius is proportional to $A^{\frac{1}{3}}$, this leads immediately to the binding energy being proportional to the volume, or nuclear mass. The coefficient is negative, i.e. it increases the binding energy, as expected.

The volume term overestimates the effect of the nuclear force because nucleons at the surface are not surrounded by other nucleons. Thus the volume term has to be corrected. This is done by the *surface* term

$$f_2(Z, A) = +a_2 A^{\frac{2}{3}}, \tag{2.49}$$

which is proportional to the surface area and decreases the binding energy. In the classical model of a real liquid drop, this term would correspond to the surface tension energy.

The *Coulomb* term accounts for the Coulomb energy of the charged nucleus, i.e. the fact that the protons repel each other. If we have a uniform charge distribution of radius proportional to $A^{\frac{1}{3}}$, then this term is

$$f_3(Z, A) = +a_3 \frac{Z(Z-1)}{A^{\frac{1}{3}}} \approx +a_3 \frac{Z^2}{A^{\frac{1}{3}}}, \tag{2.50}$$

where the approximation is sufficiently accurate for the large values of Z we will be considering. A similar effect would be present for a charged drop of a classical liquid.

The next term is the *asymmetry* term.

$$f_4(Z, A) = +a_4 \frac{(Z - A/2)^2}{A}. \tag{2.51}$$

This accounts for the observed tendency for nuclei to have $Z = N$. (There are no stable nuclei with very large neutron or proton excesses – c.f. Figure 2.7.) This term is purely quantum mechanical in origin and is due to the Pauli principle.

Part of the reason for the form (2.51) can be seen from the diagram of Figure 2.9, which shows the energy levels of a nucleus near the highest filled level in the approximation where all the energy levels are separated by the same energy Δ. Keeping A fixed and removing a proton from level 3 and adding a neutron to level 4, gives $(N - Z) = 2$ and leads to an energy increase of Δ. Repeating this for more protons, we find that the transfer of $(N - Z)/2$ nucleons decreases the binding energy by an amount $-\Delta(N - Z)^2/4$. Although we have assumed Δ is a constant, in practice it decreases like A^{-1}; hence the final form of the asymmetry term.

If we start with an even number of nucleons and progressively fill states, then the lowest energy will be when both Z and N are even. If, on the other hand, we have a

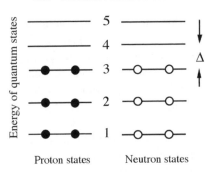

Figure 2.9 Schematic diagram of nuclear energy levels near the highest filled levels

system where both Z and N are odd and the highest filled proton state is above the highest filled neutron state, we can increase the binding energy by removing one proton from the nucleus and adding one neutron. If the highest filled proton state is below the highest filled neutron state, then we can produce the same effect by removing a neutron and adding a proton. These observations are summarized in the empirical *pairing* term, which maximizes the binding when both Z and N are even:

$$f_5(Z, A) = -f(A), \quad \text{if } Z \text{ even}, A - Z = N \text{ even}$$
$$f_5(Z, A) = 0, \qquad \text{if } Z \text{ even}, A - Z = N \text{ odd; or, } Z \text{ odd}, A - Z = N \text{ even} \quad (2.52)$$
$$f_5(Z, A) = +f(A), \quad \text{if } Z \text{ odd}, \ A - Z = N \text{ odd}$$

The exact form of the function $f(A)$ is found by fitting the data; $f(A) = a_5 A^{-\frac{1}{2}}$ is often used.

To help remember these terms, the notation VSCAP is frequently used, with

$$a_1 = a_v, \quad a_2 = a_s, \quad a_3 = a_c, \quad a_4 = a_a, \quad a_5 = a_p. \quad (2.53)$$

Precise values of the coefficients depend on the range of A fitted. One commonly used set is, in units of MeV/c^2:[12]

$$a_v = 15.56, \quad a_s = 17.23, \quad a_c = 0.697, \quad a_a = 93.14, \quad a_p = 12. \quad (2.54)$$

The fit to the binding energy data for $A > 20$ using these coefficients in the SEMF is shown in Figure 2.10. Overall the fit to the data is remarkably good for such a simple formula, but is not exact of course. For example, there are a small number of regions where the binding energy curves show enhancements that are not repro-duced. (These enhancements are due to the existence of a 'shell structure' of nucleons within the nucleus and will be discussed in Chapter 7.) Nevertheless, the SEMF gives accurate values for the binding energies for some 200 stable and many

---

[12]Note that some authors write the asymmetry term proportional to $(Z - N)^2$, which is equivalent to the form used here, but their value for the coefficient a_a will differ by a factor of four from the one in Equations (2.54).

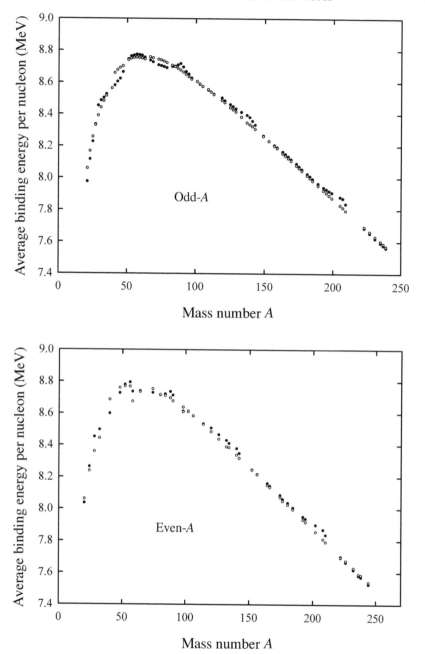

Figure 2.10   Fit to binding energy data (shown as solid circles) for odd-*A* and even-*A* nuclei using the SEMF with the coefficients given in the text; the predictions are shown as open circles and do not lie on smooth curves because *A* is not a function of *Z*

more unstable nuclei. We will use it to analyse the stability of nuclei with respect to β-decay and fission. The discussion of α-decay is deferred until Chapter 7.

Using the numerical values of Equation (2.54), the relative sizes of each of the terms in the SEMF may be calculated and for the case of odd-A are shown in Figure 2.11. In this diagram, the volume term is shown as positive and the other terms are subtracted from it to give the final SEMF curve.

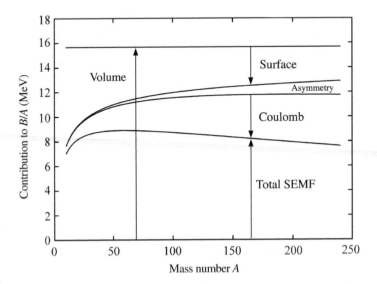

Figure 2.11   Contributions to the binding energy per nucleon as a function of mass number for odd-A from each term in the SEMF; the surface, asymmetry and Coulomb terms have been plotted so that they subtract from the volume term to give the total SEMF result in the lowest curve

Finally, from its definition, one might expect the binding energy per nucleon to be equivalent to the energy needed to remove a nucleon from the nucleus. However, to remove a neutron from a nucleus corresponds to the process

$$_{Z}^{A}Y \rightarrow {}^{A-1}_{Z}Y + n \tag{2.55a}$$

and requires an energy change

$$E_n = [M(Z, A-1) + M_n - M(Z, A)]c^2 = B(Z, A) - B(Z, A-1), \tag{2.55b}$$

whereas the removal of a proton corresponds to the process

$$_{Z}^{A}Y \rightarrow {}^{A-1}_{Z-1}X + p, \tag{2.56a}$$

where X is a different chemical species to Y, and requires an energy change

$$E_p = \left[M(Z-1, A-1) + M_p + m_e - M(Z,A) \right] c^2 = B(Z,A) - B(Z-1, A-1) + m_e c^2.$$

$$(2.56b)$$

Thus, E_p and E_n are only equal to the binding energy per nucleon in an average sense. In practice, measurements show that E_p and E_n can differ substantially from this average and from each other at certain values of (Z, A). We will see in Chapter 7 that one reason for this is the existence of a shell structure for nucleons within nuclei, similar to the shell structure of electrons in atoms, which is ignored in the liquid drop model.

2.6   β-Decay Phenomenology

By rearranging terms, the SEMF (2.46) may be written

$$M(Z, A) = \alpha A - \beta Z + \gamma Z^2 + \frac{\delta}{A^{\frac{1}{2}}}, \qquad (2.57)$$

where

$$\begin{aligned}
\alpha &= M_n - a_v + \frac{a_s}{A^{\frac{1}{3}}} + \frac{a_a}{4} \\
\beta &= a_a + (M_n - M_p - m_e) \\
\gamma &= \frac{a_a}{A} + \frac{a_c}{A^{\frac{1}{3}}} \\
\delta &= a_p
\end{aligned} \qquad (2.58)$$

$M(Z, A)$ is thus a quadratic in Z at fixed A and has a minimum at $Z = \beta/2\gamma$. For a fixed value of A, a stable nucleus will have an integer value of Z closest to the solution of this equation. For odd A, the SEMF is a single parabola, but for even A the even–even and odd–odd nuclei lie on two distinct vertically shifted parabolas, because of the pairing term. The nucleus with the smallest mass in an isobaric spectrum is stable with respect to β-decay. We will consider the two cases of odd and even A separately, using specific values of A to illustrate the main features.

2.6.1   Odd-mass nuclei

Odd-mass nuclei can arise from even-N, odd-Z, or even-Z, odd-N configurations and in practice the number of nuclei that are stable against β-decay are roughly equally distributed between these two types. The example we take is the case of

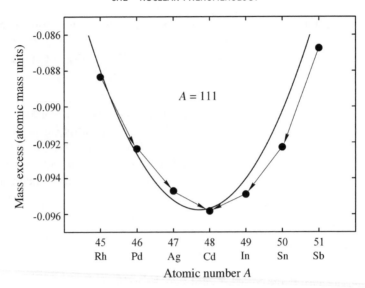

Figure 2.12 Mass parabola of the $A = 111$ isobars: the circles are experimental data and the curve is the prediction of the SEMF -- possible β-decays are indicated by arrows

the $A = 111$ isobars, which are shown in Figure 2.12. The circles show the experimental data as *mass excess* values in atomic mass units, where

$$\text{mass excess} \equiv M(Z, A) \text{ (in atomic mass units)} - A \qquad (2.59)$$

and the *atomic mass unit* (u) is defined as one twelfth of the mass of the neutral atom $^{12}_{6}C$.

The curve is the theoretical prediction from the SEMF using the numerical values of the coefficients (2.54). The exact form of the curve depends on the precise values of these coefficients. The minimum of the parabola corresponds to the isobar $^{111}_{48}Cd$ with $Z = 48$.

Isobars with more neutrons, such as $^{111}_{45}Rh$, $^{111}_{46}Pd$ and $^{111}_{47}Ag$, decay by converting a neutron to a proton, i.e.

$$n \rightarrow p + e^- + \bar{\nu}_e, \qquad (2.60)$$

so that

$$^{111}_{45}Rh \rightarrow ^{111}_{46}Pd + e^- + \bar{\nu}_e \qquad (11\,\text{s}), \qquad (2.61a)$$

$$^{111}_{46}Pd \rightarrow ^{111}_{47}Ag + e^- + \bar{\nu}_e \qquad (22.3\,\text{min}), \qquad (2.61b)$$

and

$$^{111}_{47}Ag \rightarrow ^{111}_{48}Cd + e^- + \bar{\nu}_e \qquad (7.45\,\text{days}) \qquad (2.61c)$$

This decay sequence is shown in Figure 2.12. Electron emission is energetically possible whenever the mass of the daughter atom $M(Z+1,A)$ is smaller than its isobaric neighbour, i.e.

$$M(Z,A) > M(Z+1,A). \tag{2.62}$$

Recall that we are referring here to *atoms*, so that the rest mass of the created electron is automatically taken into account.

Isobars with proton excess decay via

$$p \rightarrow n + e^+ + \nu_e, \tag{2.63}$$

i.e. positron emission, which although not possible for a free proton, *is* possible in a nucleus because of the binding energy. So for example, the nuclei $^{111}_{51}\text{Sb}$, $^{111}_{50}\text{Sn}$ and $^{111}_{49}\text{In}$ could, in principle, decay by positron emission, which is energetically possible if

$$M(Z,A) > M(Z-1,A) + 2m_e; \tag{2.64}$$

this takes account of the creation of a positron and the existence of an excess of electrons in the parent atom.

It is also theoretically possible for this sequence of transitions to occur by *electron capture*. This mainly occurs in heavy nuclei, where the electron orbits are more compact. It is usually the electron in the innermost shell (i.e. the K-shell) that is captured. Capture of such an electron gives rise to a 'hole' and causes electrons from higher levels to cascade downwards and in so doing emit characteristic X-rays. Electron capture is energetically allowed if

$$M(Z,A) > M(Z-1,A) + \varepsilon, \tag{2.65}$$

where ε is the excitation energy of the atomic shell of the daughter nucleus. The process competes with positron emission and in practice for the nuclei above this is what happens. Thus, we have

$$e^- + {}^{111}_{51}\text{Sb} \rightarrow {}^{111}_{50}\text{Sn} + \nu_e \quad (75\,\text{s}), \tag{2.66a}$$

$$e^- + {}^{111}_{50}\text{Sn} \rightarrow {}^{111}_{49}\text{In} + \nu_e \quad (35.3\,\text{min}) \tag{2.66b}$$

and

$$e^- + {}^{111}_{49}\text{In} \rightarrow {}^{111}_{48}\text{Cd} + \nu_e \quad (2.8\,\text{days}), \tag{2.66c}$$

which are manifestations of the primary reaction

$$e^- + p \rightarrow n + \nu_e. \tag{2.67}$$

So once again we arrive at the stable isobar.

2.6.2   Even-mass nuclei

Even-mass nuclei can arise from even-N, even-Z, or odd-Z, odd-N configurations, but for reasons that are explained below, nearly all even-mass nuclei that are stable against β-decay are of the even–even type, with only a handful of odd–odd types known. Consider as an example the case of $A = 102$ shown in Figure 2.13. (Recall that the plot is of mass excess, which is a very small fraction of the total mass.)

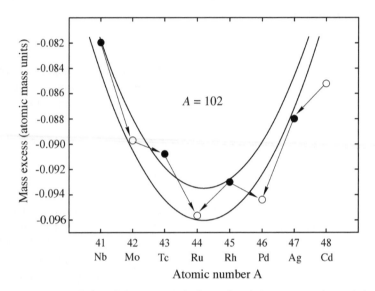

Figure 2.13   Mass parabolas of the $A = 102$ isobars: the circles are experimental data (open circles are even–even nuclei and closed circles are odd–odd nuclei); the curves are the prediction of the SEMF (upper curve is for odd--odd nuclei and lower curve for even–even nuclei) and possible β-decays are indicated by arrows

The lowest isobar is $^{102}_{44}$Ru and is β-stable. The isobar $^{102}_{46}$Pd is also stable since its two odd–odd neighbours both lie above it. In principle, the two nuclei could be connected by the reaction

$$^{102}_{46}\text{Pd} + 2e^- \rightarrow\, ^{102}_{44}\text{Ru} + 2\nu_e, \tag{2.68}$$

but this would involve a 'double electron capture' and would be heavily suppressed. The reaction has never been observed. Thus there are two β-stable isobars. This is a common situation for A-even, although no two neighbouring isobars are known to be stable. Odd–odd nuclei always have at least one more strongly bound even–even neighbour nucleus in the isobaric spectrum. They are therefore unstable. The only exceptions to this rule are a few very light nuclei.

The lifetime of a free neutron is about 887 s. The free proton is believed to be stable and can only 'decay' within a nucleus by utilizing the binding energy. Lifetimes of β emitters vary enormously from milliseconds to 10^{16} years. They depend very sensitively on the Q-value for the decay and on the properties of the nuclei involved, e.g. their spins.

2.7 Fission

Spontaneous fission has been defined as the process whereby a parent nucleus breaks into two daughter nuclei of approximately equal masses without external action. Precisely equal masses are very unlikely and in the most probable cases the daughter nuclei have mass numbers that differ by about 45, with peaks around mass numbers 95 and 140. The reason for this is unknown. The binding energy curve shows that spontaneous fission is energetically possible for nuclei with $A > 100$.[13] An example is

$$^{238}_{92}U \rightarrow {}^{145}_{57}La + {}^{90}_{35}Br + 3n, \tag{2.69}$$

with a release of about 154 MeV of energy, which is carried off as kinetic energy of the fission products. Heavy nuclei are neutron-rich and so necessarily produce neutron-rich decay products, including free neutrons. The fission products are themselves usually some way from the line of β-stability and will decay by a series of steps. For example, $^{145}_{57}La$ decays to the β-stable $^{145}_{60}Nd$ by three stages, releasing a further 8.5 MeV of energy, which in this case is carried off by the electrons and neutrinos emitted in β-decay. Although the probability of fission increases with increasing A, it is still a very rare process. For example, in $^{238}_{92}U$, the transition rate for spontaneous fission is about $3 \times 10^{-24} \, \text{s}^{-1}$ compared with about $5 \times 10^{-18} \, \text{s}^{-1}$ for α-decay, a branching fraction of 6×10^{-7}. Spontaneous emission only becomes dominant in very heavy elements with $A \geq 270$, as we shall now show.

To understand spontaneous fission we can again use the liquid drop model. In the SEMF we have assumed that the drop (i.e. the nucleus) is spherical, because this minimizes the surface area. However, if the surface is perturbed for some reason from spherical to prolate, the surface term in the SEMF will increase and the Coulomb term will decrease (assuming the volume remains the same) and the relative sizes of these two changes will determine whether the nucleus is stable against spontaneous fission.

For a fixed volume we can parametrize the deformation by the semi-major and semi-minor axes of the ellipsoid a and b, respectively as shown in Figure 2.14. One possible parametrization that preserves the volume is

$$a = R(1 + \varepsilon), \quad b = R/(1 + \varepsilon)^{\frac{1}{2}}, \tag{2.70}$$

where ε is a small parameter, so that

$$V = \frac{4}{3}\pi R^3 = \frac{4}{3}\pi ab^2. \tag{2.71}$$

---

[13]Fission in heavy nuclei was discovered by Otto Hahn, for which he received the 1944 Nobel Prize in Chemistry.

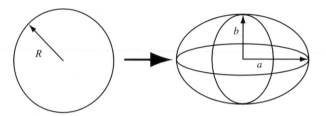

Figure 2.14   Deformation of a heavy nucleus

To find the new surface and Coulomb terms one has to find the expression for the surface of the ellipsoid in terms of a and b and expand it in a power series in ε. The algebra is unimportant and the results are:

$$E_s = a_s A^{\frac{2}{3}}\left(1 + \frac{2}{5}\varepsilon^2 + \dots\right) \qquad (2.72a)$$

and

$$E_c = a_c Z^2 A^{-\frac{1}{3}}\left(1 - \frac{1}{5}\varepsilon^2 + \dots\right). \qquad (2.72b)$$

Hence the change in the total energy is

$$\Delta E = (E_s + E_c) - (E_s + E_c)_{\text{SEMF}} = \frac{\varepsilon^2}{5}\left(2a_s A^{\frac{2}{3}} - a_c Z^2 A^{-\frac{1}{3}}\right). \qquad (2.73)$$

If $\Delta E < 0$, then the deformation is energetically favourable and fission can occur. From Equation (2.73), this happens if

$$\frac{Z^2}{A} \geq \frac{2a_s}{a_c} \approx 49, \qquad (2.74)$$

where we have used experimental values for the coefficients a_s and a_c given in Equations (2.54). The inequality is satisfied for nuclei with $Z > 116$ and $A \geq 270$.

Spontaneous fission is a potential barrier problem and this is shown in Figure 2.15. The solid line corresponds to the shape of the potential in the parent nucleus. The *activation energy* shown in Figure 2.15 determines the probability of spontaneous fission. To fission, the nucleus could in principle tunnel through the barrier, but the fragments are large and the probability for this to happen is extremely small.[14] For heavy nuclei the activation energy is about 6 MeV, but disappears for very heavy nuclei. For such nuclei, the shape of the potential corresponds closer to the dashed line and the slightest deformation will induce fission.

---

[14]The special case of α-decay will be discussed in Chapter 7. There we will show that the lifetime for such decays is expected to have an exponential dependence on the height of the fission barrier and this is observed qualitatively in fission data.

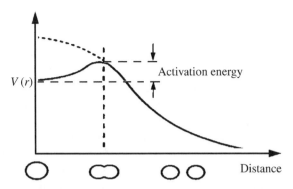

Figure 2.15 Potential energy during different stages of a fission reaction

Another possibility for fission is to supply the energy needed to overcome the barrier by a flow of neutrons. Because of the absence of a Coulomb force, a neutron can get very close to the nucleus and be captured by the strong nuclear attraction. The parent nucleus may then be excited to a state above the fission barrier and therefore split up. This process is an example of induced fission. Neutron capture by a nucleus with an odd neutron number releases not just some binding energy, but also a pairing energy. This small extra contribution makes a crucial difference to nuclear fission properties. For example, very low-energy ('thermal') neutrons can induce fission in ^{235}U, whereas only higher energy ('fast') neutrons induce fission in ^{238}U. This is because ^{235}U is an even–odd nucleus and ^{238}U is even–even. Therefore, the ground state of ^{235}U will lie higher (less tightly bound) in the potential well of its fragments than that of ^{238}U. Hence to induce fission, a smaller energy will be needed for ^{235}U than for ^{238}U. In principle, fission may be induced in ^{235}U using even zero-energy neutrons.[15]

We consider this quantitatively as follows. The capture of a neutron by ^{235}U changes an even–odd nucleus to a more tightly bound even–even (compound) nucleus of ^{236}U and releases the binding energy of the last neutron. In ^{235}U this is 6.5 MeV. As the activation energy (the energy needed to induce fission) is about 5 MeV for ^{236}U, neutron capture releases sufficient energy to fission the nucleus. The kinetic energy of the incident neutron is irrelevant and even zero-energy neutrons can induce fission in ^{235}U. In contrast, neutron capture in ^{238}U changes it from an even–even nucleus to an even–odd nucleus, i.e. changes a tightly bound nucleus to a less tightly bound one. The energy released (the binding energy of the last neutron) is about 4.8 MeV in ^{239}U and is less than the 6.5 MeV required for fission. For this reason, fast neutrons with energy of at least the difference between these two energies are required to fission ^{238}U.

---

[15]Enrico Fermi was a pioneer in the field of induced fission and received the 1938 Nobel Prize in Physics for 'demonstrations of the existence of new radioactive elements produced by neutron irradiation, and for his related discovery of nuclear reactions brought about by slow neutrons'. Fermi's citation could equally have been about his experimental discoveries and theoretical work in a wide range of areas from nuclear and particle physics to solid-state physics and astrophysics. He was probably the last 'universal physicist'.

2.8   γ-Decays

When a heavy nucleus disintegrates by either α- or β-decay, or by fission, the daughter nucleus is often left in an excited state. If this state is below the excitation energy for fission, it will de-excite, usually by emitting a high-energy photon. The energy of these photons is determined by the average energy level spacings in nuclei and ranges from a few to several MeV. They are in the gamma ray (γ) part of the electromagnetic spectrum. Because γ-decay is an electromagnetic process, we would expect the typical lifetime of an excited state to be $\sim 10^{-16}$ s. In practice, lifetimes are very sensitive to the amount of energy released in the decay and in the nuclear case other factors are also very important, particularly the quantity of angular momentum carried off by the photon. Typical lifetimes of nuclear levels are about $\sim 10^{-12}$ s.

The role of angular momentum in γ-decays is crucial. If the initial (excited) state has a total spin $\mathbf{S}_i$ and the final nucleus has a total spin $\mathbf{S}_f$, then the total angular momentum $\mathbf{J}$ of the emitted photon is given by

$$\mathbf{J} = \mathbf{S}_i - \mathbf{S}_f, \tag{2.75}$$

with

$$S_i + S_f \geq J \geq |S_i - S_f|, \tag{2.76}$$

where $S = |\mathbf{S}|$, $J = |\mathbf{J}|$. In addition,

$$m_i = M + m_f, \tag{2.77}$$

where m are the corresponding magnetic quantum numbers. Both total angular momentum and its magnetic quantum number are conserved in γ-decays.

γ-decays are further complicated because parity is conserved in these electromagnetic processes. Both the initial and final nuclear level will have an intrinsic parity, as does the photon, and in addition there is a parity associated with the angular momentum carried off by the photon, which is of the form $(-1)^J$, reflecting the symmetry of the angular part of the wavefunction (see Equation (1.14)). We will not pursue this further here, but defer a more detailed discussion until Chapter 7.

2.9   Nuclear Reactions

In Chapter 1 and earlier sections of the present chapter we discussed various aspects of reactions. In particle physics, because the projectiles and targets have relatively simple structures, this is all that is required in classifying reactions. In nuclear physics, however, because the target has a rich structure it is useful to classify reactions in more detail. In this section we do this, drawing together our previous work and also anticipating some reactions that will be encountered in later chapters.

Elastic scattering reactions were defined in Chapter 1 as those interactions where the initial and final particles are identical, i.e. $a + A \rightarrow a + A$. We also defined inelastic scattering as the situation where the final particles are the same chemical species, but one or more is in an excited state, e.g. $a + A \rightarrow a + A^*$ and in Section 2.1 we showed how the kinematics of such reactions could be used to determine the mass of the excited state. Elastic and inelastic scattering are examples of so-called *direct reactions*. These are defined as ones where the incident particle interacts in a time comparable to the time taken to transit the nucleus. They are more likely when the incident particle has an energy corresponding to a de Broglie wavelength closer to the size of a nucleon rather than that of the nucleus. The collisions are largely peripheral, with only a relatively small fraction of the available energy transferred to the target. Another direct reaction is $^{16}O(p, d)^{15}O$, i.e.

$$p + {}^{16}O \rightarrow d + {}^{15}O, \tag{2.78}$$

where we have used the notation $A(a, b)B$ for the general nuclear reaction $a + A \rightarrow b + B$. This is an example of a *pick-up reaction*, because one or more nucleons (in this case a neutron) is stripped off the target nucleus and carried away by the projectile. The 'inverse' of this reaction is $^{16}O(d, p)^{17}O$. This is an example of a *stripping reaction*, because one or more nucleons (in this case again a neutron) is stripped off the projectile and transferred to the target nucleus.

The theoretical interpretation of direct reactions is based on the assumption that the projectile experiences the average potential of the target nucleus. For example, we have seen in the optical model of Section 2.2.2 how this approach can be used to analyse differential cross sections for elastic scattering and be used to extract information about nuclear shapes and sizes. It also leads to the prediction of resonances of width typically of order 1 MeV separated by a few MeV, as observed in cross-section as functions of centre-of-mass energy for nucleon scattering from light nuclei. One way of viewing this is as a consequence of the reaction time for a direct reaction, typically 10^{-22} s , making use of the uncertainty relation between energy and time, $\Delta E \Delta t \geq \hbar$.

A second important class of interactions is where the projectile becomes loosely bound in the nucleus and shares its energy with all the nuclear constituents. This is called a *compound nucleus reaction*. The time for the system to reach statistical equilibrium depends on the nuclear species, the type of projectile and its energy, but will always be much longer than the transit time and is typically several orders of magnitude longer. An important feature of these reactions is that the properties of the compound nucleus determine its subsequent behaviour and not the mechanism by which it was formed. The compound nucleus is in an excited state and is inherently unstable. Eventually, by a statistical fluctuation, one or more nucleons will acquire sufficient energy to escape and the nucleus either emits particles or de-excites by radiating gamma rays.

If the compound nucleus is created in a region of excitation where its energy levels are well separated, the cross-section will exhibit well-defined resonances described by the Breit–Wigner formula of Section 1.6.3. These processes are depicted schematically in the energy-level diagram of Figure 2.16, which correspond to $a + A \rightarrow C^* \rightarrow b + B$, where C^* is the compound nucleus and $a + A \rightarrow C^* \rightarrow C + \gamma$, where C is the ground state corresponding to the excited state C^*. In practice, there could be many final states to which C^* could decay.

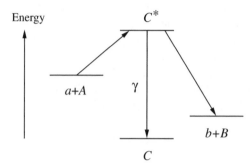

Figure 2.16 Energy-level diagram showing the excitation of a compound nucleus $C^*$and its subsequent decay

Because the time for a compound nucleus to reach statistical equilibrium is much longer than the transit time for a direct reaction, the cross-sections for a compound nucleus process can show variations on much smaller energy scales than those for direct reactions. The density of levels in the compound nucleus is high, and so a very small change in the incident energy suffices to alter completely the intermediate states, and hence the cross section. An example is shown in Figure 2.17, which gives the total cross-section for neutron scattering from ^{12}C at neutron laboratory energies of a few MeV. Peaks corresponding to resonance formation in ^{13}C are clearly identified. Their widths vary from a few tens to a few hundreds of keV, consistent with the characteristic times for compound nucleus formation and decay.

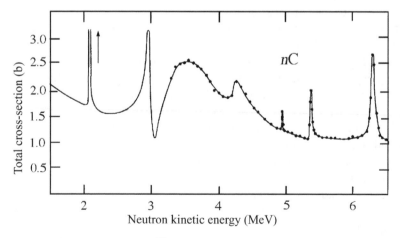

Figure 2.17   Total cross-section for n^{12}C interactions (adapted from Fo61. Copyright American Physical Society.)

The mean widths of compound nucleus excitations depend on the incident energy and the target nucleus, decreasing both with energy and rapidly with nuclear mass. Neutrons, because they are neutral, have a high probability of being captured by nuclei and their cross-sections are rich in compound nucleus effects, particularly at very low energies. This is discussed further below.

The division of reactions into direct and compound nucleus is not exhaustive and situations can occur where particles are ejected from the nucleus before full statistical equilibrium has been reached. Also, in the collisions of complex heavy ions, there is an appreciable probability for an additional reaction mechanism called *deep inelastic scattering* that is intermediate between direct and compound nucleus reactions. In this case, the probability for complete fusion of the colliding ions is small, but there can be substantial transfer of the incident kinetic energy to internal excitations of the ions. We will not discuss this or other mechanisms further, but we will encounter the concept of deep inelastic scattering again in Chapter 5 in the context of exploring the internal structure of nucleons. In practice, the various mechanisms feed the same final states as direct reactions. This is illustrated schematically in Figure 2.18 for reactions initiated using protons as the projectile.

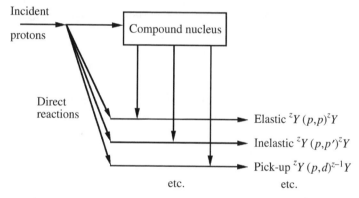

Figure 2.18 Direct and compound nucleus reactions in nuclear reactions initiated by protons

The general form of the yield $N(E)$ of secondary particles at a fixed angle as a function of the outgoing energy E, i.e. the number of particles with energy E between E and $E + dE$, is shown schematically in Figure 2.19 for the case of an incident nucleon. At the upper end of the plot (which corresponds to low-incident nucleon energies) there are a number of distinct peaks due to elastic, inelastic and transfer reactions. Then as the excitation energy is reduced, the more closely-spaced energy levels in the final nucleus are not fully resolved because of the spread in energy of the incident beam and the uncertainty in the experimental measurements of energy. At the lowest energies there is a broad continuum mainly due to the decays of compound nuclei formed by the absorption of the projectile nucleon by the target nucleus. The differential cross-sections for the two processes will be very different. Direct reactions lead to a cross-section peaked in the forward direction, falling rapidly with angle and with oscillations, as we have seen

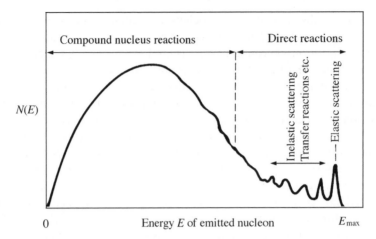

Figure 2.19   Typical spectrum of energies of the nucleons emitted at a fixed angle in inelastic nucleon--nucleus reactions

in the case of elastic scattering in Section 2.2 (Figure 2.3). On the other hand, the contribution from the compound nucleus at low energies where an isolated compound nucleus is formed is fairly isotropic and symmetric about 90°.

Many medium- and large-A nuclei can capture low-energy ($\sim (10\text{--}100)$ eV) neutrons very readily. The neutron separation energy for the final nucleus is ~ 6 MeV and thus capture leads to a compound nucleus with an excitation energy above the ground state by this separation energy. Such excitation often occurs in a region of high density of narrow states that show up as a rich resonance structure in the corresponding neutron total cross-section. An example is shown in Figure 2.20. The value of the cross-section at the resonance peaks can be many orders of

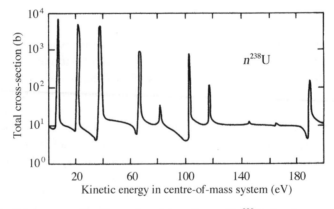

Figure 2.20   Total cross-section for neutron interactions with ^{238}U, showing many very narrow resonances (with intrinsic widths of order 10^{-2} eV) corresponding to excited states of ^{239}U (from Ga76, courtesy of Brookhaven National Laboratory)

magnitude greater than the geometrical cross-section based on the size of the nucleus. This is because the cross-section is determined dominantly by the area associated with the wavelength λ of the projectile, i.e. $\pi\lambda^2$, which is very large because λ is large.

Once formed, the compound nucleus can decay to any final state consistent with the relevant conservation laws. If this includes neutron emission, it will be the preferred decay. However, for production by very slow (thermal) neutrons with energies of the order of 0.02 eV, the available decay kinetic energy will reflect the initial energy of the projectile, which is very small. Therefore, in these cases, photon emission is often preferred. We shall see in Chapter 8 that the fact that radiative decay is the dominant mode of decay of compound nuclei formed by thermal neutrons is important in the use of nuclear fission to produce power in nuclear reactors.

Problems

2.1 Electrons with momentum 330 MeV/c are elastically scattered through an angle of 10° by a nucleus of ^{56}Fe. If the charge distribution on the nucleus is assumed to be that of a uniform hard sphere, and assuming the Born approximation is valid, by what factor would you expect the Mott cross-section to be reduced?

2.2 Show explicitly that Equation (2.28) follows from Equation (2.26).

2.3 A beam of electrons with energies 250 MeV is scattered through an angle of 10° by a heavy nucleus. It is found that the differential cross-section is 65 per cent of that expected from scattering from a point nucleus. Estimate the root mean square radius of the nucleus.

2.4 Find the form factor for a charge distribution $\rho(r) = \rho_0\exp(-r/a)/r$, where ρ_0 and a are constants.

2.5 A sample of 1 g of a radioactive isotope of atomic weight 208 decays via β-emission and 75 counts are recorded in a 24 h period. If the detector efficiency is 10 per cent, estimate the mean life of the isotope.

2.6 A 1 g sample taken from an organic artefact is found to have a β count rate of 2.1 counts per min, which are assumed to originate from the decay of ^{14}C with a mean lifetime of 8270 years. If the abundance of ^{14}C in living matter is currently 1.2×10^{-12}, what can you deduce about the approximate age of the artefact?

2.7 Nuclei of $^{212}_{86}$Rn decay by α-emission to $^{208}_{84}$Po with a mean life of 23.9 min. The $^{208}_{84}$Po nuclei in turn decay, also by α-emission, to the stable isotope $^{204}_{82}$Pb with a mean life of 2.9 years. If initially the source is pure $^{212}_{86}$Rn, how long will it take for the rate of α-emission in the final decay to reach a maximum?

2.8   Natural lanthanum has an atomic weight of 138.91 and contains 0.09 per cent of the isotope $^{138}_{57}$La. This has two decay modes: $^{138}_{57}$La $\rightarrow$ $^{138}_{58}$Ce $+ e^- + \bar{\nu}_e$ (β-decay) and $^{138}_{57}$La $+ e^- \rightarrow$ $^{138}_{56}$Ba* $+ \nu_e$ (electron capture), followed by the electromagnetic decay of the excited state $^{138}_{56}$Ba* $\rightarrow$ $^{138}_{56}$Ba $+ \gamma$ (radiative decay). There are 7.8×10^2 β-particles emitted per s per kg of natural lanthanum and there are 50 photons emitted per 100 β-particles. Estimate the mean lifetime of $^{138}_{57}$La.

2.9   Use the SEMF to estimate the energy released in the spontaneous fission reaction

$$^{235}_{92}U \rightarrow {}^{87}_{35}Br + {}^{145}_{57}La + 3n.$$

2.10   The most stable nucleus with $A = 111$ is $^{111}_{48}$Cd (see Figure 2.12). By what percentage would the fine structure constant α have to change if the most stable nucleus with $A = 111$ were to be $^{111}_{47}$Ag? Assume that altering α does not change particle masses.

2.11   The transuranic isotope $^{269}_{108}$Hs decays 100 per cent via α-emission with a lifetime of 27 s, i.e. $^{269}_{108}$Hs $\rightarrow$ $^{265}_{106}$Sg $+ \alpha$, where the kinetic energy of the α-particle is $E_\alpha = 9.23$ MeV. Calculate the mass of the $^{269}_{108}$Hs nucleus in atomic mass units.

2.12   The isotope $^{238}_{94}$Pu decays via α-emission to the essentially stable isotope $^{234}_{92}$U with a lifetime of 126.7 years and a release of 5.49 MeV of kinetic energy. This energy is converted to electrical power in a space probe designed to reach planet X in a journey planned to last 4 years. If the efficiency of power conversion is 5 per cent and on reaching planet X the probe requires at least 200 W of power to perform its landing tasks, how much $^{238}_{94}$Pu would be needed at launch?

2.13   On planet X it is found that the isotopes ^{205}Pb($\tau = 1.53 \times 10^7$y) and ^{204}Pb (stable) are present with abundances n_{205} and n_{204}, with $n_{205}/n_{204} = 2 \times 10^{-7}$. If at the time of the formation of planet X both isotopes were present in equal amounts, how old is the planet?

2.14   The reaction $^{45}_{21}$Sc$(d, p)^{46}_{21}$Sc has a Q-value of 6.54 MeV and a resonance when the incident deuteron laboratory kinetic energy is 2.76 MeV. Would you expect the same resonance to be excited in the reaction $^{43}_{20}$Ca$(\alpha, n)^{46}_{22}$Ti and if so at what value of the laboratory kinetic energy of the alpha particle? You may use the fact that the β-decay $^{46}_{21}$Sc $\rightarrow$ $^{46}_{22}$Ti $+ e^- + \bar{\nu}_e$ has a Q-value of 2.37 MeV and the mass difference between the neutron and a hydrogen atom is 0.78 MeV/c^2.

2.15   A radioisotope with decay constant λ is produced at a constant rate P. Show that the number of atoms at time t is $N(t) = P[1 - \exp(-\lambda t)]/\lambda$.

2.16   Radioactive ^{36}Cl (half-life 3×10^5 years) is produced by irradiating 1 g of natural nickel chloride (NiCl$_2$, molecular weight 129.6) in a neutron beam of flux $F = 10^{14}$cm^{-2}s^{-1}. If the neutron absorption cross-section ^{35}Cl$(n, \gamma)^{36}$Cl is $\sigma = 43.6$ b and 75.8 per cent of natural chlorine is ^{35}Cl, use the result of Problem 2.15 to estimate the time it would take to produce a 3×10^5 Bq source of ^{36}Cl.

2.17 Consider the total cross-section data for the n^{238}U interaction shown in Figure 2.20. There is a resonance R at the centre-of-mass neutron kinetic energy $E_n = 10$ eV with width $\Gamma = 10^{-2}$ eV and the total cross-section there is $\sigma_{max} = 9 \times 10^3$ b. Use this information to find the partial widths $\Gamma_{n,\gamma}$ for the decays $R \rightarrow n + {}^{238}$U and $R \rightarrow \gamma + {}^{238}$U, if these are the only two significant decay modes. The spin of the ground state of ^{238}U is zero.

3

Particle Phenomenology

In this chapter we shall look at some of the basic phenomena of particle physics – the properties of leptons and quarks, and the bound states of the latter, the hadrons. In later chapters we will discuss theories and models that attempt to explain these and other particle data.

3.1 Leptons

We have seen that the spin-$\frac{1}{2}$ leptons are one of the three classes of elementary particles in the standard model and we shall start with a discussion of their basic properties. Then we shall look in more detail at the neutral leptons, the neutrinos and, amongst other things, examine an interesting property they can exhibit, based on simple quantum mechanics, if they have non-zero masses.

3.1.1 Leptons multiplets and lepton numbers

There are six known leptons and they occur in pairs, called *generations*, which we write, for reasons that will become clear presently, as:

$$\begin{pmatrix} \nu_e \\ e^- \end{pmatrix}, \qquad \begin{pmatrix} \nu_\mu \\ \mu^- \end{pmatrix}, \qquad \begin{pmatrix} \nu_\tau \\ \tau^- \end{pmatrix}. \tag{3.1}$$

Each generation comprises a *charged lepton* with electric charge $-e$, and a *neutral neutrino*. The three charged leptons (e^-, μ^-, τ^-) are the familiar electron, together with two heavier particles, the *mu lepton* (usually called the *muon,* or just *mu*) and the *tau lepton* (usually called the *tauon,* or just *tau*). The associated neutrinos are called the *electron neutrino, mu neutrino,* and *tau*

Nuclear and Particle Physics   B. R. Martin
© 2006 John Wiley & Sons, Ltd

neutrino, respectively.[1] In addition to the leptons, there are six corresponding antileptons:

$$\begin{pmatrix} e^+ \\ \bar{\nu}_e \end{pmatrix}, \quad \begin{pmatrix} \mu^+ \\ \bar{\nu}_\mu \end{pmatrix}, \quad \begin{pmatrix} \tau^+ \\ \bar{\nu}_\tau \end{pmatrix}. \tag{3.2}$$

Ignoring gravity, the charged leptons interact only via electromagnetic and weak forces, whereas for the neutrinos, only weak interactions have been observed.[2] Because of this, neutrinos, which are all believed to have extremely small masses, can be detected only with considerable difficulty.

The masses and lifetimes of the leptons are listed in Table 3.1. The electron and the neutrinos are stable, for reasons that will become clear shortly. The muons decay by the weak interaction processes

$$\mu^+ \rightarrow e^+ + \nu_e + \bar{\nu}_\mu \quad \text{and} \quad \mu^- \rightarrow e^- + \bar{\nu}_e + \nu_\mu, \tag{3.3a}$$

Table 3.1 Properties of leptons: all have spin $\frac{1}{2}$ and masses are given units of MeV/c²; the antiparticles (not shown) have the same masses as their associated particles, but the electric charges (Q) and lepton numbers (L_ℓ , $\ell = e$, μ, τ) are reversed in sign

Name and symbol	Mass	Q	L_e	L_μ	L_τ	Lifetime (s)	Major decays
Electron e^-	0.511	-1	1	0	0	Stable	None
Electron neutrino ν_e	$<2.2\,\text{eV}/c^2$	0	1	0	0	Stable	None
Muon (mu) μ^-	105.7	-1	0	1	0	2.197×10^{-6}	$e^- \bar{\nu}_e \nu_\mu$ (100%)
Muon neutrino ν_μ	<0.19	0	0	1	0	Stable	None
Tauon (tau) τ^-	1777.0	-1	0	0	1	2.906×10^{-13}	$\mu^- \bar{\nu}_\mu \nu_\tau$ (17.4%)
							$e^- \bar{\nu}_e \nu_\tau$ (17.8%)
							ν_τ+hadrons ($\sim$64%)
Tauon neutrino ν_τ	<18.2	0	0	0	1	Stable	None

with lifetimes $(2.19703 \pm 0.00004) \times 10^{-6}$ s. The tau also decays by the weak interaction, but with a much shorter lifetime $(2.906 \pm 0.011) \times 10^{-13}$ s. (This illustrates what we have already seen in nuclear physics, that lifetimes depend sensitively on the energy released in the decay, i.e. the Q-value.) Because it is heavier than the muon, the tau has sufficient energy to decay to many different final states, which can include both hadrons and leptons. However, about 35 per cent

---

[1]Leon Lederman, Melvin Schwartz and Jack Steinberger shared the 1988 Nobel Prize in Physics for their use of neutrino beams and the discovery of the muon neutrino. Martin Perl shared the 1995 Nobel Prize in Physics for his pioneering work in lepton physics and in particular for the discovery of the tau lepton.
[2]Although neutrinos have zero electric charge they could, in principle, have a charge *distribution* that would give rise to a magnetic moment (like neutrons) and hence electromagnetic interactions. This would of course be forbidden in the standard model because the neutrinos are defined to be point-like.

of decays again lead to purely leptonic final states, via reactions which are very similar to muon decay, for example:

$$\tau^+ \rightarrow \mu^+ + \nu_\mu + \bar{\nu}_\tau \quad \text{and} \quad \tau^- \rightarrow e^- + \bar{\nu}_e + \nu_\tau. \tag{3.3b}$$

Associated with each generation of leptons is a quantum number called a *lepton number*. The first of these lepton numbers is the *electron number*, defined for any state by

$$L_e \equiv N(e^-) - N(e^+) + N(\nu_e) - N(\bar{\nu}_e), \tag{3.4}$$

where $N(e^-)$ is the number of electrons present, $N(e^+)$ is the number of positrons present and so on. For single-particle states, $L_e = 1$ for e^- and ν_e, $L_e = -1$ for e^+ and $\bar{\nu}_e$, and $L_e = 0$ for all other particles. The *muon* and *tauon numbers* are defined in a similar way and their values for all single particle states are summarized in Table 3.1. They are zero for all particles other than leptons. For multiparticle states, the lepton numbers of the individual particles are simply added. For example, the final state in neutron β-decay (i.e. $n \rightarrow p + e^- + \bar{\nu}_e$) has

$$L_e = L_e(p) + L_e(e^-) + L_e(\bar{\nu}_e) = (0) + (1) + (-1) = 0, \tag{3.5}$$

like the initial state, which has $L_e(n) = 0$.

In the standard model, the value of each lepton number is postulated to be conserved in *any* reaction. The decays (3.3) illustrate this principle of *lepton number conservation*. Until recently this was considered an absolute conservation law, but in Section 3.1.4 we will discuss growing evidence that neutrinos are not strictly massless, which would imply that conservation of individual lepton numbers is not an exact law. However, for the present we will assume lepton numbers are conserved, as in the standard model. In electromagnetic interactions, this reduces to the conservation of $N(e^-) - N(e^+)$, $N(\mu^-) - N(\mu^+)$ and $N(\tau^-) - N(\tau^+)$, since neutrinos are not involved. This implies that the charged leptons can only be created or annihilated in particle–antiparticle pairs. For example, in the electromagnetic reaction

$$e^+ + e^- \rightarrow \mu^+ + \mu^- \tag{3.6}$$

an electron pair is annihilated and a muon pair is created by the mechanism of Figure 3.1.

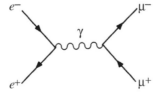

Figure 3.1 Single-photon exchange in the reaction $e^+e^- \rightarrow \mu^+\mu^-$

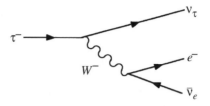

Figure 3.2   Dominant Feynman diagram for the decay $\tau^- \to e^- \bar{\nu}_e \nu_\tau$

In weak interactions more general possibilities are allowed, which still conserve lepton numbers. For example, in the tau-decay process $\tau^- \to e^- + \bar{\nu}_e + \nu_\tau$, a tau converts to a tau neutrino and an electron is created together with an antineutrino, rather than a positron. The dominant Feynman graph corresponding to this process is shown in Figure 3.2.

Lepton number conservation, like electric charge conservation, plays an important role in understanding reactions involving leptons. Observed reactions conserve lepton numbers, while reactions that do not conserve lepton numbers are 'forbidden' and are not observed. For example, the neutrino scattering reaction

$$\nu_\mu + n \to \mu^- + p \tag{3.7}$$

is observed experimentally, while the apparently similar reaction

$$\nu_\mu + n \to e^- + p, \tag{3.8}$$

which violates both L_e and L_μ conservation, is not. Another example which violates both L_e and L_μ conservation is $\mu^- \to e^- + \gamma$. If this reaction were allowed, the dominant decay of the muon would be electromagnetic and the muon lifetime would be much shorter than its observed value. This is very strong evidence that even if lepton numbers are not absolutely conserved, they are conserved to a high degree of accuracy.

Finally, conservation laws explain the stability of the electron and the neutrinos. The electron is stable because electric charge is conserved in all interactions and the electron is the lightest charged particle. Hence decays to lighter particles that satisfy all other conservation laws, like $e^- \to \nu_e + \gamma$, are necessarily forbidden by electric charge conservation. In the same way, lepton number conservation implies that the lightest particles with non-zero values of the three lepton numbers – the three neutrinos – are stable, whether they have zero masses or not. Of course, if lepton numbers are not conserved, then the latter argument is invalid.

3.1.2   Neutrinos

As we mentioned in Chapter 1, the existence of the *electron neutrino* ν_e was first postulated by Pauli in 1930. He did this in order to understand the observed nuclear β-decays

$$(Z, N) \rightarrow (Z+1, N-1) + e^- + \bar{\nu}_e \qquad (3.9)$$

and

$$(Z, N) \rightarrow (Z-1, N+1) + e^+ + \nu_e \qquad (3.10)$$

that were discussed in Chapter 2. The neutrinos and antineutrinos emitted in these decays are not observed experimentally, but are inferred from energy and angular momentum conservation. In the case of energy, if the antineutrino were not present in the first of the reactions, the energy E_e of the emitted electron would be a unique value equal to the difference in rest energies of the two nuclei, i.e.

$$E_e = \Delta M c^2 = [M(Z, N) - M(Z+1, N-1)]c^2, \qquad (3.11)$$

where for simplicity we have neglected the extremely small kinetic energy of the recoiling nucleus. However, if the antineutrino is present, the electron energy would not be unique, but would lie in the range

$$m_e c^2 \leq E_e \leq (\Delta M - m_{\bar{\nu}_e})c^2, \qquad (3.12)$$

depending on how much of the kinetic energy released in the decay is carried away by the neutrino. Experimentally, the observed energies span the whole of the above range and in principle a measurement of the energy of the electron near its maximum value of $E_e = (\Delta M - m_{\bar{\nu}_e})c^2$ determines the neutrino mass. The most accurate results come from tritium decay and are compatible with zero mass for electron antineutrinos. When experimental errors are taken into account, the experimentally allowed range is

$$0 \leq m_{\bar{\nu}_e} < 2.2\,\text{eV}/c^2 \approx 4.3 \times 10^{-6} m_e. \qquad (3.13)$$

We will discuss this experiment in more detail in Chapter 7, after we have considered the theory of β-decay.

The masses of both ν_μ and ν_τ can similarly be directly inferred from the e^- and μ^- energy spectra in the leptonic decays of muons and tauons, using energy conservation. The results from these and other decays show that the neutrino masses are very small compared with the masses of the associated charged leptons. The present limits are given in Table 3.1.

Small neutrino masses, compatible with the above limits, can be ignored in most circumstances, and there are theoretical attractions in assuming neutrino masses are precisely zero, as is done in the standard model. However, we will show in the following section that there is now strong evidence for physical phenomena that could not occur if the neutrinos had exactly zero mass. The consequences of neutrinos having small masses have therefore to be taken seriously.

Because neutrinos only have weak interactions, they can only be detected with extreme difficulty. For example, electron neutrinos and antineutrinos of sufficient energy can in principle be detected by observing the *inverse β-decay* processes

$$\nu_e + n \rightarrow e^- + p \tag{3.14}$$

and

$$\bar{\nu}_e + p \rightarrow e^+ + n. \tag{3.15}$$

However, the probability for these and other processes to occur is extremely small. In particular, the neutrinos and antineutrinos emitted in β-decays, with energies of the order of 1 MeV, have mean free paths in matter of the order of 10^6 km.[3] Nevertheless, if the neutrino flux is intense enough and the detector is large enough, the reactions can be observed. In particular, uranium fission fragments are neutron rich, and decay by electron emission to give an antineutrino flux that can be of the order of 10^{17} m^{-2} s^{-1} or more in the vicinity of a nuclear reactor, which derives its energy from the decay of nuclei. These antineutrinos will occasionally interact with protons in a large detector, enabling examples of the inverse β-decay reaction to be observed. As mentioned in Chapter 1 (Footnote 12), electron neutrinos were first detected in this way by Reines and Cowan in 1956, and their interactions have been studied in considerable detail since.

The mu neutrino, ν_μ, has been detected using the reaction $\nu_\mu + n \rightarrow \mu^- + p$ and other reactions. In this case, well-defined high-energy ν_μ beams can be created in the laboratory by exploiting the decay properties of *pions*, which are particles we have mentioned briefly in Chapter 1 and which we will meet in more detail presently. The probability of neutrinos interacting with matter increases rapidly with energy (this will be demonstrated in Section 6.5.2) and, for large detectors, events initiated by such beams are so copious that they have become an indispensable tool in studying both the fundamental properties of weak interactions and the internal structure of the proton. Finally, in 2000, a few examples of tau neutrinos were reported, so that almost 70 years after Pauli first suggested the existence of a neutrino, all three types had been directly detected.

3.1.3   Neutrino mixing and oscillations

Neutrinos are assumed to have zero mass in the standard model. However, as mentioned above, data from the β-decay of tritium are compatible with a non-zero

---

[3]The mean free path is the distance a particle would have to travel in a medium for there to be a significant probability of an interaction. A formal definition is given in Chapter 4.

mass. A phenomenon that can occur if neutrinos have non-zero masses is *neutrino mixing*. This arises if we assume that the observed neutrino states ν_e, ν_μ and ν_τ which take part in weak interactions, i.e. the states that couple to electrons, muons and tauons, respectively, are not eigenstates of mass, but instead are linear combinations of three other states ν_1, ν_2 and ν_3 which do have definite masses m_1, m_2 and m_3, i.e. *are* eigenstates of mass. For algebraic simplicity we will consider the case of mixing between just two states, one of which we will assume is ν_μ and the other we will denote by ν_x. Then, in order to preserve the orthonormality of the states, we can write

$$\nu_\mu = \nu_1 \cos\alpha + \nu_2 \sin\alpha \tag{3.16}$$

and

$$\nu_x = -\nu_1 \sin\alpha + \nu_2 \cos\alpha. \tag{3.17}$$

Here α is a *mixing angle* which must be determined from experiment. If $\alpha \neq 0$ then some interesting predictions follow.

Measurement of the mixing angle may be made in principle by studying the phenomenon of *neutrino oscillation*. When, for example, a muon neutrino is produced with momentum **p** at time $t = 0$, the ν_1 and ν_2 components will have slightly different energies E_1 and E_2 due to their slightly different masses. In quantum mechanics, their associated waves will therefore have slightly different frequencies, giving rise to a phenomenon somewhat akin to the 'beats' heard when two sound waves of slightly different frequency are superimposed. As a result of this, one finds that the original beam of muon neutrinos develops a ν_x component whose intensity oscillates as it travels through space, while the intensity of the muon neutrino beam itself is correspondingly reduced, i.e. muon neutrinos will 'disappear'.

This effect follows from simple quantum mechanics. To illustrate this we will consider a muon neutrino produced with momentum **p** at time $t = 0$. The initial state is therefore

$$\left|\nu_\mu, \mathbf{p}\right\rangle = \left|\nu_1, \mathbf{p}\right\rangle \cos\alpha + \left|\nu_2, \mathbf{p}\right\rangle \sin\alpha, \tag{3.18}$$

where we use the notation $\left|P, \mathbf{p}\right\rangle$ to denote a state of a particle P having momentum **p**. After time t this will become

$$a_1(t)\left|\nu_1, \mathbf{p}\right\rangle \cos\alpha + a_2(t)\left|\nu_2, \mathbf{p}\right\rangle \sin\alpha, \tag{3.19}$$

where

$$a_i(t) = e^{-iE_i t/\hbar} \qquad (i = 1, 2) \tag{3.20}$$

are the usual oscillating energy factors associated with any quantum mechanical stationary state.[4] For $t \neq 0$, the linear combination (3.19) does not correspond to a pure muon neutrino state, but can be written as a linear combination

$$A(t)|\nu_\mu, \mathbf{p}\rangle + B(t)|\nu_x, \mathbf{p}\rangle, \qquad (3.21)$$

of ν_μ and ν_x states, where the latter is

$$|\nu_x, \mathbf{p}\rangle = -|\nu_1, \mathbf{p}\rangle \sin \alpha + |\nu_2, \mathbf{p}\rangle \cos \alpha. \qquad (3.22)$$

The functions $A(t)$ and $B(t)$ are found by solving Equations (3.18) and (3.22) for $|\nu_1, \mathbf{p}\rangle$ and $|\nu_2, \mathbf{p}\rangle$, then substituting the results into (3.19) and comparing at with (3.21). This gives,

$$A(t) = a_1(t) \cos^2 \alpha + a_2(t) \sin^2 \alpha \qquad (3.23)$$

and

$$B(t) = \sin \alpha \cos \alpha \, [a_2(t) - a_1(t)]. \qquad (3.24)$$

The probability of finding a ν_x state is therefore

$$P(\nu_\mu \to \nu_x) = |B(t)|^2 = \sin^2(2\alpha) \sin^2[(E_2 - E_1) t/2\hbar] \qquad (3.25)$$

and thus oscillates with time, while the probability of finding a muon neutrino is reduced by a corresponding oscillating factor. Similar effects are predicted if instead we start from electron or tau neutrinos. In each case the oscillations vanish if the mixing angle is zero, or if the neutrinos have equal masses, and hence equal energies, as can be seen explicitly from Equation (3.25). In particular, such oscillations are not possible if the neutrinos both have zero masses.

Returning to Equation (3.25), since neutrino masses are very small, $E_{1,2} \gg m_i c^2$ $(i = 1, 2)$ and we can write

$$E_2 - E_1 = \left(m_2^2 c^4 + p^2 c^2\right)^{1/2} - \left(m_1^2 c^4 + p^2 c^2\right)^{1/2} \approx \frac{m_2^2 c^4 - m_1^2 c^4}{2pc}. \qquad (3.26)$$

Also, $E \approx pc$ and $t \approx |\mathbf{x}|/c \equiv L/c$, where L is the distance from the point of production. Thus Equation (3.25) may be written

$$P(\nu_\mu \to \nu_x) \approx \sin^2(2\alpha) \sin^2\left[\frac{\Delta(m^2 c^4)L}{4\hbar c E}\right], \qquad (3.27a)$$

---

[4]See, for example, Chapter 1 of Ma92.

with

$$P(\nu_\mu \to \nu_\mu) = 1 - P(\nu_\mu \to \nu_x), \qquad (3.27b)$$

where $\Delta(m^2c^4) \equiv m_2^2c^4 - m_1^2c^4$. These formulae assume that the neutrinos are propagating in a vacuum, whereas in real experiments they will be passing through matter and the situation is more complicated than these simple results suggest. This formalism can be extended to the general case of mixing between all three neutrino species, but at the expense of additional free parameters.[5]

Attempts to establish neutrino oscillations rest on using the inverse β-decay reactions (3.14) and (3.15) to produce electrons and the analogous reactions for muon neutrinos to produce muons, which are then detected. In addition, the time t is determined by the distance of the neutrino detector from the source of the neutrinos, since their energies are always much greater than their possible masses, and they travel at approximately the speed of light. Hence, for example, if we start with a source of muon neutrinos, the flux of muons observed in a detector should vary with its distance from the source of the neutrinos, if appreciable oscillations occur. In practice, oscillations at the few per cent level are very difficult to detect for experimental reasons that we will not discuss here.

3.1.4   Neutrino masses

There are a number of different types of experiment that can explore neutrino oscillations and hence neutrino masses. The first of these to produce definitive evidence for oscillations was that of a Japanese group in 1998 using the giant *Super Kamiokande* detector to study *atmospheric neutrinos* produced by the action of cosmic rays.[6] (Neutrinos of each generation are often referred to as having a different *flavour* and so the observations are evidence for flavour oscillation.)

The *Super Kamiokande* detector is shown in Figure 3.3. (Detectors will be discussed in detail in Chapter 4, so the description here will be brief.) It consists of a stainless steel cylindrical tank of roughly 40 m diameter and 40 m height, containing about 50 000 metric tons of very pure water. The detector is situated deep underground in a mountain in Japan, at a depth equivalent to 2700 m of water. This is to use the rocks above to shield the detector from cosmic ray muons. The volume is separated into a large inner region, the walls of which are lined with 11 200 light-sensitive devices called photomultipliers (the physics of these will be discussed in Chapter 4). These register the presence of electrons or muons

---

[5]See, for example, the Review of Particle Properties published biannually by the Particle Data Group (2004 edition: Ei04). The PDG Review is also available at http://pdg.lbl.gov. This publication contains a wealth of useful data about elementary particles and their interactions and we will refer to it in future simply as PDG04.
[6]Cosmic neutrinos were first detected (independently) by Raymond Davis Jr. and Masatoshi Koshiba, for which they were jointly awarded the 2002 Nobel Prize in Physics.

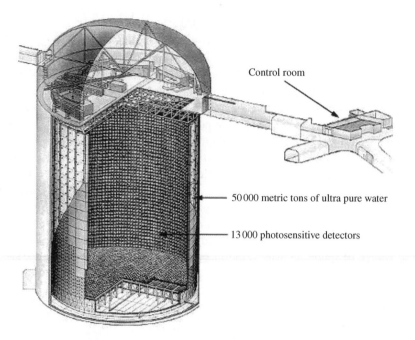

Figure 3.3 A schematic diagram of the *Super Kamiokande* detector (adapted from an original University of Hawaii, Manoa, illustration -- with permission)

indirectly by detecting the light (the so-called Čerenkov radiation – again, see Chapter 4) emitted by relativistic charged particles (the electrons or muons) that are created in, or pass through, the water. The outer region of water acts as a shield against low-energy particles entering the detector from outside. An additional 1200 photomultipliers are located there to detect muons that enter or exit the detector.

When cosmic ray protons collide with atoms in the upper atmosphere they create many pions, which in turn create neutrinos mainly by the decay sequences

$$\pi^- \rightarrow \mu^- + \bar{\nu}_\mu, \quad \pi^+ \rightarrow \mu^+ + \nu_\mu \tag{3.28}$$

and

$$\mu^- \rightarrow e^- + \bar{\nu}_e + \nu_\mu, \quad \mu^+ \rightarrow e^+ + \nu_e + \bar{\nu}_\mu. \tag{3.29}$$

From this, one would naively expect to detect two muon neutrinos for every electron neutrino. However, the ratio was observed to be about 1.3 to 1 on average, suggesting that the muon neutrinos produced might be oscillating into other species.

Clear confirmation for this was found by exploiting the fact that the detector could measure the direction of the detected neutrinos to study the azimuthal dependence of the effect. Since the flux of cosmic rays that lead to neutrinos with energies above about 1 GeV is isotropic, the production rate for neutrinos should be the same all around the Earth. In particular, one can compare the measured flux from neutrinos produced in the atmosphere directly above the detector, which have a short flight path before detection, with those incident from directly below, which have travelled a long way through the Earth before detection, and so have had plenty of time to oscillate (perhaps several cycles). Experimentally, it was found that the yield of electron neutrinos from above and below were the same within errors and consistent with the expectation for no oscillations. However, while the yield of muon neutrinos from above accorded with the expectation for no significant oscillations, the flux of muon neutrinos from below was a factor of about two lower. This is clear evidence for muon neutrino oscillations.

In a later development of the experiment, the flux of muon neutrinos was measured as a function of L/E by estimating L from the reconstructed neutrino direction. Values of L range from 15 km to 13000 km. The results are shown in Figure 3.4 in the form of the ratio of observed number of events to the theoretical expectation if there were no oscillations. The data show clear evidence for a deviation of this ratio from unity, particularly at large values of L/E.

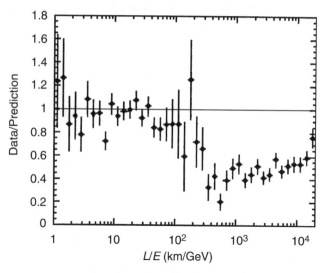

Figure 3.4 Data from the *Super Kamiokande* detector showing evidence for neutrino oscillations in atmospheric neutrinos (adapted from As04, copyright American Physical Society)

Other experiments also set limits on $P(\nu_\mu \to \nu_e)$ and taking these into account the most plausible hypothesis is that muon neutrinos are changing into tau neutrinos, which for the neutrino energies concerned could not be detected by Super Kamiokande. The data are consistent with this hypothesis and yield the values

$$1.9 \times 10^{-3} \le \Delta(m^2 c^4) \le 3.0 \times 10^{-3} \, (\text{eV})^2, \quad \sin^2(2\alpha) > 0.9 \qquad (3.30)$$

at 90 per cent confidence level. This conclusion is supported by preliminary results from laboratory-based experiments that start with a beam of ν_μ and measure the flux at a large distance (250 km) from the origin. Analysis of the data yields similar parameters to those above.

A second piece of evidence for neutrino oscillations comes from our knowledge of the Sun. We shall see in Chapter 8 that the energy of the Sun is due to various nuclear reactions and these produce a huge flux of electron neutrinos that can be detected at the surface of the Earth. Since the astrophysics of the Sun and nuclear production processes are well understood, this flux can be calculated with some confidence by what is known as the standard solar model.[7] However, the measured count rate is about a factor of two lower than the theoretical expectation. This is the so-called *solar neutrino problem*. It was first investigated by Davis and co-workers in the late 1960s who studied the reaction

$$\nu_e + {}^{37}\text{Cl} \rightarrow {}^{37}\text{Ar} + e^-, \tag{3.31}$$

to detect the neutrinos. (This required sensitive radiochemical analysis to confirm the production of ${}^{37}\text{Ar}$.) This reaction has a threshold of 0.81 MeV and is therefore only sensitive to relatively high-energy neutrinos from the Sun. Such neutrinos come predominantly from the weak interaction decay

$$^{8}\text{B} \rightarrow {}^{8}\text{Be} + e^+ + \nu_e, \tag{3.32}$$

where the neutrinos have an average energy $\sim 7\,\text{MeV}$. More recent experiments have studied the same process using the reactions

$$\text{(a) } \nu_x + d \rightarrow e^- + p + p, \quad \text{(b) } \nu_x + d \rightarrow \nu_x + p + n, \quad \text{(c) } \nu_x + e^- \rightarrow \nu_x + e^-, \tag{3.33}$$

to detect the neutrinos, where d is a deuteron. The first of these reactions clearly can be initiated with electron neutrinos only, whereas the other two can be initiated with neutrinos of any flavour. The measured flux of ν_e from reaction (a) agrees well with the standard solar model, but the ratio of the flux for ν_e to that for ν_x, where x could be a combination of μ and τ, obtained by using data from all three reactions, is less than unity. For example, the Sudbury Neutrino Observatory (SNO) experiment finds a ratio of about 0.3. Thus there is a flux of neutrinos of a type that did not come from the original decay process. The observations are further clear evidence for flavour oscillation.

Although the neutrinos from (3.32) have been extensively studied, this decay contributes only about 10^{-4} of the total solar neutrino flux. It is therefore important

---

[7]See, for example, Chapter 4 of Ph94.

to detect neutrinos from other reactions and in particular from the reaction

$$p + p \rightarrow d + e^+ + \nu_e, \tag{3.34}$$

which is the primary reaction that produces the energy of the Sun and contributes approximately 90 per cent of the solar neutrino flux. (It will be discussed in more detail in Chapter 8.) The neutrinos in this reaction have average energies of ~ 0.26 MeV and so cannot be detected by reaction (3.31). Instead, the reaction

$$\nu_e + {}^{71}\mathrm{Ga} \rightarrow {}^{71}\mathrm{Ge} + e^- \tag{3.35}$$

has been used, which has a threshold of 233 keV. (The experiments can also detect neutrinos from the solar reaction $e^- + {}^7\mathrm{Be} \rightarrow {}^7\mathrm{Li} + \nu_e$.) Just as for the original experiments of Davis et al., there are formidable problems in identifying the radioactive products from this reaction, which produces only about 1 atom of ${}^{71}\mathrm{Ge}$ per day in a target of 30 tons of gallium. Nevertheless, results from these experiments confirm the deficit of electron neutrinos and find between 60 and 70 per cent of the flux expected from the standard solar model without flavour changing.

These solar neutrino results require that interactions with matter play a significant role in flavour changing and imply, for example, that a substantial fraction of a beam of $\bar{\nu}_e$ would change to antineutrinos of other flavours after travelling a distance of the order of 100 km from its source. This prediction has been tested by the KamLAND group in Japan. They have studied the $\bar{\nu}_e$ flux from more than 60 reactors in Japan and South Korea after the neutrinos have travelled distances of between 150 and 200 km. They found that the $\bar{\nu}_e$ flux was only about 60 per cent of that expected from the known characteristics of the reactors. A simultaneous analysis of the data from this experiment and the solar neutrino data yields the result:

$$7.6 \times 10^{-5} \leq \Delta(m^2 c^4) \leq 8.8 \times 10^{-5} \, (\mathrm{eV})^2, \quad 0.32 \leq \tan^2(\alpha) \leq 0.48. \tag{3.36}$$

The existence of neutrino oscillations (flavour changing), and by implication non-zero neutrino masses, is now generally accepted on the basis of the above set of experiments.

What are the consequences of these results for the standard model? The observation of oscillations does not lead to a measurement of the neutrino masses, only (squared) mass differences, but combined with the tritium β-decay experiment, it would be natural to assume that neutrinos all had very small masses, with the mass differences being of the same order-of-magnitude as the masses themselves. The standard model can be modified to accommodate small masses, although methods for doing this are not without their own problems.[8] Unfortunately, the various

---

[8]One possibility will be mentioned briefly in Chapter 9 as part of a discussion of the general question of how masses arise in the standard model.

experiments – although producing compatible values for the mixing angle $\alpha \sim 40°$ – yield wildly different values for the mass difference, as can be seen from Equations (3.30) and (3.36). However, the analyses have been made in the framework of a two-component mixing model, whereas there are of course three neutrinos. Thus it could be, for example, that two of the neutrino states are separated by a small mass difference given by Equation (3.36) and the third is separated from them by a relatively large mass difference given by Equation (3.30). Progress will have to await experiments currently being planned to detect oscillations directly using prepared neutrino beams and which will make measurements at great distances from their origin. These experiments are expected to produce data in the next few years and should yield definitive values of the neutrino mass differences and the various mixing angles involved.[9]

The consequences for lepton number conservation are unclear. In the simple mixing model above, the total lepton number could still be conserved, but individual lepton numbers would not. However, there are other theoretical descriptions of neutrino oscillations and this is an open question. A definitive answer would be to detect *neutrinoless double β-decay*, such as

$$^{76}\text{Ge} \rightarrow {}^{76}\text{Se} + 2e^-, \tag{3.37}$$

where the final state contains two electrons, but no antineutrinos. This could occur if the neutrino emitted by the parent nucleus was internally absorbed by the daughter nucleus (i.e. it never appears as a real particle) which is possible only if $\nu_e \equiv \bar{\nu}_e$. A very recent experiment claims to have detected this decay, but the result is not universally accepted and at present 'the jury is out'. Experiments planned for the next few years should settle important questions about lepton number conservation and the nature of neutrinos.

3.1.5 Universal lepton interactions – the number of neutrinos

The three neutrinos have similar properties, but the three charged leptons are strikingly different. For example: the mass of the muon is roughly 200 times greater than that of the electron and consequently its magnetic moment is 200 times smaller; high-energy electrons are stopped by modest thicknesses of a centimetre or so of lead, while muons are the most penetrating form of radiation known, apart from neutrinos; and the tauon lifetime is many orders of magnitude smaller than the muon lifetime, while the electron is stable. It is therefore a remarkable fact that all experimental data are consistent with the assumption that the interactions of the electron and its associated neutrino are identical to those of the muon and its associated neutrino and of the tauon and its neutrino, *provided the*

---

[9]For a review of these experiments see, for example, http://www.hep.anl.gov/ndk/hypertext/nuindustry.html.

mass differences are taken into account. This property, called *lepton universality*, can be verified with great precision, because we have a precise theory of electromagnetic and weak interactions (to be discussed in Chapter 6), which enables predictions to be made of the mass dependence of all observables.

For example, when we discuss experimental methods in Chapter 4, we will show that the *radiation length*, which is a measure of how far a charged particle travels through matter before losing a certain fraction of its energy by radiation, is proportional to the squared mass of the radiating particle. Hence it is about 4×10^4 times greater for muons than for electrons, explaining their much greater penetrating power in matter. As another example, we have seen that the rates for weak β-decays are extremely sensitive to the kinetic energy released in the decay (recall the enormous variation in the lifetimes of nuclei decaying via β-decay). From dimensional arguments and the fact that they are weak interactions, the rates for muon and tau leptonic decays are predicted to be proportional to the fifth power of the relevant Q-values multiplied by G_F^2, the square of the Fermi coupling.[10] Thus, from universality, the ratio of the decay rates Γ is given approximately by

$$\frac{\Gamma(\tau^- \to e^- + \bar{\nu}_e + \nu_\tau)}{\Gamma(\mu^- \to e^- + \bar{\nu}_e + \nu_\mu)} \approx \left(\frac{Q_\tau}{Q_\mu}\right)^5 = 1.37 \times 10^6. \tag{3.38}$$

This is in excellent agreement with the experimental value of 1.35×10^6 (and is even closer in a full calculation) and accounts very well for the huge difference between the tau and muon lifetimes. The above are just some of the most striking manifestations of the universality of lepton interactions.

A question that arises naturally is whether there are more generations of leptons, with identical interactions, waiting to be discovered. This question has been answered, under reasonable assumptions, by an experimental study of the decays of the Z^0 boson. This particle, one of the two gauge bosons associated with the weak interaction, has a mass of 91 GeV/c^2. It decays, among other final states, to neutrino pairs

$$Z^0 \to \nu_\ell + \bar{\nu}_\ell \quad (\ell = e, \mu, \tau). \tag{3.39}$$

If we assume universal lepton interactions and neutrino masses which are small compared with the mass of the Z^0,[11] the decay rates to a given neutrino pair will all be equal and thus

$$\Gamma_{\text{neutrinos}} \equiv \Gamma_{\nu_e} + \Gamma_{\nu_\mu} + \Gamma_{\nu_\tau} + \cdots = N_\nu \Gamma_\nu, \tag{3.40}$$

---

[10]The increase of the decay rate as the fifth power of Q is known as *Sargent's Rule*.
[11]More precisely, we assume $m_\nu \leq M_Z/2$, so that the decays $Z \to \nu\bar{\nu}$ are not forbidden by energy conservation.

where N_ν is the number of neutrino species and Γ_ν is the decay rate to any given pair of neutrinos. The measured total decay rate may then be written

$$\Gamma_{\text{total}} = \Gamma_{\text{hadrons}} + \Gamma_{\text{leptons}} + \Gamma_{\text{neutrinos}}, \tag{3.41}$$

where the first two terms on the right are the measured decay rates to hadrons and charged leptons, respectively. Although the rate to neutrinos Γ_ν is not directly measured, it can be calculated in the standard model and combining this with experimental data for the other decay modes, a value of N_ν may be found. The best value using all available data is $N_\nu = 3.00 \pm 0.08$, which is consistent with the expectation for three neutrino species, but not four. The conclusion is that only three generations (flavours) of leptons can exist, if we assume universal lepton interactions and exclude very large neutrino masses.

Why there are just three generations of leptons remains a mystery, particularly as the extra two generations seem to tell us nothing fundamental that cannot be deduced from the interaction of the first generation.

3.2   Quarks

We turn now to the strongly interacting particles – the quarks and their bound states, the hadrons. These also interact by the weak and electromagnetic interactions, although such effects can often be neglected compared with the strong interactions. To this extent we are entering the realm of 'strong interaction physics'.

3.2.1   Evidence for quarks

Several hundred hadrons (not including nuclei) have been observed since pions were first produced in the laboratory in the early 1950s and all have zero or integer electric charges: $0, \pm 1,$ or ± 2 in units of e. They are all bound states of the fundamental spin-$\frac{1}{2}$ quarks, whose electric charges are either $+\frac{2}{3}$ or $-\frac{1}{3}$, and/or antiquarks, with charges $-\frac{2}{3}$ or $+\frac{1}{3}$. The quarks themselves have never been directly observed as single, free particles and, as remarked earlier, this fact initially made it difficult for quarks to be accepted as anything other than convenient mathematical quantities for performing calculations. Only later, when the fundamental reason for this was realized (it will be discussed in Chapter 6), were quarks universally accepted as physical entities. Nevertheless, there is compelling experimental evidence for their existence. The evidence comes from three main areas: *hadron spectroscopy, lepton scattering* and *jet production*.

Hadron spectroscopy

This is the study of the static properties of hadrons: their masses, lifetimes and decay modes, and especially the values of their quantum numbers, including spin, electric charge and several more that we define in Section 3.2.2. As mentioned in Chapter 1, the existence and properties of quarks were first inferred from hadron spectroscopy by Gell-Mann and independently by Zweig in 1964 and the close correspondence between the experimentally observed hadrons and those predicted by the quark model, which we will examine in more detail later, remains one of the strongest reasons for our belief in the existence of quarks.

Lepton scattering

It was mentioned in earlier chapters that in the early 1960s experiments were first performed where electrons were scattered from protons and neutrons. These strongly suggested that nucleons were not elementary. By the late 1960s this work had been extended to higher energies and with projectiles that included muons and neutrinos. In much the same way as Rutherford deduced the existence of the nucleus in atoms, high-energy lepton scattering, particularly at large momentum transfers, revealed the existence of point-like entities within the nucleons, which we now identify as quarks.

Jet production

High-energy collisions can cause the quarks within hadrons, or newly created quark–antiquark pairs, to fly apart from each other with very high energies. Before they can be observed, these quarks are converted into 'jets' of hadrons (a process referred to as *fragmentation*) whose production rates and angular distributions reflect those of the quarks from which they originated. They were first clearly identified in experiments at the DESY laboratory in Hamburg in 1979, where electrons and positrons were arranged to collide 'head-on' in a magnetic field. An example of a 'two-jet' event is shown in Figure 3.5. The picture is a computer reconstruction of an end view along the beam direction; the solid lines indicate the reconstructed charged particle trajectories taking into account the known magnetic field, which is also parallel to the beam direction; the dotted lines indicate the reconstructed trajectories of neutral particles, which were detected outside this device by other means.

The production rate and angular distribution of the observed jets closely matches that of quarks produced in the reaction

$$e^+ + e^- \rightarrow q + \bar{q}, \tag{3.42}$$

by the mechanism of Figure 3.6. Such jets have now been observed in many reactions, and are strong evidence for the existence of quarks within hadrons.

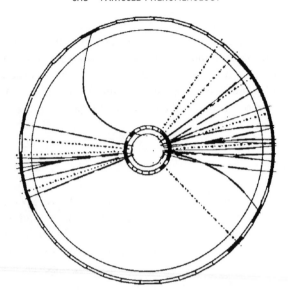

Figure 3.5   Two-jet event in e^+e^- collisions

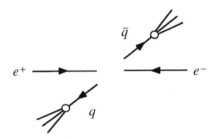

Figure 3.6   Mechanism for two-jet production in e^+e^- annihilation reaction

The failure to detect free quarks is not an experimental problem. Firstly, free quarks would be easily distinguished from other particles by their fractional charges and their resulting ionization properties.[12] Secondly, electric charge conservation implies that a fractionally charged particle cannot decay to a final state composed entirely of particles with integer electric charges. Hence the lightest fractionally charged particle, i.e. the lightest free quark, would be stable and so presumably easy to observe. Finally, some of the quarks are not very massive (see below) and because they interact by the strong interaction, one would expect free quarks to be copiously produced in, for example, high-energy proton–proton collisions. However, despite careful and exhaustive searches in ordinary matter, in cosmic rays and in high-energy collision products, free quarks have

---

[12]We will see in Chapter 4 that energy losses in matter due to ionization are proportional to the square of the charge and thus would be 'anomalously' small for quarks.

never been observed. The conclusion – that quarks exist solely within hadrons and not as isolated free particles – is called *confinement*. It is for this reason that we are forced to study the properties of hadrons, the bound states of quarks.

The modern theory of strong interactions, called *quantum chromodynamics* (*QCD*), which is discussed in Chapter 5, offers at least a qualitative account of confinement, although much of the detail eludes us due to the difficulty of performing accurate calculations. In what follows, we shall assume confinement and use the properties of quarks to interpret the properties of hadrons.

3.2.2 Quark generations and quark numbers

Six distinct types, or *flavours*, of spin-$\frac{1}{2}$ quarks are now known to exist. Like the leptons, they occur in pairs, or *generations*, denoted

$$\begin{pmatrix} u \\ d \end{pmatrix}, \quad \begin{pmatrix} c \\ s \end{pmatrix}, \quad \begin{pmatrix} t \\ b \end{pmatrix}. \tag{3.43}$$

Each generation consists of a quark with charge $+\frac{2}{3}$ (u, c, or t) together with a quark of charge $-\frac{1}{3}$ (d, s, or b), in units of e. They are called the *down* (d), *up* (u), *strange* (s), *charmed* (c), *bottom* (b) and *top* (t) quarks. The quantum numbers associated with the s, c, b and t quarks are called *strangeness, charm, beauty* and *truth*, respectively. The antiquarks are denoted

$$\begin{pmatrix} \bar{d} \\ \bar{u} \end{pmatrix}, \quad \begin{pmatrix} \bar{s} \\ \bar{c} \end{pmatrix}, \quad \begin{pmatrix} \bar{b} \\ \bar{t} \end{pmatrix} \tag{3.44}$$

with charges $+\frac{1}{3}$ ($\bar{d}$, $\bar{s}$, or $\bar{b}$) and $-\frac{2}{3}$ ($\bar{u}$, $\bar{c}$, $\bar{t}$).

Approximate quark masses are given in Table 3.2. Except for the top quark, these masses are inferred indirectly from the observed masses of their hadron

Table 3.2 Properties of quarks: all have spin $\frac{1}{2}$ and masses are given units of GeV/c²; the antiparticles (not shown) have the same masses as their associated particles, but the electric charges (Q) are reversed in sign (in the major decay modes, X denotes other particles)

Name	Symbol	Mass	Q	Lifetime (s)	Major decays
Down	d	$m_d \approx 0.3$	$-1/3$		
Up	u	$m_u \approx m_d$	$2/3$		
Strange	s	$m_s \approx 0.5$	$-1/3$	10^{-8}–10^{-10}	$s \rightarrow u + X$
Charmed	c	$m_c \approx 1.5$	$2/3$	10^{-12}–10^{-13}	$c \rightarrow s + X$
					$c \rightarrow d + X$
Bottom	b	$m_b \approx 4.5$	$-1/3$	10^{-12}–10^{-13}	$b \rightarrow c + X$
Top	t	$m_t = 180 \pm 12$	$2/3$	$\sim 10^{-25}$	$t \rightarrow b + X$

bound states, together with models of quark binding.[13] In this context they are also referred to as *constituent* quark masses.

The stability of quarks in hadrons – like the stability of protons and neutrons in atomic nuclei – is influenced by their interaction energies. However, for the s, c and b quarks these effects are small enough for them to be assigned approximate lifetimes of 10^{-8}–10^{-10} s for the s quark and 10^{-12}–10^{-13} s for both the c and b quarks. The top quark is much heavier than the other quarks and its lifetime is of the order of 10^{-25} s. This lifetime is so short that when top quarks are created they decay too quickly to form observable hadrons. In contrast to the other quarks, our knowledge of the top quark is based entirely on observations of its decay products.

When we talk about 'the decay of quarks' we always mean that the decay takes place within a hadron, with the other bound quarks acting as 'spectators', i.e. not taking part in the interaction. Thus, for example, in this picture neutron decay at the quark level is given by the Feynman diagram of Figure 3.7 and no free quarks are observed. Note that it is assumed that the exchanged particle interacts with only one constituent quark in the nucleons. This is the essence of the *spectator model*. (This is not dissimilar to the idea of a single nucleon decaying within a radioactive nucleus.)

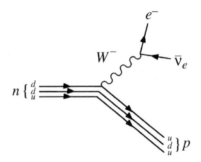

Figure 3.7 Spectator model quark Feynman diagram for the decay $n \rightarrow pe^-\bar{\nu}_e$

In strong and electromagnetic interactions, quarks can only be created or destroyed as particle–antiparticle pairs, just like electrons as we discussed in Section 3.1.1. This implies, for example, that in electromagnetic processes corresponding to the Feynman diagram of Figure 3.8, the reaction $e^+ + e^- \rightarrow c + \bar{c}$, which creates a $c\bar{c}$ pair, is allowed, but the reaction $e^+ + e^- \rightarrow c + \bar{u}$ producing a $c\bar{u}$ pair, is forbidden.[14]

More generally, it implies conservation of each of the six *quark numbers*

$$N_f \equiv N(f) - N(\bar{f}) \qquad (f = u, \ d, \ s, \ c, \ b, \ t) \qquad (3.45)$$

where $N(f)$ is the number of quarks of flavour f present and $N(\bar{f})$ is the number of antiquarks of flavour $\bar{f}$ present. For example, for single-particle states; $N_c = 1$ for the

---

[13]An analogy would be to deduce the mass of nucleons from the masses of nuclei via a model of the nucleus.
[14]Again, these reactions and associated Feynman diagrams do not imply that free quarks are created. Spectator quarks are implicitly present to form hadrons in the final state.

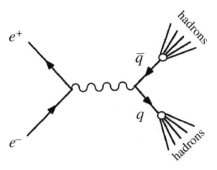

Figure 3.8   Production mechanism for the reaction $e^+e^- \to q\bar{q}$

c quark; $N_c = -1$ for the $\bar{c}$ antiquark; and $N_c = 0$ for all other particles. Similar results apply for the other quark numbers N_f, and for multi-particle states the quark numbers of the individual particles are simply added. Thus a state containing the particles u, u, d, has $N_u = 2$, $N_d = 1$ and $N_f = 0$ for the other quark numbers with $f = s, c, b, t$.

In weak interactions, more general possibilities are allowed, and only the total quark number

$$N_q \equiv N(q) - N(\bar{q}) \tag{3.46}$$

is conserved, where $N(q)$ and $N(\bar{q})$ are the total number of quarks and antiquarks present, irrespective of their flavour. This is illustrated by the decay modes of the quarks themselves, some of which are listed in Table 3.2, which are all weak inter-action processes, and we have seen it also in the decay of the neutron in Figure 3.7. Another example is the *main* decay mode of the charmed quark, which is

$$c \to s + u + \bar{d}, \tag{3.47}$$

in which a c quark is replaced by an s quark and a u quark is created together with a $\bar{d}$ antiquark. This clearly violates conservation of the individual quark numbers N_c, N_s, N_u and N_d, but the total quark number N_q is conserved.

In practice, it is convenient to replace the total quark number N_q in analyses by the *baryon number*, defined by

$$B \equiv N_q/3 = [N(q) - N(\bar{q})]/3. \tag{3.48}$$

Like the electric charge and the lepton numbers introduced in the last section, the baryon number is conserved in *all known interactions*, and unlike the lepton number, there are no experiments that suggest otherwise.[15]

---

[15]However, there are *theories* beyond the standard model that predict baryon number non-conservation, although there is no experimental evidence to support this prediction. These will be discussed briefly in Chapter 9.

3.3 Hadrons

In principle, the properties of atoms and nuclei can be explained in terms of their proton, neutron and electron constituents, although in practice many details are too complicated to be accurately calculated. However, the properties of these constituents can be determined without reference to atoms and nuclei by studying them directly as free particles in the laboratory. In this sense atomic and nuclear physics are no longer fundamental, although they are still very interesting and important if we want to understand the world we live in.

In the case of hadrons, the situation is more complicated. Their properties are explained in terms of a few fundamental quark constituents; but the properties of the quarks themselves can only be studied experimentally by appropriate measurements on hadrons. Whether we like it or not, studying quarks without hadrons is not an option.

3.3.1 Flavour independence and charge multiplets

One of the most fundamental properties of the strong interaction is *flavour independence*. This is the statement that the strong force between two quarks at a fixed distance apart is independent of which quark flavours u, d, s, c, b, t are involved. Thus, for example, the strong forces between us and ds pairs are identical. The same principle applies to quark–antiquark forces which are, however, *not* identical to quark–quark forces, because in the former case annihilations can occur. Flavour independence does not apply to the electromagnetic interaction, since the quarks have different electric charges, but compared with the strong force between quarks, the electromagnetic force is a small correction. In addition, in applying flavour independence one must take proper account of the quark mass differences, which can be non-trivial. However, there are cases where these corrections are small or easily estimated, and the phenomenon of flavour independence is plain to see.

One consequence of flavour independence is the striking observation that hadrons occur in families of particles with approximately the same masses, called *charge multiplets*. Within a given family, all particles have the same spin-parity and the same baryon number, strangeness, charm and beauty, but differ in their electric charges. Examples are the triplet of pions, (π^+, π^0, π^-) and the nucleon doublet (p, n). The latter behaviour reflects an approximate symmetry between u and d quarks. This arises because, as we shall see in Section 3.3.2, these two quarks have only a very small mass difference

$$m_d - m_u = (3 \pm 1)\text{MeV}/c^2, \tag{3.49}$$

so that in this case mass corrections can to a good approximation be neglected. For example, consider the proton and neutron. We shall see in the next section that their quark content is $p(938) = uud$ and $n(940) = udd$. If we neglect the small mass difference between the *u* and *d* quarks and also the electromagnetic interactions, which is equivalent to setting all electric charges to zero, so that the forces acting on the *u* and *d* quarks are exactly equal, then replacing the *u* quark by a *d* quark in the proton would produce a 'neutron' which would be essentially identical to the proton. Of course the symmetry is not exact because of the small mass difference between the *u* and *d* quarks and because of the electromagnetic forces, and it is these that give rise to the small differences in mass within multiplets.

Flavour independence of the strong forces between *u* and *d* quarks also leads directly to the *charge independence of nuclear forces*, e.g. the equality of the force between any pair of nucleons, provided the two particles are in the same spin state. Subsumed in the idea of charge independence is the idea of *charge symmetry*, i.e. the equality of the proton–proton and neutron–neutron forces, again provided the two particles are in the same spin state. Evidence for the latter is found in studies of nuclei with the same value of *A*, but the values of *N* and *Z* interchanged (*mirror nuclei*). An example is shown in Figure 3.9. The two nuclei $^{11}_{5}$B and $^{11}_{6}$C have the same number of *np* pairs, but $^{11}_{5}$B has 10 *pp* pairs and 15 *nn* pairs, whereas $^{11}_{6}$C has 15 *pp* pairs and 10 *nn* pairs. Thus, allowing for the Coulomb interaction, the approximate equality of the level structures of these two nuclei, as seen in Figure 3.9, means *charge symmetry* is approximately verified. To test charge independence in a nuclear context we would have to look at the level structure in three related nuclei such as $^{11}_{4}$Be, $^{11}_{5}$B and $^{11}_{6}$C.

Here the test is not so clear-cut because an *np* pair is not subject to the restrictions of the Pauli principle like *pp* and *nn* pairs and there is evidence (to be discussed briefly in Chapter 7) that the *np* force is stronger in the $S = 1$ state than in the $S = 0$ state. Nevertheless, the measured energy levels in such triplets of nuclei support the idea of approximate charge independence of nuclear forces.

The symmetry between *u* and *d* quarks is called *isospin symmetry* and greatly simplifies the interpretation of hadron physics. It is described by the same mathematics as ordinary spin, hence the name. For example, the proton and neutron are viewed as the 'up' and 'down' components of a single particle, the nucleon *N*, that has an isospin quantum number $\mathbf{I} = \frac{1}{2}$, with I_3 values $\frac{1}{2}$ and $-\frac{1}{2}$, assigned to the proton and neutron, where I_3 is analogous to the magnetic quantum number in the case of ordinary spin. Likewise, the three pions π^+, π^- and π^0 are part of a triplet π with $\mathbf{I} = 1$ corresponding to I_3 values 1, 0 and -1, respectively. In discussing the strong interactions between pions and nucleons, it is then only necessary to consider the πN interaction with total isospin either $\frac{1}{2}$ or $\frac{3}{2}$.

As an example, we will consider some predictions for the hadronic resonance $\Delta(1232)$. The $\Delta(1232)$ has $\mathbf{I} = \frac{3}{2}$ and four charge states Δ^{++}, Δ^+, Δ^0 and Δ^- (see Table 3.3) corresponding to $I_3 = \frac{3}{2}, \frac{1}{2}, -\frac{1}{2}, -\frac{3}{2}$, respectively. If we use the notation $\left| \pi N; I, I_3 \right\rangle$ for a πN state, then $\left| \pi N; \frac{3}{2}, \frac{3}{2} \right\rangle$ is the unique state $\pi^+ p$ and may be written

$$\left| \pi N; \frac{3}{2}, \frac{3}{2} \right\rangle = \left| \pi; 1, 1 \right\rangle \left| N; \frac{1}{2}, \frac{1}{2} \right\rangle. \qquad (3.50)$$

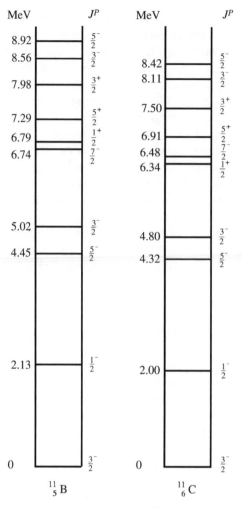

Figure 3.9 Low-lying energy levels with spin-parity J^P of the mirror nuclei $^{11}_5$B and $^{11}_6$C. (data from Aj90)

The other πN states may then be obtained by applying quantum mechanical shift (ladder) operators to Equation (3.50), as is done when constructing ordinary spin states. This gives[16]

$$\left|\pi N; \frac{3}{2}, \frac{1}{2}\right\rangle = -\sqrt{\frac{1}{3}}\left|\pi^+ n\right\rangle + \sqrt{\frac{2}{3}}\left|\pi^0 p\right\rangle \tag{3.51}$$

and hence isospin invariance predicts

$$\frac{\Gamma(\Delta^+ \rightarrow \pi^+ n)}{\Gamma(\Delta^+ \rightarrow \pi^0 p)} = \frac{1}{2}, \tag{3.52}$$

which is in good agreement with experiment.

---

[16]The reason for the minus sign and other details are given in, for example, Appendix D of Ma97.

Secondly, by constructing all the πN isospin states by analogy with Equations (3.50) and (3.51) we can show that

$$\left|\pi^- p\right\rangle = \frac{1}{\sqrt{3}}\left|\pi N; \frac{3}{2}, -\frac{1}{2}\right\rangle - \sqrt{\frac{2}{3}}\left|\pi N; \frac{1}{2}, -\frac{1}{2}\right\rangle \tag{3.53a}$$

and

$$\left|\pi^0 n\right\rangle = \sqrt{\frac{2}{3}}\left|\pi N; \frac{3}{2}, -\frac{1}{2}\right\rangle + \frac{1}{\sqrt{3}}\left|\pi N; \frac{1}{2}, -\frac{1}{2}\right\rangle. \tag{3.53b}$$

Then, if M_I is the amplitude for scattering in a pure isospin state I,

$$M(\pi^- p \to \pi^- p) = \frac{1}{3}M_3 + \frac{2}{3}M_1 \tag{3.54a}$$

and

$$M(\pi^- p \to \pi^0 n) = \frac{\sqrt{2}}{3}M_3 - \frac{\sqrt{2}}{3}M_1. \tag{3.54b}$$

At the $\Delta(1232)$, the available energy is such that the total cross-section is dominated by the elastic ($\pi^- p \to \pi^- p$) and charge-exchange ($\pi^- p \to \pi^0 n$) reactions. In addition, because the $\Delta(1232)$ has $\mathbf{I} = \frac{3}{2}$, $M_3 \gg M_1$, so

$$\sigma_{\text{total}}(\pi^- p) = \sigma(\pi^- p \to \pi^- p) + \sigma(\pi^- p \to \pi^0 n) \propto \frac{1}{3}|M_3|^2 \tag{3.55a}$$

and

$$\sigma_{\text{total}}(\pi^+ p) \propto |M_3|^2. \tag{3.55b}$$

Thus, neglecting small kinematic corrections due to mass differences (phase space corrections), isospin symmetry predicts

$$\frac{\sigma_{\text{total}}(\pi^+ p)}{\sigma_{\text{total}}(\pi^- p)} = 3. \tag{3.56}$$

Figure 3.10 shows the two total cross-sections at low energies. There are clear peaks with Breit–Wigner forms at a mass of 1232 MeV corresponding to the production of the $\Delta(1232)$ and the ratio of the peaks is in good agreement with the prediction of Equation (3.56).

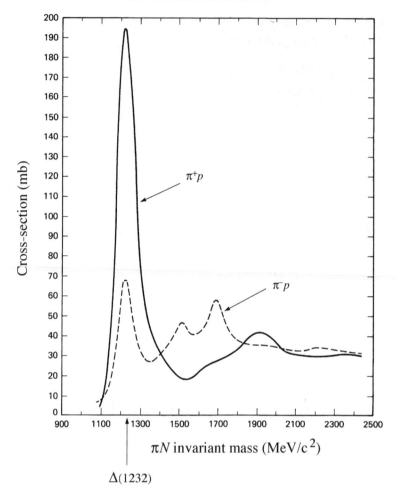

Figure 3.10 Total cross-sections for π^-p and π^+p scattering

3.3.2 Quark model spectroscopy

The observed hadrons are of three types. There are *baryons* and their antiparticles *antibaryons*, which have half-integral spin, and *mesons*, which have integral spin. In the *quark model of hadrons* the baryons are assumed to be bound states of three quarks ($3q$), *antibaryons* are assumed to be bound states of three antiquarks ($3\bar{q}$) and *mesons* are assumed to be bound states of a quark and an antiquark ($q\bar{q}$).[17] The

---

[17]In addition to these so-called 'valence' quarks there could also, in principle, be other constituent quarks present in the form of a cloud of virtual quarks and antiquarks – the so-called 'sea' quarks – the origin of which we will discuss in Chapter 5. In this chapter we consider only the valence quarks which determine the static properties of hadrons. The masses of the constituent quarks could be quite different from those that appear in the fundamental strong interaction Hamiltonian for quark–quark interactions via gluon exchange (i.e. QCD), because those quarks are free of the dynamical effects experienced in hadrons. The latter are referred to as 'current' quarks.

baryons and antibaryons have baryon numbers 1 and -1 respectively, while the mesons have baryon number 0. Hence the baryons and antibaryons can annihilate each other in reactions which conserve baryon number to give mesons or, more rarely, photons or lepton–antilepton pairs, in the final state.

The lightest known baryons are the proton and neutron, with the quark compositions given in Section 3.3.1:

$$p = uud \quad \text{and} \quad n = udd. \tag{3.57}$$

These particles have been familiar as constituents of atomic nuclei since the 1930s. The birth of particle physics as a new subject, distinct from atomic and nuclear physics, dates from 1947, when hadrons other than the neutron and proton were first detected. These were the *pions*, already mentioned, and the *kaons*, discovered in cosmic rays by groups in Bristol and Manchester Universities, UK, respectively.

The discovery of the pions was not totally unexpected, since Yukawa had famously predicted their existence and their approximate masses in 1935, in order to explain the observed range of nuclear forces (recall the discussion in Section 1.5.2). This consisted of finding what mass was needed in the Yukawa potential to give the observed range of the strong nuclear force (which was poorly known at the time). After some false signals, a particle with the right mass and suitable properties was discovered – this was the pion. Here and in what follows we will give the hadron masses in brackets in units of MeV/c^2 and use a superscript to indicate the electric charge in units of e. Thus the pions are $\pi^{\pm}(140)$, $\pi^{0}(135)$. Pions are the lightest known mesons and have the quark compositions

$$\pi^{+} = u\bar{d}, \quad \pi^{0} = u\bar{u}, \, d\bar{d}, \quad \pi^{-} = d\bar{u}. \tag{3.58}$$

While the charged pions have a unique composition, the neutral pion is composed of both $u\bar{u}$ and $d\bar{d}$ pairs in equal amounts. Pions are copiously produced in high-energy collisions by strong interaction processes such as $p + p \rightarrow p + n + \pi^{+}$.

In contrast to the discovery of the pions, the discovery of the kaons was totally unexpected, and they were almost immediately recognized as a completely new form of matter, because they had supposedly 'strange' properties. Eventually, after several years, it was realized that these properties were precisely what would be expected if kaons had non-zero values of a hitherto unknown quantum number, given the name *strangeness*, which was conserved in strong and electromagnetic interactions, but not necessarily conserved in weak interaction. Particles with non-zero strangeness were named *strange particles*, and with the advent of the quark model in 1964, it was realized that strangeness S was, apart from a sign, the strangeness quark number introduced earlier, i.e.

$$S = -N_s. \tag{3.59}$$

Kaons are the lightest strange mesons, with the quark compositions:

$$K^+(494) = u\bar{s} \quad \text{and} \quad K^0(498) = d\bar{s}, \tag{3.60}$$

where K^+ and K^0 have $S = +1$ and their antiparticles K^- and $\bar{K}^0$ have $S = -1$, while the lightest strange baryon is the *lambda*, with the quark composition $\Lambda = uds$. Subsequently, hadrons containing c and b quarks have also been discovered, with non-zero values of the *charm* and *beauty* quantum numbers defined by

$$C \equiv N_c \equiv N(c) - N(\bar{c}) \quad \text{and} \quad \tilde{B} \equiv -N_b \equiv -N(b) - N(\bar{b}). \tag{3.61}$$

The above examples illustrate just some of the many different combinations of quarks that form baryons or mesons. These and some further examples are shown in Table 3.3 and a complete listing is given in the PDG Tables.

Table 3.3 Some examples of baryons and mesons, with their major decay modes; masses are in MeV/c^2

Particle	Mass	Lifetime (s)	Major decays
$\pi^+(u\bar{d})$	140	2.6×10^{-8}	$\mu^+\nu_\mu$ ($\sim$100%)
$\pi^0(u\bar{u}, d\bar{d})$	135	8.4×10^{-17}	$\gamma\gamma$ ($\sim$100%)
$K^+(u\bar{s})$	494	1.2×10^{-8}	$\mu^+\nu_\mu$ (64%)
			$\pi^+\pi^0$ (21%)
$K^{*+}(u\bar{s})$	892	$\sim 1.3 \times 10^{-23}$	$K^+\pi^0$, $K^0\pi^+$ ($\sim$100%)
$D^-(d\bar{c})$	1869	1.1×10^{-12}	Several seen
$B^-(b\bar{u})$	5278	1.6×10^{-12}	Several seen
$p(uud)$	938	Stable	None
$n(udd)$	940	887	$pe^-\bar{\nu}_e$ (100%)
$\Lambda(uds)$	1116	2.6×10^{-10}	$p\pi^-$ (64%)
			$n\pi^0$ (36%)
$\Delta^{++}(uuu)$	1232	$\sim 0.6 \times 10^{-23}$	$p\pi^+$ (100%)
$\Omega^-(sss)$	1672	0.8×10^{-10}	ΛK^- (68%)
			$\Xi^0\pi^-$ (24%)
$\Lambda_c^+(udc)$	2285	2.1×10^{-13}	Several seen

To proceed more systematically one could, for example, construct all the mesons states of the form $q\bar{q}$, where q can be any of the six quark flavours. Each of these is labelled by its spin and its intrinsic parity P. The simplest such states would have the spins of the two quarks antiparallel with no orbital angular momentum between them and so have spin-parity $J^P = 0^-$. (Recall from Chapter 1 that quarks and antiquarks have opposite parities.) If, for simplicity, we consider those states composed of just u, d and s quarks, there will be nine such mesons and they have quantum numbers which may be identified with the observed mesons (K^0, K^+), $(\bar{K}^0, K^-)$, $(\pi^\pm, \pi^0)$ and two neutral particles, which are called η and η'. This *supermultiplet* is shown Figure 3.11(a) as a plot of Y, the *hypercharge*, defined as

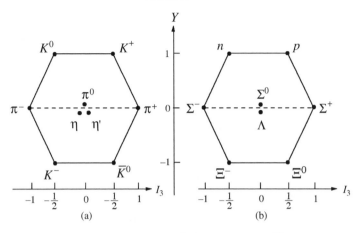

Figure 3.11   The lowest-lying states with (a) $J^P = 0^-$ and (b) $J = \frac{1}{2}^+$ that are composed of u, d and s quarks

$Y \equiv B + S + C + \tilde{B} + T$, against I_3, the third component of isospin. This can be extended to the lowest-lying qqq states and the lowest-lying supermultiplet consists of the eight $J^P = \frac{1}{2}^+$ baryons shown in Figure 3.11(b).[18]

It is a remarkable fact that the states observed experimentally agree with those predicted by the simple combinations qqq, $\bar{q}\bar{q}\bar{q}$ and $q\bar{q}$ and until very recently there was no evidence for states corresponding to any other combinations. However, some recent experiments have claimed evidence for the existence of a few states outside this scheme, possibly ones involving five quarks, although other experiments have failed to confirm this. Nevertheless, it is still a fact that hadron states are overwhelmingly composed of the simplest quark combinations of the basic quark model. This was one of the original pieces of evidence for the existence of quarks and remains one of the strongest today.

The scheme may also be extended to more quark flavours, although the diagrams become increasingly complex. For example, Figure 3.12 shows the predicted $J^P = \frac{3}{2}^+$ baryon states formed from u, d, s and c quarks when all three quarks have their spins aligned, but still with zero orbital angular momentum between them. All the states in the bottom plane have been detected as well as many in the higher planes and with the possible exception of the five-quark states mentioned previously, no states have been found that are outside this scheme. The latest situation may be found in the PDG Tables.

For many quark combinations there exist not one, but several states. For example, the lowest-lying state of the $u\bar{d}$ system has spin-parity 0^- and is the π^+ meson. It can be regarded as the 'ground state' of the $u\bar{d}$ system. Here the spins of the quark

---

[18]If you try to try to verify Figure 3.11, you will find that it is necessary to assume that the overall hadronic wavefunctions $\Psi = \psi_{\text{space}}\psi_{\text{spin}}$ are *symmetric* under the exchange of identical quarks, i.e. opposite to the symmetry required by the Pauli principle. This apparent contradiction will be resolved in Chapter 5.

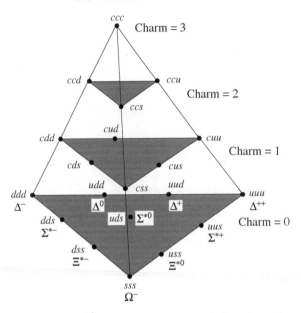

Figure 3.12   The $J = \frac{3}{2}^+$ baryon states composed of u, d, s and c quarks

constituents are anti-aligned to give a total spin $\mathbf{S} = \mathbf{0}$ and there is no orbital angular momentum $\mathbf{L}$ between the two quarks, so that the total angular momentum, which we identify as the spin of the hadron, is $\mathbf{J} = \mathbf{L} + \mathbf{S} = \mathbf{0}$. Other 'excited' states can have different spin-parities depending on the different states of motion of the quarks within the hadron.

An example is the $K^{*+}(890)$ meson shown in Table 3.3 with $J^P = 1^-$. In this state the u and $\bar{s}$ quarks have their spins aligned so that $\mathbf{S} = \mathbf{1}$ and there is no orbital angular momentum between them, i.e. $\mathbf{L} = \mathbf{0}$, so that the spin of the K^{*+} is $\mathbf{J} = \mathbf{L} + \mathbf{S} = \mathbf{1}$. This is a *resonance* and such states usually decay by the strong interaction, with very short lifetimes, of order 10^{-23} s. The mass distribution of their decay products is described by the Breit–Wigner formula we met in Section 1.6.3. The spin of a resonance may be found from an analysis of the angular distributions of its decay products. This is because the distribution will be determined by the wavefunction of the decaying particle and this will contain an angular part proportional to a spherical harmonic labelled by the orbital angular momentum between the decay products. Thus from a measurement of the angular distribution of the decay products, the angular momentum may be found, and hence the spin of the resonance. It is part of the triumph of the quark model that it successfully accounts for the excited states of the various quark systems, as well as their ground states, when the internal motion of the quarks is properly taken into account.

From experiments such as electron scattering we know that hadrons have typical radii r of the order of 1 fm and hence associated time scales r/c of the order of 10^{-23} s. The vast majority are highly unstable resonances, corresponding to excited states of the various quark systems, and decay to lighter hadrons by the strong interaction, with lifetimes of this order. The $K^{*+}(890) = u\bar{s}$ resonance, mentioned above, is an example. It decays to $K^+\pi^0$ and $K^0\pi^+$ final states with a lifetime of 1.3×10^{-23} s. The quark description of the process $K^{*+} \to K^0 + \pi^+$, for example, is

$$u\bar{s} \to d\bar{s} + u\bar{d}. \tag{3.62}$$

From this we see that the final state contains the same quarks as the initial state, plus an additional $d\bar{d}$ pair, so that the quark numbers N_u and N_d are separately conserved. This is characteristic of strong and electromagnetic processes, which are only allowed if each of the quark numbers N_u, N_d, N_s, N_c and N_b is separately conserved.

Since leptons and photons do not have strong interactions, hadrons can only decay by the strong interaction if lighter states composed solely of other hadrons exist with the same quantum numbers. While this is possible for the majority of hadrons, it is not in general possible for the lightest state corresponding to any given quark combination. These hadrons, which cannot decay by strong interactions, are long-lived on a timescale of the order of 10^{-23} s and are often called *stable particles*. It is more accurate to call them *long-lived particles*, because except for the proton they are not absolutely stable, but decay by either the electromagnetic or weak interaction.

The proton is stable because it is the lightest particle with non-zero baryon number and baryon number is conserved in all known interactions. A few of the other long-lived hadrons decay by electromagnetic interactions to final states that include photons. These decays, like the strong interaction, conserve all the individual quark numbers. An example of this is the neutral pion, which has $N_u = N_d = N_s = N_c = N_b = 0$ and decays by the reaction

$$\pi^0(u\bar{u}, d\bar{d}) \to \gamma + \gamma, \tag{3.63}$$

with a lifetime of 0.8×10^{-16} s. However, most of the long-lived hadrons have non-zero values for at least one of the quark numbers, and can only decay by the weak interaction, in which quark numbers do not have to be conserved. For example, the positive pion decays with a lifetime of 2.6×10^{-8} s by the reaction

$$\pi^+ \to \mu^+ + \nu_\mu, \tag{3.64}$$

while the $\Lambda(1116) = uds$ baryon decays mainly by the reactions

$$\Lambda \to p + \pi^- \quad \text{and} \quad n + \pi^0, \tag{3.65}$$

with a lifetime of 2.6×10^{-10} s. The quark interpretations of these reactions are

$$(u\bar{d}) \to \mu^+ + \nu_\mu, \tag{3.66}$$

in which a u quark annihilates with a $\bar{d}$ antiquark, violating both N_u and N_d conservation; and for lambda decay to charged pions,

$$sud \rightarrow uud + d\bar{u}, \tag{3.67}$$

in which an s quark turns into a u quark and a $u\bar{d}$ pair is created, violating N_d and N_s conservation.

We see from the above that the strong, electromagnetic or weak nature of a given hadron decay can be determined by inspecting quark numbers. The resulting lifetimes can then be summarized as follows. Strong decays lead to lifetimes that are typically of the order of 10^{-23} s. Electromagnetic decay rates are suppressed by powers of the fine structure constant α relative to strong decays, leading to observed lifetimes in the range 10^{-16}–10^{-21} s. Finally, weak decays give longer lifetimes, which depend sensitively on the characteristic energy of the decay.

A useful measure of the decay energy is the Q-value, the kinetic energy released in the decay of the particle at rest, which we metioned before in Section 2.3. In the weak interactions of hadrons, Q-values of the order of $10^2 - 10^3$ MeV are typical, leading to lifetimes in the range 10^{-7}–10^{-13} s, but there are some exceptions, notably neutron decay, $n \rightarrow p + e^- + \bar{\nu}_e$, for which

$$Q = m_n - m_p - m_e - m_{\bar{\nu}_e} = 0.79\,\text{MeV} \tag{3.68}$$

is unusually small, leading to a lifetime of about 10^3 s. Thus hadron decay lifetimes are reasonably well understood and span some 27 orders of magnitude, from about 10^{-24} s to about 10^3 s. The typical ranges corresponding to each interaction are summarized in Table 3.4.

Table 3.4   Typical lifetimes of hadrons decaying by the three interactions

Interaction	Lifetimes (s)
Strong	10^{-22}–10^{-24}
Electromagnetic	10^{-16}–10^{-21}
Weak	10^{-7}–10^{-13}

3.3.3   Hadron masses and magnetic moments

The quark model can make predictions for hadronic magnetic moments and masses in a way that is analogous to the semi-empirical mass formula for nuclear masses, i.e. the formulae have a theoretical basis, but contain parameters that have to be determined from experiment. We start by examining the case of baryon magnetic moments.

These have been measured only for the $\frac{1}{2}^+$ octet of states composed of u, d and s quarks and so we will consider only these. In this supermultiplet, the quarks have zero orbital angular momentum and so the hadron magnetic moments are just the sums of contributions from the constituent quark magnetic moments, which we will assume are of the Dirac form, i.e.

$$\mu_q \equiv \left\langle q, S_z = \frac{1}{2} \middle| \hat{\mu}_z \middle| q, S_z = \frac{1}{2} \right\rangle = e_q e\hbar/2m_q = (e_q M_p/m_q)\mu_N, \qquad (3.69)$$

where e_q is the quark charge in units of e and $\mu_N \equiv e\hbar/2M_p$ is the nuclear magneton. Thus

$$\mu_u = \frac{2M_p}{3m_u}\mu_N, \quad \mu_d = -\frac{M_p}{3m_d}\mu_N \quad \text{and} \quad \mu_s = -\frac{M_p}{3m_s}\mu_N. \qquad (3.70)$$

Consider, for example, the case of the $\Lambda(1116) = uds$. It is straightforward to show that the configuration that ensures that the predicted quantum numbers of the supermultiplet agree with experiment is to have the ud pair in a spin-0 state. Hence it makes no contribution to the Λ spin or magnetic moment. Thus we have the immediate prediction

$$\mu_\Lambda = \mu_s = -\frac{M_p}{3m_s}\mu_N. \qquad (3.71)$$

For $\frac{1}{2}^+$ baryons B with quark configuration aab, the aa pair is in the symmetric spin-1 state with parallel spins (again this is to ensure that the predicted quantum numbers of the supermultiplet agree with experiment) and magnetic moment $2\mu_a$. The 'spin-up' baryon state is given by

$$\left| B; S = \frac{1}{2}, S_z = \frac{1}{2} \right\rangle = \sqrt{\frac{2}{3}} \left| b; S = \frac{1}{2}, S_z = -\frac{1}{2} \right\rangle \left| aa; S = 1, S_z = 1 \right\rangle$$
$$- \sqrt{\frac{1}{3}} \left| b; S = \frac{1}{2}, S_z = \frac{1}{2} \right\rangle \left| aa; S = 1, S_z = 0 \right\rangle \qquad (3.72)$$

The first term corresponds to a state with magnetic moment $2\mu_a - \mu_b$, since the b quark has $S_z = -\frac{1}{2}$; the second term corresponds to a state with magnetic moment μ_b, since the aa pair has $S_z = 0$ and does not contribute. Hence the magnetic moment of B is given by

$$\mu_B = \frac{2}{3}(2\mu_a - \mu_b) + \frac{1}{3}\mu_b = \frac{4}{3}\mu_a - \frac{1}{3}\mu_b. \qquad (3.73)$$

For example, the magnetic moment of the proton is

$$\mu_p = \frac{4}{3}\mu_u - \frac{1}{3}\mu_d = \frac{M_p}{m}\mu_N, \tag{3.74}$$

where we have neglected the mass difference between the u and d quarks, as suggested by isospin symmetry, and set $m_u \approx m_d \equiv m$. The predictions for the magnetic moments of all the other members of the $\frac{1}{2}^+$ octet may be found in a similar way in terms of just two parameters, the masses m and m_s. A best fit to the measured magnetic moments (but not taking account of the errors on the data[19]) yields the values $m = 0.344\,\text{GeV/c}^2$ and $m_s = 0.539\,\text{GeV/c}^2$. The predicted moments are shown in Table 3.5. The agreement is good, but by no means perfect and suggests that the assumption that baryons are pure three-quark states with zero orbital angular momentum between them is not exact. For example, there could be small admixtures of states with non-zero orbital angular momentum.

Table 3.5 Magnetic moments of the $\frac{1}{2}^+$ baryon octet as predicted by the constituent quark model, compared with experiment in units of μ_N, the nuclear magneton; these have been obtained using $m = 0.344$ GeV/c^2 and $m_s = 0.539$ GeV/c^2 -- errors on the nucleon moments are of the order of 10^{-7}

Particle	Moment	Prediction	Experiment
$p(938)$	$\frac{4}{3}\mu_u - \frac{1}{3}\mu_d$	2.73	2.793
$n(940)$	$\frac{4}{3}\mu_d - \frac{1}{3}\mu_u$	-1.82	-1.913
$\Lambda(1116)$	μ_s	-0.58	-0.613 ± 0.004
$\Sigma^+(1189)$	$\frac{4}{3}\mu_u - \frac{1}{3}\mu_s$	2.62	2.458 ± 0.010
$\Sigma^-(1197)$	$\frac{4}{3}\mu_d - \frac{1}{3}\mu_s$	-1.02	-1.160 ± 0.025
$\Xi^0(1315)$	$\frac{4}{3}\mu_s - \frac{1}{3}\mu_u$	-1.38	-1.250 ± 0.014
$\Xi^-(1321)$	$\frac{4}{3}\mu_s - \frac{1}{3}\mu_d$	-0.47	-0.651 ± 0.003

We now turn to the prediction of hadron masses. The mass differences between members of a given supermulitplet are conveniently separated into the small mass differences between members of the same isospin multiplet and the much larger mass differences between members of different isospin multiplets. The size of the former suggests that they have their origin in electromagnetic effects, and if we neglect them then a first approximation would be to assume that the mass differences are due solely to differences in the constituent quark masses. If we concentrate on hadrons with quark structures composed of u, d and s quarks, since

---

[19]If we had fitted taking account of the errors, the fit would be dominated by the proton and neutron moments because they have very small errors.

their masses are the best known from experiment, this assumption leads directly to the relations

$$M_\Xi - M_\Sigma = M_\Xi - M_\Lambda = M_\Lambda - M_N = m_s - m_{u,d} \tag{3.75}$$

for the $\frac{1}{2}^+$ baryon octet and

$$M_\Omega - M_{\Xi^*} = M_{\Xi^*} - M_{\Sigma^*} = M_{\Sigma^*} - M_\Delta = m_s - m_{u,d} \tag{3.76}$$

for the $\frac{3}{2}^+$ decuplet. These give numerical estimates for $m_s - m_{u,d}$ in the range 120 to 200 MeV/c^2, which are consistent with the estimate from magnetic moments above.

These results support the suggestion that baryon mass differences (and by analogy meson mass differences) are dominantly due to the mass differences of their constituent quarks. However, this cannot be the complete explanation, because if it were then the $\frac{1}{2}^+$ nucleon would have the same mass as the $\frac{3}{2}^+$ $\Delta(1232)$, as they have the same quark constituents, and similarly for other related particles in the $\frac{1}{2}^+$ octet and $\frac{3}{2}^+$ decuplet. The absence of orbital angular momentum in these states means that there is nothing equivalent to the 'fine structure' of atomic physics. The difference lies in the spin structures of these states.

If we take the case of two spin-$\frac{1}{2}$ particles with magnetic moments $\boldsymbol{\mu}_i$ and $\boldsymbol{\mu}_j$ separated by a distance r_{ij} then the interaction energy is proportional to $\boldsymbol{\mu}_i \cdot \boldsymbol{\mu}_j / r_{ij}^3$. If, in addition, the particles are point-like and have charges e_i and e_j, the moments will be of the Dirac form $\boldsymbol{\mu}_i = (e_i/m_i)\mathbf{S}_i$. Then for two particles in a relative S-state it can be shown that the interaction energy is given by

$$\Delta E = \frac{8\pi}{3} \frac{e_i e_j}{m_i m_j} |\psi(0)|^2 \mathbf{S}_i \cdot \mathbf{S}_j, \tag{3.77}$$

where $\psi(0)$ is the wavefunction at the origin, $r_{ij} = 0$. (When averaged over all space, the interaction is zero except at the origin.) In atomic physics this is known as the *hyperfine interaction* and causes very small splittings in atomic energy levels. In the hadron case, the electric charges must be replaced by their strong interaction equivalents with appropriate changes to the overall numerical factor. The resulting interaction is called (for reasons that will be clear in Chapter 5) the *chromomagnetic interaction*. As we cannot calculate the equivalent quark–quark wavefunction, for the purposes of a phenomenological analysis we will write the contribution to the hadron mass as

$$\Delta M \propto \frac{\mathbf{S}_1 \cdot \mathbf{S}_2}{m_1 m_2}. \tag{3.78}$$

This of course assumes that $|\psi(0)|^2$ is the same for all states, which will not be exactly true.

Consider first the case of mesons. By writing the total spin squared as

$$\mathbf{S}^2 \equiv (\mathbf{S}_1 + \mathbf{S}_2)^2 = \mathbf{S}_1^2 + \mathbf{S}_2^2 + 2\mathbf{S}_1 \cdot \mathbf{S}_2, \qquad (3.79)$$

we easily find the expect values of $\mathbf{S}_1 \cdot \mathbf{S}_2$ are $-\frac{3}{4}\hbar^2$ for the $\mathbf{S} = \mathbf{0}$ (pseudoscalar) mesons and $\frac{1}{4}\hbar^2$ for the $\mathbf{S} = \mathbf{1}$ (vector) mesons. Then the masses may be written

$$M(\text{meson}) = m_1 + m_2 + \Delta M, \qquad (3.80)$$

where $m_{1,2}$ are the masses of the constituent quarks and

$$\Delta M(J^P = 0^- \text{ meson}) = -\frac{3a}{4}\frac{1}{m_1 m_2}, \quad \Delta M(J^P = 1^- \text{ meson}) = \frac{a}{4}\frac{1}{m_1 m_2} \qquad (3.81)$$

and a is a constant to be found from experiment. The masses of the members of the 0^- and 1^- meson supermultiplets then follow from a knowledge of their quark compositions. For example, the K-mesons have one u or d quark and one s quark and so

$$M_K = m + m_s - \frac{3a}{8}\left(\frac{1}{m^2} + \frac{1}{m_s^2}\right). \qquad (3.82)$$

Predictions for the masses of all the mesons are shown in Table 3.6, which also gives the best fit to the measured masses (again ignoring the relative errors on the latter) using these formulae. The predictions correspond to the values

$$m = 0.308 \,\text{GeV}/c^2, \quad m_s = 0.482 \text{GeV}/c^2, \quad a = 0.0588 \,(\text{GeV}/c^2)^3. \qquad (3.83)$$

Note that the quark mass values are smaller than those obtained from fitting the baryon magnetic moments. There is no contradiction in this, because there is no reason that quarks should have the same effective masses in mesons as in baryons.

Table 3.6   Meson masses (in Gev/c^2) in the constituent quark model compared with experimental values

Particle	Mass	Prediction	Experiment
π	$2m - \dfrac{3a}{4m^2}$	0.15	0.137
K	$m + m_s - \dfrac{3a}{8}\left(\dfrac{1}{m^2} + \dfrac{1}{m_s^2}\right)$	0.46	0.496
η	$\dfrac{2}{3}m + \dfrac{4}{3}m_s - \dfrac{a}{4}\left(\dfrac{1}{m^2} + \dfrac{2}{m_s^2}\right)$	0.57	0.549
ρ	$2m + \dfrac{a}{4m^2}$	0.77	0.770
ω	$2m + \dfrac{a}{4m^2}$	0.77	0.782
K^*	$m + m_s + \dfrac{a}{8}\left(\dfrac{1}{m^2} + \dfrac{1}{m_s^2}\right)$	0.87	0.892
ϕ	$2m_s + \dfrac{a}{4m_s^2}$	1.03	1.020

The comparison with the measured values is very reasonable, but omitted from the fit is the η' state where the fit is very poor indeed. Unlike the atomic case, the spin–spin interaction in the strong interaction case leads to substantial corrections to the meson masses.

The baryons are somewhat more complicated, because in this case we have three pairs of spin–spin couplings to consider. In general the spin–spin contribution to the mass is

$$\Delta M \propto \sum_{i<j} \frac{\mathbf{S}_i \cdot \mathbf{S}_j}{m_i m_j}, \quad i,j = 1,3. \tag{3.84}$$

In the case of the $\frac{3}{2}^+$ decuplet, all three quarks have their spins aligned and every pair therefore combines to make spin-1. Thus for example,

$$(\mathbf{S}_1 + \mathbf{S}_2)^2 = \mathbf{S}_1^2 + \mathbf{S}_2^2 + 2\mathbf{S}_1 \cdot \mathbf{S}_2 = 2\hbar^2, \tag{3.85}$$

giving $\mathbf{S}_1 \cdot \mathbf{S}_2 = \hbar^2/4$ and in general

$$\mathbf{S}_1 \cdot \mathbf{S}_2 = \mathbf{S}_1 \cdot \mathbf{S}_3 = \mathbf{S}_2 \cdot \mathbf{S}_3 = \hbar^2/4. \tag{3.86}$$

Using this result, the mass of the $\Sigma^*(1385)$, for example, may be written

$$M_{\Sigma^*} = 2m + m_s + \frac{b}{4}\left(\frac{1}{m^2} + \frac{2}{mm_s}\right), \tag{3.87}$$

where b is a constant to be determined from experiment. (There is no reason for b to be equal to the constant a used in the meson case because the quark wavefunctions and numerical factors in the baryonic equivalent of Equation (3.77) will be different in the two cases.)

In the case of the $\frac{1}{2}^+$ octet, we have

$$(\mathbf{S}_1^2 + \mathbf{S}_2^2 + \mathbf{S}_3^2) = \mathbf{S}_1^2 + \mathbf{S}_2^2 + \mathbf{S}_3^2 + 2(\mathbf{S}_1 \cdot \mathbf{S}_2 + \mathbf{S}_1 \cdot \mathbf{S}_3 + \mathbf{S}_2 \cdot \mathbf{S}_3) = 3\hbar^2/4 \tag{3.88}$$

and hence

$$\mathbf{S}_1 \cdot \mathbf{S}_2 + \mathbf{S}_1 \cdot \mathbf{S}_3 + \mathbf{S}_2 \cdot \mathbf{S}_3 = -3\hbar^2/4. \tag{3.89}$$

In addition, we have to consider the symmetry of the spin wavefunctions of individual hadrons. For example, without proof (this will be given in Chapter 5), the spins of the u and d pair in the Λ must combine to give $\mathbf{S} = 0$. Thus, $(\mathbf{S}_u + \mathbf{S}_d)^2 = 0$, so that $\mathbf{S}_u \cdot \mathbf{S}_d = -3\hbar^2/4$. Then,

$$M_\Lambda = m_u + m_d + m_s + b\left[\frac{\mathbf{S}_u \cdot \mathbf{S}_d}{m_u m_d} + \frac{\mathbf{S}_u \cdot \mathbf{S}_s}{m_u m_s} + \frac{\mathbf{S}_d \cdot \mathbf{S}_s}{m_d m_s}\right]. \tag{3.90}$$

Table 3.7   Baryon masses (in Gev/c^2) in the constituent quark model compared with experimental values

Particle	Mass	Prediction	Experiment
N	$3m - \dfrac{3b}{4m^2}$	0.89	0.939
Λ	$2m + m_s - \dfrac{3b}{4}\left(\dfrac{1}{m^2}\right)$	1.08	1.116
Σ	$2m + m_s + \dfrac{b}{4}\left(\dfrac{1}{m^2} - \dfrac{4}{mm_s}\right)$	1.15	1.193
Ξ	$m + 2m_s + \dfrac{b}{4}\left(\dfrac{1}{m_s^2} - \dfrac{4}{mm_s}\right)$	1.32	1.318
Δ	$3m + \dfrac{3b}{4m^2}$	1.07	1.232
Σ^*	$2m + m_s + \dfrac{b}{4}\left(\dfrac{1}{m} + \dfrac{2}{mm_s}\right)$	1.34	1.385
Ξ^*	$m + 2m_s + \dfrac{b}{4}\left(\dfrac{2}{mm_s} + \dfrac{1}{m_s^2}\right)$	1.50	1.533
Ω	$3m_s + \dfrac{3b}{m_s^2}$	1.68	1.673

Finally, setting $m_u = m_d = m$ and absorbing factors of $\hbar^2$ into the constant b, gives

$$M_\Lambda = 2m + m_s + b\left[\frac{\mathbf{S}_u \cdot \mathbf{S}_d}{m^2} + \frac{(\mathbf{S}_1 \cdot \mathbf{S}_2 + \mathbf{S}_1 \cdot \mathbf{S}_3 + \mathbf{S}_2 \cdot \mathbf{S}_3 - \mathbf{S}_u \cdot \mathbf{S}_d)}{mm_s}\right] = 2m + m_s - \frac{3b}{4m^2},$$

$$(3.91)$$

where we have used Equation (3.89). The resulting formulae for all the $\frac{1}{2}^+$ octet and $\frac{3}{2}^+$ decuplet masses are shown in Table 3.7. Also shown are the predicted masses, where for consistency we have used the same quark mass values obtained earlier in fitting baryon magnetic moments, i.e. $m = 0.308\,\text{GeV}/c^2$ and $m_s = 0.482\,\text{GeV}/c^2$, and varied only the parameter b, giving a value $0.0225\,(\text{GeV}/c^2)^3$. The fit is quite reasonable and although better fits can be obtained by allowing the masses to vary there is little justification for this, given the approximations of the analysis.

Overall, what we learn from the above is that the constituent quark model is capable of giving a reasonably consistent account of hadron masses and magnetic moments, at least for the low-lying states (the η' is an exception), provided a few parameters are allowed to be found from experiment.

Problems

3.1   Which of the following reactions are allowed and which are forbidden by the conservation laws appropriate to weak interactions?

(a) $\nu_\mu + p \rightarrow \mu^+ + n$;

(b) $\nu_e + p \rightarrow n + e^- + \pi^+$;

(c) $\Lambda \rightarrow \pi^+ + e^- + \bar{\nu}_e$;

(d) $K^+ \rightarrow \pi^0 + \mu^+ + \nu_\mu$.

3.2 Draw the lowest-order Feynman diagram at the quark level for the following decays:

(a) $D^- \rightarrow K^0 + \pi^-$;

(b) $\Lambda \rightarrow p + e^- + \bar{\nu}_e$.

3.3 Consider the following combinations of quantum numbers $(Q, B, S, C, \tilde{B})$ where $Q =$ electric charge, $B =$ baryon number, $S =$ strangeness, $C =$ charm and $\tilde{B} =$ beauty:

(a) $(-1, 1, -2, 0, -1)$;

(b) $(0, 0, 1, 0, 1)$.

Which of these possible states are compatible with the postulates of the quark model?

3.4 Consider a scenario where overall hadronic wavefunctions Ψ consist of just spin and space parts, i.e. $\Psi = \psi_{space} \psi_{spin}$. What would be the resulting multiplet structure of the lowest-lying baryon states composed of u, d and s quarks?

3.5 Draw Feynman diagrams at the quark level for the reactions:

(a) $e^+ + e^- \rightarrow \bar{B}^0 + B^0$, where B is a meson containing a b-quark;

(b) $\pi^- + p \rightarrow K^0 + \Lambda^0$.

3.6 Find the parity P and charge conjugation C values for the ground-state $(J = 0)$ meson π and its first excited $(J = 1)$ state ρ. Why does the charged pion have a longer lifetime than the ρ? Explain also why the decay $\rho^0 \rightarrow \pi^+\pi^-$ has been observed, but not the decay $\rho^0 \rightarrow \pi^0\pi^0$.

3.7 The particle Y^- can be produced in the strong interaction process $K^- + p \rightarrow K^+ + Y^-$. Deduce its baryon number, strangeness, charm and beauty, and using these, its quark content. The $Y^-(1311)$ decays by the reaction $Y^- \rightarrow \Lambda + \pi^-$. Give a rough estimate of its lifetime.

3.8 Verify the expression in Table 3.7 for the mass of the $\frac{1}{2}^+ \Sigma$ baryon, given that the spins of the two non-strange quarks combine to give $\mathbf{S} = 1$.

3.9 Consider the reaction $K^- + p \rightarrow \Omega^- + K^+ + K^0$ followed by the sequence of decays

$$\Omega^- \rightarrow \Xi^0 + \pi^-$$
$$\quad \hookrightarrow \pi^0 + \Lambda, \qquad K^+ \rightarrow \pi^+ + \pi^0 \qquad \text{and} \quad K^0 \rightarrow \pi^+ + \pi^- + \pi^0$$
$$\quad \hookrightarrow \gamma + \gamma \qquad\qquad \hookrightarrow \mu^+ + \nu_\mu$$

Classify each process as strong, weak or electromagnetic and give your reasons.

3.10 Draw the lowest-order Feynman diagram for the decay $K^+ \to \mu^+ + \nu_\mu + \gamma$ and hence deduce the form of the overall effective coupling.

3.11 A KamLAND-type experiment detects $\bar{\nu}_e$ neutrinos at a distance of 200 m from a nuclear reactor and finds that the flux is (90 ± 10) per cent of that expected if there were no oscillations. Assuming maximal mixing and a mean neutrino energy of 3 MeV, use this result to estimate upper and lower bounds on the squared mass of the $\bar{\nu}_e$.

3.12 Comment on the feasibility of the following reactions:

(a) $p + \bar{p} \to \pi^+ + \pi^-$;

(b) $p \to e^+ + \gamma$;

(c) $\Sigma^0 \to \Lambda + \gamma$;

(d) $p + p \to \Sigma^+ + n + K^0 + \pi^+$;

(e) $\Xi^- \to \Lambda + \pi^-$;

(f) $\Delta^+ \to p + \pi^0$.

3.13 Use the results of Section 3.3.1 to deduce a relation between the total cross-sections for the reactions $\pi^- p \to K^0 \Sigma^0$, $\pi^- p \to K^+ \Sigma^-$ and $\pi^+ p \to K^+ \Sigma^+$ at a fixed energy.

3.14 At a certain energy $\sigma(\pi^+ n) \approx \sigma(\pi^- p)$, whereas $\sigma(K^+ n) \neq \sigma(K^- p)$. Explain this.

4
Experimental Methods

In earlier chapters we have discussed the results of a number of experiments, but said almost nothing about how such experiments are done. In this chapter we will take a brief look at experimental methods. This is a very extensive subject and the aim will not be to give a comprehensive review, but rather to emphasize the physical principles behind the methods. More details may be found in specialized texts.[1]

4.1 Overview

To explore the structure of nuclei (nuclear physics) or hadrons (particle physics) requires projectiles whose wavelengths are at least as small as the effective radii of the nuclei or hadrons. This determines the minimum value of the momentum $p = h/\lambda$ and hence the energy required. The majority of experiments are conducted using beams of particles produced by machines called *accelerators*. This has the great advantage that the projectiles are of a single type, and have energies that may be controlled by the experimenter.[2] For example, beams that are essentially mono-energetic may be prepared, and can be used to study the energy dependence of interactions. The beam, once established, is directed onto a target so that interactions may be produced. In a *fixed-target* experiment the target is stationary in the laboratory. Nuclear physics experiments are almost invariably of this type, as are many experiments in particle physics.

In particle physics, high energies are also required to produce new and unstable particles and this reveals a disadvantage of fixed-target experiments when large

---

[1]See, for example, Fe86 and Kl86.
[2]Nevertheless, important experiments are still performed without using accelerators, for example some of those described in Chapter 3 on neutrino oscillations used cosmic rays and nuclear reactors. In fact cosmic rays are still the source of the very highest-energy particles.

Nuclear and Particle Physics   B. R. Martin
© 2006 John Wiley & Sons, Ltd

centre-of-mass energies are required. The centre-of-mass energy is important because it is a measure of the energy available to create new particles. In the laboratory frame at least some of the final-state particles must be in motion to conserve momentum. Consequently, at least some of the initial beam energy must reappear as kinetic energy of the final-state particles, and is therefore unavailable for particle production. In contrast, in the centre-of-mass frame the total momentum is zero and, in principle, all the energy is available for particle production.

To find the centre-of-mass energy we use the expression

$$E_{CM}^2 = (P_t + P_b)^2 c^2, \tag{4.1}$$

where P is the particle's four-momentum and the subscripts t and b refer to target and beam, respectively.[3] For a fixed-target experiment in the laboratory we have

$$P_t = (m_t c, \ \mathbf{0}); \quad P_b = (E_L/c, \ \mathbf{p_b}). \tag{4.2}$$

Expanding Equation (4.1) gives

$$E_{CM}^2 = \left(P_t^2 + P_b^2 + 2P_t P_b\right)c^2 \tag{4.3}$$

and using $P_t^2 = m_t^2 c^2$ etc., together with the general result

$$P_i P_j = E_i E_j / c^2 - \mathbf{p}_i \cdot \mathbf{p}_j , \tag{4.4}$$

we have

$$E_{CM} = \left[m_b^2 c^4 + m_t^2 c^4 + 2m_t c^2 E_L\right]^{1/2}. \tag{4.5}$$

At high energies this increases only as $(E_L)^{\frac{1}{2}}$ and so an increasingly smaller fraction of the beam energy is available for particle production, most going to impart kinetic energy to the target.

In a *colliding-beam* accelerator, two beams of particles travelling in almost opposite directions are made to collide at a small or zero crossing angle. If for simplicity we assume the particles in the two beams have the same mass and laboratory energy E_L and collide at zero crossing angle, then the total centre-of-mass energy is

$$E_{CM} = 2E_L. \tag{4.6}$$

This increases linearly with the energy of the accelerated particles, and hence is a significant improvement on the fixed-target result. Colliding-beam experiments are not, however, without their own disadvantages. The colliding particles have to be stable, which limits the interactions that can be studied, and the collision rate in the intersection region is smaller than that achieved in fixed-target experiments, because the particle densities in the beams are low compared with a solid or liquid target.

---

[3]A brief summary of relativistic kinematics is given in Appendix B.

Finally, details of the particles produced in the collision (e.g. their momenta) are deduced by observing their interactions with the material of *detectors*, which are placed in the vicinity of the interaction region. A wide range of detectors is available. Some have a very specific characteristic, others serve more than one purpose. Modern experiments, particularly in particle physics, typically use several types in a single experiment.

In this chapter we start by describing some of the different types of accelerator that have been built, the beams that they can produce and also how beams of neutral and unstable particles can be prepared. Then we discuss the ways in which particles interact with matter, and finally review how these mechanisms are exploited in the construction of a range of particle detectors.

4.2   Accelerators and Beams

All accelerators use electromagnetic forces to boost the energy of stable charged particles. These are injected into the machine from a device that provides a high-intensity source of low-energy particles, for example an electron gun (a hot filament), or a proton ion source. The accelerators used for nuclear structure studies may be classified into those that develop a steady accelerating field (DC machines) and those in which radio frequency electric fields are used (AC machines). All accelerators for particle physics are of the latter type. We start with a brief description of DC machines.

4.2.1   DC accelerators

The earliest type of DC accelerator was the *Cockcroft–Walton machine*, in which ions pass through sets of aligned electrodes that are operated at successively higher potentials. These machines are limited to energies of about 1 MeV, but are still sometimes used as injectors as part of the multistage process of accelerating particles to higher energies.[4]

The most important DC machine in current use is the *van de Graaff accelerator* and an ingenious version of this, known as the *tandem van de Graaff*, that doubles the energy of the simple machine, is shown schematically in Figure 4.1. The key to this type of device is to establish a very high voltage. The van de Graaff accelerator achieves this by using the fact that the charge on a conductor resides on its outermost surface and hence if a conductor carrying charge touches another conductor it will transfer its charge to the outer surface of the second conductor.

---

[4]Sir John Cockcroft and Ernest Walton received the 1951 Nobel Prize in Physics for the development of their accelerator and the subsequent nuclear physics experiments they did using it.

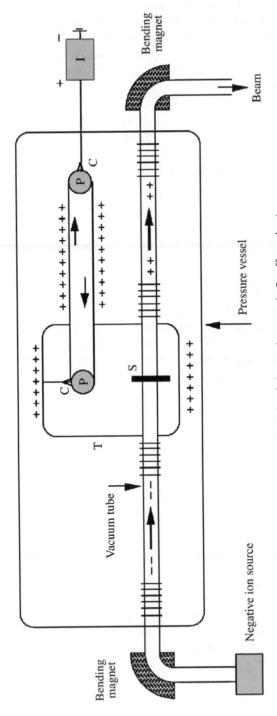

Figure 4.1 Principle of the tandem van de Graaff accelerator

In Figure 4.1, a high voltage source at I passes positive ions to a belt via a comb arrangement at C. The belt is motor driven via the pulleys at P and the ions are carried on the belt to a second pulley where they are collected by another comb located within a metal vessel T. The charges are then transferred to the outer surface of the vessel, which acts as an extended terminal. In this way a high voltage is established on T. Singly-charged negative ions are injected from a source and accelerated along a vacuum tube towards T. Within T there is a stripper S (for example a thin carbon foil) that removes two or more electrons from the projectiles to produce positive ions. The latter then continue to accelerate through the second half of the accelerator increasing their energy still further and finally may be bent and collimated to produce a beam of positive ions. This brief account ignores many technical details. For example, an inert gas at high pressure is used to minimize electrical breakdown by the high voltage. The highest energy van de Graaff accelerator can achieve a potential of about 30–40 MeV for singly-charged ions and greater if more than one electron is removed by the stripper. It has been a mainstay of nuclear research.

4.2.2   AC accelerators

Accelerators using radio frequency (r.f.) electric fields may conveniently be divided into *linear* and *cyclic* varieties.

Linear accelerators

In a linear accelerator (or *linac*) for acclerating ions, particles pass through a series of metal pipes called *drift tubes*, that are located in a vacuum vessel and connected successively to alternate terminals of an r.f. oscillator, as shown in Figure 4.2. Positive ions accelerated by the field move towards the first drift tube. If the

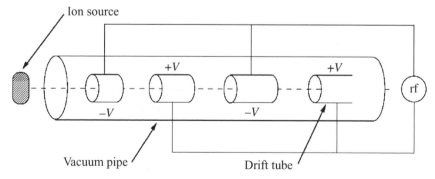

Figure 4.2   Acceleration in a linear ion accelerator

alternator can change its direction before the ions passes through that tube, then they will be accelerated again on their way between the exit of the first and entry to the second tube, and so on. Thus the particles will form bunches. Because the particles are accelerating, their speed is increasing and hence the lengths of the drift tubes have to increase to ensure continuous acceleration. To produce a useful beam the particles must keep in phase with the r.f. field and remain focused.

In the case of electrons, their velocity very rapidly approaches the speed of light. In this case a variation of the linac method that is more efficient is used to accelerate them. Bunches of particles pass through a straight evacuated waveguide with a periodic array of gaps, similar to the ion accelerator. Radio frequency oscillations in the gaps are used to establish a moving electromagnetic wave in the structure, with a longitudinal component of the electric field moving in phase with the particles. As long as this phase relationship can be maintained, the particles will be continuously accelerated. Radio frequency power is pumped into the waveguide at intervals to compensate for resistive losses and gives energy to the electrons. The largest electron linac is at the SLAC laboratory in Stanford, USA, and has a maximum energy of 50 GeV. It is over 3 km long.

An ingenious way of reducing the enormous lengths of high-energy linacs has been developed at the Continuous Electron Beam Accelerator Facility (CEBAF) at the Jefferson Laboratory in the USA. This utilizes the fact that above about 50 MeV, electron velocities are very close to the speed of light and thus electrons of very different energies can be accelerated in the same drift tube. Instead of a single long linac, the CEBAF machine consists of two much shorter linacs and the beam from one is bent and passed through the other. This can be repeated for up to four cycles. Even with the radiation losses inherent in bending the beams, very intense beams can be produced with energies between 0.5 and 6.0 GeV. CEBAF is proving to be an important machine in the energy region where nuclear physics and particle physics descriptions overlap.

Cyclic accelerators

Cyclic accelerators used for low-energy nuclear physics experiments are of a type called *cyclotrons*. They are also used to produce beams of particles for medical applications, including proton beams for radiation therapy. Cyclotrons operate in a somewhat different way to cyclic accelerators used in particle physics, which are called *synchrotrons*. In a cyclotron,[5] charged particles are constrained to move in near-circular orbits by a magnetic field during the acceleration process. There are several types of cyclotron; we will describe just one. This is illustrated

---

[5]The cyclotron was invented by Ernest Lawrence, who received the 1939 Nobel Prize in Physics for this and the experimental work he did using it.

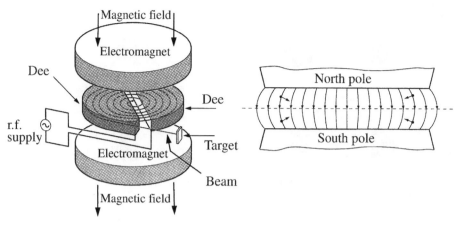

Figure 4.3 Schematic diagram of a cyclotron (adapted from Kr88, copyright John Wiley & Sons)

schematically in Figure 4.3. The accelerator consists of two 'dee'-shaped sections across which an r.f. electric field is established. Charged ions are injected into the machine near its centre and are constrained to traverse outward in spiral trajectories by a magnetic field. The ions are accelerated each time they pass across the gap between the dees. At the maximum radius, which corresponds to the maximum energy, the beam is extracted. The shape of the magnetic field, which is also shown in Figure 4.3, ensures that forces act on particles not orbiting in the medium plane to move them closer to this plane. This brief description ignores the considerable problems that have to be overcome to ensure that the beam remains focused during the acceleration.

The principle of a *synchrotron* is analogous to that of a linear accelerator, but where the acceleration takes place in a near circular orbit rather than in a straight line. The beam of particles travels in an evacuated tube called the *beam pipe* and is constrained in a circular or near circular path by an array of dipole magnets called bending magnets (Figure 4.4). Acceleration is achieved as the beam repeatedly traverses one or more cavities placed in the ring where energy is given to the particles. Since the particles travel in a circular orbit they continuously emit radiation, called in this context *synchrotron radiation*. The amount of energy radiated per turn by a relativistic particle of mass m is proportional to $1/m^4$. For electrons the losses are thus very severe, and the need to compensate for these by the input of large amounts of r.f. power limits the energies of electron synchrotrons.

The momentum in GeV/c of an orbiting particle assumed to have unit charge is given by $p = 0.3B\rho$, where B is the magnetic field in Tesla and ρ, the radius of curvature, is measured in metres. Because p is increased during acceleration, B must also be steadily increased if ρ is to remain constant, and the final momentum is limited both by the maximum field available and by the size of the ring. With

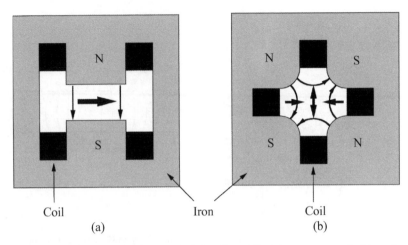

Figure 4.4 Cross-section of (a) a typical bending (dipole) magnet, and (b) a focusing (quadrupole) magnet; the thin arrows indicate field directions; the thick arrows indicate the force on a negative particle travelling into the paper

conventional electromagnets, the largest field attainable over an adequate region is about 1.5 T, and even with superconducting coils it is only of the order of 10 T. Hence the radius of the ring must be very large to achieve very high energies. For example, the Tevatron accelerator, located at the Fermi National Laboratory, Chicago, USA, which accelerates protons to an energy of 1 TeV, has a radius of 1 km. A large radius is also important to limit synchrotron radiation losses in electron machines.

In the course of its acceleration, a beam may make typically 10^5 traversals of its orbit before reaching its maximum energy. Consequently stability of the orbit is vital, both to ensure that the particles continue to be accelerated, and that they do not strike the sides of the vacuum tube. In practice, the particles are accelerated in bunches each being synchronized with the r.f. field. In equilibrium, a particle increases its momentum just enough to keep the radius of curvature constant as the field B is increased during one rotation, and the circulation frequency of the particle is in step with the r.f. of the field. This is illustrated in Figure 4.5. With obvious changes, a similar principle is used in linear accelerators.

In practice, the particles remain in the bunch, but their trajectories oscillate about the stable orbits. These oscillations are controlled by a series of focusing magnets, usually of the quadrupole type, which are placed at intervals around the beam and act like optical lenses. A schematic diagram of one of these is shown in Figure 4.4. Each focuses the beam in one direction and so alternate magnets have their field directions reversed.

In addition to the energy of the beam, one is also concerned to produce a beam of high intensity, so that interactions will be plentiful. The intensity is ultimately limited by defocussing effects, e.g. the mutual repulsion of the particles in the

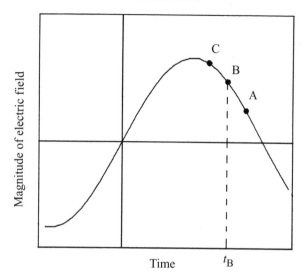

Figure 4.5 Magnitude of the electric field as a function of time at a fixed point in the rf cavity: particle B is synchronous with the field and arrives at time t_B; particle A (C) is behind (ahead of) B and receives an increase (decrease) in its rotational frequency -- thus particles oscillate about the equilibrium orbit

beam, and a number of technical problems have to be overcome which are outside the scope of this brief account.[6]

Fixed-target machines and colliders

Both linear and cyclic accelerators can be divided into *fixed-target* and *colliding-beam* machines. The latter are also known as *colliders*, or sometimes in the case of cyclic machines, *storage rings*.[7] In fixed-target machines, particles are accelerated to the highest operating energy and then the beam is extracted from the machine and directed onto a stationary target, which is usually a solid or liquid. Much higher energies have been achieved for protons than electrons, because of the large radiation losses inherent in electron machines mentioned earlier. The intensity of the beam is such that large numbers of interactions can be produced, which can either be studied in their own right or used to produce secondary beams.

The main disadvantage of fixed-target machines for particle physics has been mentioned earlier: the need to achieve large centre-of-mass energies to produce

---

[6]Very recently (2005) significant progress has been made on an 'induction synchrotron' in which a changing magnetic field produces the electric field that accelerates the particles. This device has the potential to overcome certain effects that limit the intensity achievable in conventional synchrotrons.

[7]The use of the terms *storage rings* and *colliders* as synonymous is not strictly correct, because we will see that the former can also describe a machine that stores a single beam for use on both internal and external fixed targets.

new particles. Almost all new machines for particle physics are therefore colliders, although some fixed-target machines for specialized purposes are still constructed. The largest collider currently under construction is the Large Hadron Collider (LHC), which is being built at CERN, Geneva, Switzerland. This is a massive *pp* accelerator of circumference 27 km, with each beam having an energy of 7 TeV. A schematic diagram of the CERN site showing the LHC and some of its other accelerators is shown in Figure 4.6. The acceleration process starts with a linac

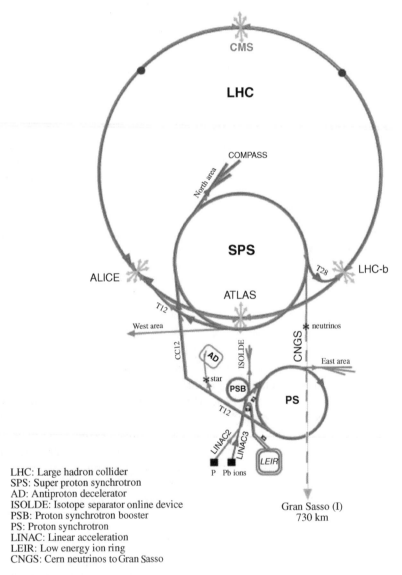

LHC: Large hadron collider
SPS: Super proton synchrotron
AD: Antiproton decelerator
ISOLDE: Isotope separator online device
PSB: Proton synchrotron booster
PS: Proton synchrotron
LINAC: Linear acceleration
LEIR: Low energy ion ring
CNGS: Cern neutrinos to Gran Sasso

Figure 4.6 A schematic diagram of the CERN site showing the LHC and some of its other accelerators (CERN photo, reproduced with permission)

whose beam is boosted in energy in the Proton Synchrotron Booster (PSB) and passed to the Proton Synchrotron (PS), a machine that is still the source of beams for lower-energy experiments. The beam energy is increased still further in the Super Proton Synchrotron (SPS) that also provides beams for a range of experiments as well as the injection beams for the LHC itself. Four beam intersection points are shown in the LHC and experiments (ALICE, CMS, LHC-b and ATLAS) will be located at each of these. The extracted neutrino beam shown at the bottom of the diagram is sent to the Gran Sasso laboratory 730 km away and is used, amongst other things, for experiments on neutrino oscillations of the type discussed in Chapter 3.

Another very large collider we should mention is the Relativistic Heavy Ion Collider (RHIC), located at Brookhaven National Laboratory, USA. This unique machine, which began operation in 2000 following 10 years of development and construction, is the first collider in the world capable of accelerating heavy ions. Like the LHC above, there are several stages, involving a linac, a tandem van de Graaff and a synchrotron, before the ions are injected into the main machine. There they form two counter-circulating beams controlled by two 4-km rings of superconducting magnets and are accelerated to an energy of 100 GeV/nucleon. Thus the total centre-of-mass energy is 200 GeV/nucleon. Collisions occur at six intersection points, where major experiments can be sited. RHIC primarily accelerates ions of gold and is used to study matter at extreme energy-densities, where a new state of matter called a 'quark–gluon plasma' is predicted to occur. We will return to this briefly in Chapter 9.

The performance of a collider is characterized by its luminosity, which was defined in Chapter 1. The general formula for luminosity given there is shown in Problem 1.10 to reduce in the case of a collider to the useful form

$$L = n\frac{N_1 N_2}{A} f, \tag{4.7}$$

where $N_i (i = 1, 2)$ are the numbers of particles in the n colliding bunches, A is the cross-sectional area of the beam and f is the frequency, i.e. $f = 1/T$, where T is the time taken for the particles to make one traversal of the ring.

An interesting proton synchrotron for nuclear physics studies is the COSY facility located at the Research Centre Jülich, Germany. Low-energy protons are pre-accelerated in a cyclotron, then cooled to reduce their transverse momentum and injected into a synchrotron, where they are further accelerated to momenta in the range 600–3700 MeV/c (corresponding to energies of 175–2880 MeV). The protons can be stored in the ring for appreciable times and are available for experiments not only in the usual way by extracting the beam, but also by using the circulating beam to interact with a very thin internal target. Thus we have a mixture of storage rings and fixed targets. The fact that the circulating beam may make as many as 10^{10} traversals through the target compensates to some extent for its low particle density.

4.2.3   Neutral and unstable particle beams

The particles used in accelerators must be stable and charged, but one is also interested in the interaction of neutral particles, e.g. photons and neutrons, as well as those of unstable particles, such as charged pions. Beams appropriate for performing such experiments are produced in a number of ways.

We have seen that neutrons are the natural product of radioactive decays and we will see in Chapter 8 that a large flux of neutrons is present in a nuclear reactor. Typically these will have a spectrum concentrated at low energies of 1–2 MeV, but extending as high as 5–6 MeV. Purpose-built reactors exist for research purposes, such as the ILL reactor at the Institut Laue-Langevin, France. Another source of neutrons is via the spallation process. The most important neutron spallation source at present is ISIS located at the Rutherford Appleton Laboratory, UK. Protons which have been accelerated in a linac to 70 MeV are injected into a synchrotron that further accelerates them to 800 MeV, where they collide with a heavy metal target of tantalum. The interaction drives out neutrons from the target and provides an intense pulsed source. In each case, if beams of lower-energy neutrons are required these are produced by slowing down faster neutrons in moderators, which are materials with a large cross-section for elastic scattering, but a small cross-section for absorption. In Chapter 8 we will see that moderators are vital for the successful extraction of power from fission nuclear reactors.

Beams of unstable particles can be formed provided their constituents live long enough to travel appreciable distances in the laboratory. One way of doing this is to direct an extracted primary beam onto a heavy target. In the resulting interactions with the target nuclei, many new particles are produced which may then be analysed into secondary beams of well-defined momentum. Such beams will ideally consist predominantly of particles of one type, but if this cannot be achieved, then the wanted species may have to be identified by other means. In addition, if these secondary beams are composed of unstable particles, they can themselves be used to produce further beams formed from their decay products. Two examples will illustrate how, in principle, such secondary particle beams can be formed.

Consider firstly the construction of a K^- beam from a primary beam of protons. By allowing the protons to interact with a heavy target, secondary particles will be produced. Most of these will be pions, but a few will be kaons (that have to be produced with a hyperon to conserve strangeness – this an example of so-called *associated production*). A collimator can be used to select particles in a particular direction, and the K^- component can subsequently be removed and focused into a mono-energetic beam by selective use of electrostatic fields and bending and focusing magnets.

The pion beam may also be used to produce a beam of neutrinos. For example, the π^- is unstable and as we have seen, one of its weak interaction decays modes is $\pi^- \rightarrow \mu^- + \bar{\nu}_\mu$. So if the pions are passed down a long vacuum pipe, many will decay in flight to give muons and anti-neutrinos, which will mostly travel in essentially the same direction as the initial beam. The muons and any remaining pions can then be removed by passing the beam through a very long absorber, leaving the neutrinos. In this case the final neutrino beam will have a momentum spectrum reflecting the initial momentum spectrum of the pions and, since neutrinos are neutral, no further momentum selection using magnets is possible.

4.3 Particle Interactions with Matter

In order to be detected, a particle must undergo an interaction with the material of a detector. In this section we discuss these interactions, but only in sufficient detail to be able to understand the detectors themselves.

The first possibility is that the particle interacts with an atomic nucleus. For example, this could be via the strong nuclear interaction if it is a hadron, or by the weak interaction if it is a neutrino. We know from the work of Chapter 1 that both are *short-range interactions*. If the energy is sufficiently high, new particles may be produced, and such reactions are often the first step in the detection process. In addition to these short-range interactions, a charged particle will also excite and ionize atoms along its path, giving rise to *ionization energy losses*, and emit radiation, leading to *radiation energy losses*. Both of these processes are due to the long-range electromagnetic interaction. They are important because they form the basis of most detectors for charged particles. Photons are also directly detected by electromagnetic interactions, and at high energies their interactions with matter lead predominantly to the production of e^+e^- pairs via the *pair production* process $\gamma \rightarrow e^+ + e^-$, which has to occur in the vicinity of a nucleus to conserve energy and momentum. (Recall the discussion in Chapter 1 on the range of forces.) All these types of interactions are described in the following sections.

4.3.1 Short-range interactions with nuclei

For hadrons, the most important short-range interactions with nuclei are due to the strong nuclear force which, unlike the electromagnetic interaction, is as important for neutral particles as for charged ones, because of the charge-independence of the strong interaction. Both elastic scattering and inelastic reactions may result. At high energies, many inelastic reactions are possible, most of them involving the production of several particles in the final state.

Many hadronic cross-sections show considerable structure at low energies due to the production of hadronic resonances, but at energies above about 3 GeV, total cross-sections are usually slowly varying in the range 10–50 mb and are much larger than the elastic cross-section. (The example of $\pi^- p$ scattering is shown in Figure 4.7.) This is of the same order-of-magnitude as the 'geometrical'

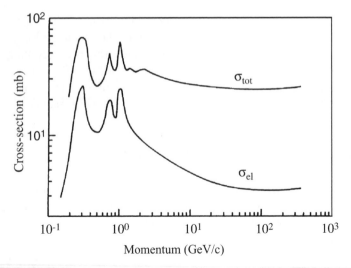

Figure 4.7 Total and elastic cross-sections for $\pi^- p$ scattering as functions of the pion laboratory momentum

cross-section $\pi r^2 \approx 30$ mb, where $r \approx 1$ fm is the approximate range of the strong interaction between hadrons. Total cross-sections on nuclei are much larger see for example Figure (2.17), increasing roughly as the square of the nuclear radius, i.e. as $A^{2/3}$.

A special case is the detection of *thermal* neutrons (defined as those with kinetic energies energies below about 0.02 eV). We have seen in Chapter 2 that neutrons in this region have very large cross-sections for being absorbed, leading to the production of a compound nucleus which decays by delayed emission of a γ-ray. Examples of these so-called neutron activation reactions are $^{63}\text{Cu}(n,\gamma)^{64}\text{Cu}$ and $^{55}\text{Mn}(n,\gamma)^{56}\text{Mn}$.

The probability of a hadron-nucleus interaction occurring as the hadron traverses a small thickness dx of material is given by $n\sigma_{\text{tot}}\text{d}x$, where n is the number of nuclei per unit volume in the material. Consequently, the mean distance travelled before an interaction occurs is given by

$$\ell_c = 1/n\sigma_{\text{tot}}. \tag{4.8}$$

This is called the *collision length*. An analogous quantity is the *absorption length*, defined by

$$\ell_a = 1/n\sigma_{\text{inel}}, \tag{4.9}$$

that governs the probability of an inelastic collision. In practice, $\ell_c \approx \ell_a$ at high energies. As examples, the interaction lengths are between 10 and 40 cm for nucleons of energy in the range 100–300 GeV interacting with metals such as iron.

Neutrinos and antineutrinos can also be absorbed by nuclei, leading to reactions of the type

$$\bar{\nu}_\ell + p \rightarrow \ell^+ + X, \tag{4.10}$$

where ℓ is a lepton and X denotes any hadron or set of hadrons allowed by the conservation laws. Such processes are weak interactions (because they involve neutrinos) and the associated cross-sections are extremely small compared with the cross-sections for strong interaction processes. The corresponding interaction lengths are therefore enormous. Nonetheless, in the absence of other possibilities such reactions are the basis for detecting neutrinos. Finally, photons can be absorbed by nuclei, giving *photoproduction* reactions such as $\gamma + p \rightarrow X$. However, these electromagnetic interactions are only used to detect photons at low energies, because at higher energies there is a far larger probability for e^+e^- pair production in the Coulomb field of the nucleus. We will return to this in Section 4.3.4.

4.3.2 Ionization energy losses

Ionization energy losses are important for all charged particles, and for particles other than electrons and positrons they dominate over radiation energy losses at all but the highest attainable energies. The theory of such losses, which are due dominantly to Coulomb scattering from the atomic electrons, was worked out by Bethe, Bloch and others in the 1930s. The result is called the Bethe–Bloch formula, and for spin-0 bosons with charge $\pm q$ (in units of e), mass M and velocity v, it takes the approximate form (neglecting small corrections for highly relativistic particles)

$$-\frac{dE}{dx} = \frac{D q^2 n_e}{\beta^2} \left[\ln\left(\frac{2m_e c^2 \beta^2 \gamma^2}{I} \right) - \beta^2 \right], \tag{4.11}$$

where x is the distance travelled through the medium;

$$D = \frac{4\pi\alpha^2 \hbar^2}{m_e} = 5.1 \times 10^{-25} \text{MeV cm}^2, \tag{4.12}$$

m_e is the electron mass, $\beta = v/c$ and $\gamma = (1 - \beta^2)^{-1/2}$. The other constants refer to the properties of the medium: n_e is the electron density; I is the mean ionization potential of the atoms averaged over all electrons, which is given approximately by $I = 10 Z$ eV for Z greater than 20. The corresponding formula for spin-$\frac{1}{2}$ particles differs from this, but in practice the differences are small and may be neglected in discussing the main features of ionization energy loses.

Examples of the behaviour of $-dE/dx$ for muons, pions and protons traversing a range of materials is shown in Figure 4.8. It is common practice to absorb the density ρ of the medium by dividing by ρ and expressing dE/dx in terms of an

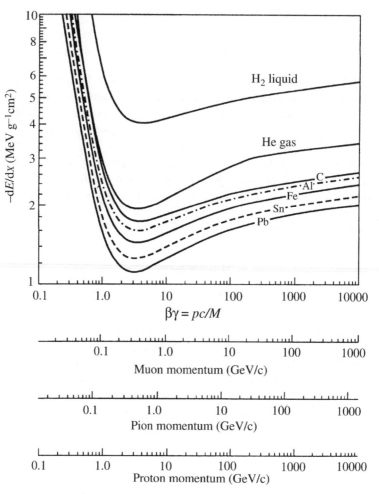

Figure 4.8 Ionization energy loss for muons, pions and protons on a variety of materials (reprinted from Ei04, copyright Elsevier, with permission)

equivalent thickness of gm cm^{-2} – hence the units in Figure 4.8. As can be seen, $-dE/dx$ falls rapidly as the velocity increases from zero because of the $1/\beta^2$ factor in the Bethe–Bloch equation. All particles have a region of 'minimum ionization' for $\beta\gamma$ in the range 3–4. Beyond this, β tends to unity, and the logarithmic factor in the Bethe–Bloch formula gives a 'relativistic rise' in $-dE/dx$.

The magnitude of the energy loss depends on the medium. The electron density is given by $n_e = \rho N_A Z/A$, where N_A is Avogadro's number, and ρ and A are the mass density and atomic weight of the medium, so the mean energy loss is proportional to the density of the medium. The remaining dependence on the medium is relatively weak because $Z/A \approx 0.5$ for all atoms except the very light and the very heavy elements, and because the ionization energy I only enters the Bethe–Bloch formula logarithmically. In the 'minimum ionization' region where $\beta\gamma \approx 3$–4, the minimum value of $-dE/dx$ can be calculated from Equation (4.11) and for a particle with unit charge is given approximately by

$$\left(-\frac{dE}{dx}\right)_{min} \approx 3.5\frac{Z}{A}\,\mathrm{MeVg^{-1}cm^2}. \tag{4.13}$$

Ionization losses are proportional to the squared charge of the particle, so that a fractionally charged particle with $\beta\gamma \geq 3$ would have a much lower rate of energy loss than the minimum energy loss of any integrally charged particle. This has been used as a means of identifying possible free quarks, but without success.

From the knowledge of the rate of energy loss, we can calculate the attenuation as a function of distance travelled in the medium. This is called the *Bragg curve*. Most of the ionization loss occurs near the end of the path where the speed is smallest and the curve has a pronounced peak (the *Bragg peak*) close to the end point before falling rapidly to zero at the end of the particle's path length. The *range R*, i.e. the mean distance a particle travels before it comes to rest is defined as

$$R \equiv \int_0^{x_{max}} dx(\beta), \tag{4.14}$$

which, using Equation (4.11), may be written

$$R = \int_0^{\beta_{initial}} \left[-\frac{dE}{dx}\right]^{-1}\frac{dE}{d\beta}\,d\beta = \frac{M}{q^2 n_e}F(\beta_{initial}), \tag{4.15}$$

where F is a function of the initial velocity and we have used the relation $E = \gamma Mc^2$ to show the dependence on the projectile mass M.

The range as given by Equation (4.15) is actually an average value because scattering is a statistical process and there will therefore be a spread of values for individual particles. The spread will be greater for light particles and smaller for heavier particles such as α-particles. These properties have implications for the use of radiation in therapeutic situations, where it may be necessary to deposit energy within a small region at a specific depth of tissue, for example to precisely target a cancer. The biological effects of radiation will be discussed in Chapter 8.

Because neutrons are uncharged, direct detection is not possible by ionization methods. However, they can be detected via the action of the charged products of induced direct nuclear reactions. Commonly used reactions are ${}^6\mathrm{Li}(n,\alpha){}^3\mathrm{H}$, ${}^{10}\mathrm{B}(n,\alpha){}^7\mathrm{Li}$ and ${}^3\mathrm{He}(n,p){}^3\mathrm{H}$. All these reactions are exothermic and so are very suitable for detecting neutrons with energies below about 20 MeV. Moreover, as nuclear cross-sections tend to increase as v^{-1} at low energies, detection becomes more efficient the slower the neutron.

4.3.3   Radiation energy losses

When a charged particle traverses matter it can also lose energy by radiative collisions, especially with nuclei. The electric field of a nucleus will accelerate and decelerate the particles as they pass, causing them to radiate photons, and hence lose energy. This process is called *bremsstrahlung* (literally 'braking radiation' in German) and is a particularly important contribution to the energy loss for electrons and positrons.

The dominant Feynman diagrams for electron bremsstrahlung in the field of a nucleus, i.e.

$$e^- + (Z,A) \rightarrow e^- + \gamma + (Z,A), \tag{4.16}$$

are shown in Figure 4.9 and are of the order of $Z^2\alpha^3$. The function of the nucleus is to absorb the recoil energy and so ensure that energy and momentum are simultaneously conserved (recall the discussion of Feynman diagrams in Chapter 1).

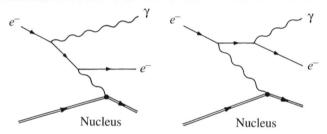

Figure 4.9   Dominant Feynman diagrams for the bremsstrahlung process $e^- + (Z,A) \rightarrow e^- + \gamma + (Z,A)$

There are also contributions from bremsstrahlung in the fields of the atomic electrons, each of the order of α^3. Since there are Z atomic electrons for each nucleus, these give a total contribution of the order of $Z\alpha^3$, which is small compared with the contribution from the nucleus for all but the lightest elements. A detailed calculation shows that for relativistic electrons with $E \gg mc^2/\alpha Z^{1/3}$, the average rate of energy loss is given by

$$-\mathrm{d}E/\mathrm{d}x = E/L_R. \tag{4.17}$$

The constant L_R is called the *radiation length* and is a function of Z and n_a, the number density of atoms/cm^3 in the medium. Integrating Equation (4.17) gives

$$E = -E_0 \exp(-x/L_R), \tag{4.18}$$

where E_0 is the initial energy. It follows that the radiation length is the average thickness of material that reduces the mean energy of an electron or positron by a factor e. For example, the radiation length in lead is 0.566 cm.

From these results, we see that at high energies the radiation losses are proportional to E/m_p^2 for an arbitrary charged particle of mass m_p. On the other hand, the ionization energy losses are only weakly dependent on the projectile mass and energy at very high energies. Consequently, radiation losses completely dominate the energy losses for electrons and positrons at high enough energies, but are much smaller than ionization losses for all particles other than electrons and positrons at all but the highest energies.

Taking into account the above and the results of Section 4.3.2, we see that at low energies, particles with the same kinetic energy but different masses can have substantially different ranges. Thus, for example, an electron of 5 MeV has a range that is several hundred times that of an α-particle of the same kinetic energy.

4.3.4   Interactions of photons in matter

In contrast to heavy charged particles, photons have a high probability of being absorbed or scattered through large angles by the atoms in matter. Consequently, a collimated monoenergetic beam of I photons per second traversing a thickness dx of matter will lose

$$dI = -I\frac{dx}{\lambda} \qquad (4.19)$$

photons per second, where

$$\lambda = (n_a \sigma_\gamma)^{-1} \qquad (4.20)$$

is the mean free path before absorption or scattering out of the beam, and σ_γ is the total photon interaction cross-section with an atom. The mean free path λ is analogous to the collision length for hadronic reactions. Integrating Equation (4.19) gives

$$I(x) = I_0 e^{-x/\lambda} \qquad (4.21)$$

for the intensity of the beam as a function of distance, where I_0 is the initial intensity.

The main processes contributing to σ_γ are: *Rayleigh scattering*, in which the photon scatters coherently from the atom, the *photoelectric effect*, in which the photon is absorbed by the atom as a whole with the emission of an electron; *Compton scattering*,[8] where the photon scatters from an atomic electron; and *electron–positron pair production* in the field of a nucleus or of an atomic electron. The corresponding cross-sections on carbon and lead are shown in Figure 4.10, where it can be seen that above a few MeV the cross-section is dominated by pair production from the nucleus. The pair production process is closely related to

---

[8]Arthur Compton shared the 1927 Nobel Prize in Physics for the discovery of the increase in wavelength that occurs when photons with energies of around 0.5–3.5 MeV interact with electrons in a material – the original *Compton effect*.

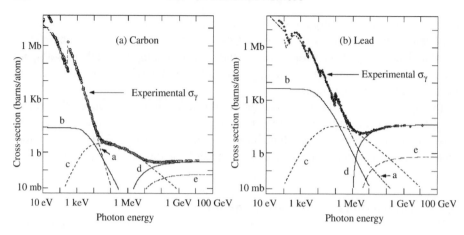

Figure 4.10   Total experimental photon cross-section σ_γ on (a) a carbon atom, and (b) a lead atom, together with the contributions from (a) the photoelectric effect, (b) Rayleigh (coherent atomic) scattering, (c) Compton scattering, (d) pair production in the field of the nucleus, and (e) pair production in the field of the atomic electrons (adapted from Ei04, copyright Elsevier, with permission)

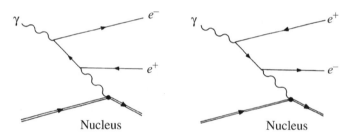

Figure 4.11   The pair production process $\gamma + (Z,A) \rightarrow e^- + e^+ + (Z,A)$

electron bremsstrahlung, as can be seen by comparing the Feynman diagrams shown in Figures 4.9 and 4.11.

The cross-section for pair production rises rapidly from threshold, and is given to a good approximation by

$$\sigma_{\text{pair}} = \frac{7}{9}\frac{1}{n_a L_R},$$ (4.22)

for $E_\gamma \gg mc^2/\alpha Z^{1/3}$, where L_R is the radiation length. Substituting these results into Equation (4.21), gives

$$I(x) = I_0 \exp(-7x/9L_R),$$ (4.23)

so that at high energies, photon absorption, like electron radiation loss, is characterized by the radiation length L_R.

4.4 Particle Detectors

The detection of a particle means more than simply its localization. To be useful this must be done with a resolution sufficient to enable particles to be separated in both space and time in order to determine which are associated with a particular event. We also need to be able to identify each particle and measure its energy and momentum. No single detector is optimal with respect to all these requirements, although some are multifunctional. For example, calorimeters, primarily used for making energy measurements, can also have very good space and time resolution. Many of the devices discussed below are commonly used both in nuclear and particle physics, but in the former a small number of types of detector is often sufficient, whereas in particle physics, both at fixed-target machines and colliders, modern experiments commonly use very large multi-component detectors which integrate many different sub-detectors in a single device. Such systems rely heavily on fast electronics and computers to monitor and control the sub-detectors, and to coordinate, classify and record the vast amount of information flowing in from different parts of the apparatus. In this section we will briefly introduce some of the most important detectors currently available, but detector development is a rapidly-moving major area of research and new devices are frequently developed, so the list below is by no means exhaustive.[9]

4.4.1 Gas detectors

Most gas detectors detect the ionization produced by the passage of a charged particle through a gas, typically an inert one such as argon, either by collecting the ionization products or induced charges onto electrodes, or (historically) by making the ionization track visible in some form. The average energy needed to produce an electron–ion pair is $30 \pm 10\,\text{eV}$, with a weak dependence on the gas used and the energy of the incident particle. In practice, the output is a pulse at the anode (which is amplified by electronic means), with the bulk of the signal being due to the positive ions because of their longer drift distance. For a certain range of applied voltages – the so-called 'proportional region' (see below) – these devices are primarily used to provide accurate measurements of a particle's position. As position detectors, gas detectors largely replaced earlier detectors which used visual techniques, such as cloud chambers, bubble chambers and stacks of photographic emulsions, although the latter are still an ingredient in some neutrino experiments. Although historically important, none of these visual devices are now

---

[9]For more detailed discussions of particle detectors see, for example, Gr96 and the references in Footnote 1. There are also useful reviews in Chapter 5 of Ho97 and Ei04.

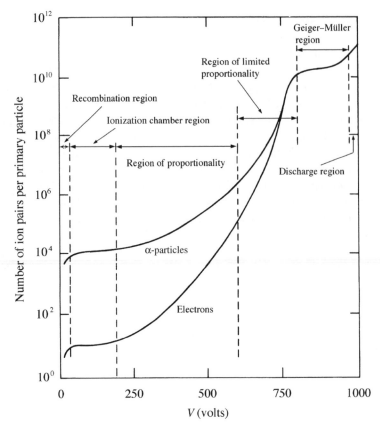

Figure 4.12   Gas amplification factor as a function of voltage V applied in a single-wire gas detector, with a wire radius typically 20 μm, for a strongly ionizing particle (α) and a weakly ionizing particle (electron)

in general use and they have been superceded by electronic detectors.[10] In particle physics experiments being planned at the new accelerators currently being built, gas detectors themselves are being replaced by a new generation of solid-state detectors based on silicon.

To understand the principles of gas detectors we refer to Figure 4.12, which shows the number of ion pairs produced per incident charged particle (the *gas*

---

[10]These early detector techniques produced many notable discoveries and their importance has been recognized by the award of no less than five Nobel Prizes in Physics: a share of the 1927 Prize to Charles Wilson for the invention and use of the cloud chamber; the 1948 Prize to Patrick Blackett for further developments of the cloud chamber and discoveries made with it; the 1950 Prize to Cecil Powell for development of the photographic emulsion technique and its use to discover pions; the 1960 Prize to Donald Glaser for the invention of the bubble chamber; and the 1968 Prize to Luis Alvarez for developing the bubble chamber and associated data analysis techniques resulting in the discovery of a large number of hadronic resonances.

amplification factor) as a function of the applied voltage *V* for two cases: a heavily ionizing particle (e.g. an alpha particle – upper curve) and a lightly ionizing particle (e.g. an electron – lower curve).

Ionization chamber

At low applied voltages, the output signal is very small because electron–ion pairs recombine before reaching the electrodes, but as the voltage increases the number of pairs increases to a saturation level representing complete collection. This is the region of the *ionization chamber*. The simplest type of chamber is a parallel plate condenser filled with an inert gas and having an electric field $E = V/d$, where d is the distance between the plates. In practice the gas mixture must contain at lease one 'quenching' component that absorbs ultraviolet light and stops a plasma forming and spreading throughout the gas.

Another arrangement is cylindrical with an inner anode of radius r_a and an outer cathode of radius r_c, giving an electric field

$$E(r) = \frac{V}{r\ln(r_c/r_a)} \tag{4.24}$$

at a radial distance r from the centre of the anode wire. The output signal is proportional to the number of ions formed and hence the energy deposited by the radiation, but is independent of the applied voltage. However, the signal is very small compared with the noise of all but the slowest electronic circuits and requires considerable amplification to be useful. Overall, the energy resolution and the time resolution of the chamber are relatively poor and ionization chambers are of very limited use in recording individual pulses. They are used, for example, as beam monitors, where the particle flux is very large, and in medical environments to calibrate radioactive sources.

As mentioned previously, neutrons cannot be directly detected by ionization methods, but neutron flux measurements can be made with ionization chambers (or proportional chambers – see below) filled with BF_3 by utilizing the neutron activation reactions of Section 4.3.1.

Proportional counters

If the voltage is increase beyond the region of operation of the ionization chamber, we move into the *proportional region*. In this region, a cylindrical arrangement as used in the ionization chamber will produce electric field strengths of the order of 10^4–10^5 V/cm near the wire and this is strong enough for electron–ion pairs released in the primary ionization to gain sufficient energy to cause secondary ionization. The rapid increase in amplification due to secondary ionization is called a *Townsend avalanche*. The output signal at the anode is still proportional to the energy lost by the original particle. There are a number of different types of device working in the proportional region and they are sometimes generically referred to as *track chambers* or *wire chambers*.

The earliest detector using this idea was the *proportional counter*, which consists of a cylindrical tube filled with gas (again a quenching component in the gas is required) and maintained at a negative potential, and a fine central anode wire at a positive potential. Again, neutrons can be detected indirectly by using the direct nuclear reaction ^{3}He$(n, p)^3$H mentioned in Section 4.3.2 in a proportional chamber filled with a mixture of ^{3}He and krypton. Subsequently, the resolution of proportional counters was greatly improved as a result of the discovery that if many anode wires were arranged in a plane between a common pair of cathode plates, each wire acts as an independent detector. This device is called a *multiwire proportional chamber* (MWPC), and was introduced in 1968.[11] An MWPC can achieve spatial resolutions of 200 μm or less, and has a typical time resolution of about 3 ns.

A schematic diagram of an MWPC is shown in Figure 4.13. The planes (a) have anode wires into the page and those in plane (b) are at right angles. The wire spacings are typically 2 mm. The cathodes are the faces of the chambers. A positive voltage applied to the anode wires generates a field as shown in the upper corner. A particle crossing the chamber ionizes the gas and the electrons drift along the field lines to the anode wires. In this particular example, there would be signals from one wire in the upper (a) chamber and two in the lower (a) chamber.

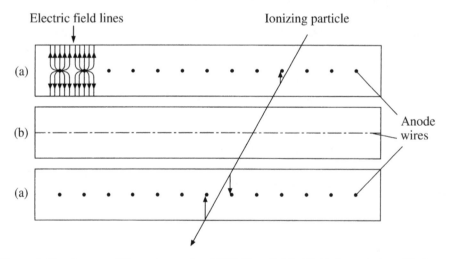

Figure 4.13 A group of three planes of an MWPC (from Po00 with kind permission of Springer Science and Business Media)

---

[11]The MWPC was invented by Georges Charpak and for this and other developments in particle detectors he was awarded the 1992 Nobel Prize in Physics.

Even better spatial resolutions are obtained in a related device called a *drift chamber*, which has now largely replaced the MWPC as a general detector.[12] This uses the fact that the liberated electrons take time to drift from their point of production to the anode. Thus the time delay between the passage of a charged particle through the chamber and the creation of a pulse at the anode is related to the distance between the particle trajectory and the anode wire. In practice, additional wires are incorporated to provide a relatively constant electric field in each cell in a direction transverse to normal incidence. A reference time has to be defined, which, for example, could be done by allowing the particle to pass through a scintillator positioned elsewhere in the experiment (scintillation counters are discussed in Section 4.4.2). The electrons drift for a time and are then collected at the anode, thus providing a signal that the particle has passed. If the drift time can be measured accurately (to within a few ns) and if the drift velocity is known, then spatial resolutions of 100–200 µm can easily be achieved, and specialized detectors can reduce this still further.

Drift chambers are constructed in a variety of geometries to suit the nature of the experiment, and arrangements where the wires are in planar, radial or cylindrical configurations have all been used. The latter type is also called a 'jet chamber' and a two-jet event in a jet chamber was shown in Figure 3.5 as evidence for the existence of quarks.

One of the most advanced applications of proportional and drift chamber principles is embodied in the *time projection chamber* (TPC) illustrated schematically in Figure 4.14. This device consists of a cylindrical barrel, typically

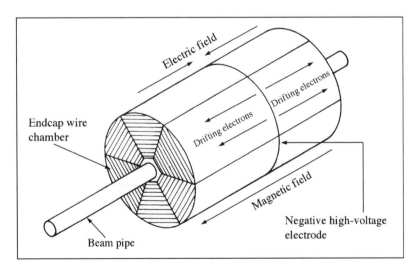

Figure 4.14 Schematic diagram of a time projection chamber (TPC) (adapted from Kl86, copyright Cambridge University Press)

---

[12]In the new generation of colliders, drift chambers are largely being replaced by detectors based on silicon.

2 m long and 1 m in diameter, surrounding the beam pipe of a collider. At each end of the chamber is a segmented layer of proportional counters. The electric drift field E, due to a negative high-voltage electrode plane at the centre of the chamber, and a strong magnetic field B are aligned parallel and anti-parallel to the axis of the cylinder. Because of this, the Lorentz forces on the drifting electrons vanish and electrons formed along the track of an ionizing particle emerging from the interaction point at the centre of the barrel, drift towards one of the endcaps along helical trajectories whose direction is parallel to the axis of the barrel. Their locations are measured by a set of anode wires located between rectangular cathodes in the endcaps. The remaining third coordinate necessary to reconstruct the position of a point on the track is found from the time it takes for the electrons to drift from the point of production to the endcaps where they are detected. The TPC has excellent spatial resolution.

Recently a more robust form of chamber has evolved, in which the wires are replaced by conductive metal strips on a printed circuit board. This is called a *microstrip gas chamber* (MSGC) and is being incorporated in experiments being designed for the new generation of accelerators currently planned or under construction.

Beyond the region of proportionality

Referring again to Figure 4.12, by increasing the external voltage still further one moves into a region where the output signal ceases to be proportional to the number of ion pairs produced and hence the incident energy. This is the region of *limited proportionality*. In this region a type of gas detector called a streamer tube operates, but this will not be discussed here. Eventually the process runs out of control and we enter the *Geiger–Müller region* where the output signal is independent of the energy lost by the incident particle. In this region a quenching agent is not used. Detectors working in this region are called Geiger–Müller counters. Physically they are similar to the simple cylindrical proportional counter and are widely used as portable radiation monitors in the context of safety regulations.

For completeness, we can mention that if the gas amplification factor is taken beyond the Geiger–Müller region, the avalanche develops moving plasmas or streamers. Recombination of ions then leads to visible light which can be made to generate an electrical output. Eventually complete breakdown occurs and a spark is emitted as the incident particle traverses the gas. Detectors in this region, called streamer and spark chambers (these were of parallel plate construction, rather than cylindrical), were widely used in the 1970s and 1980s and played an important role in hadron physics, but are no longer in general use.

4.4.2 Scintillation counters

For charged particles we have seen that energy losses occur due to excitation and ionization of atomic electrons in the medium of the detector. In suitable materials, called *scintillators*, a small fraction of the excitation energy re-emerges as visible light (or sometimes in the UV region) during de-excitation. In a scintillation counter this light passes down the scintillator and onto the face of a *photodetector* – a device that converts a weak photon signal to a detectable electric impulse. An important example of a photodetector is the *photomultiplier tube*, a schematic diagram of which is shown in Figure 4.15.

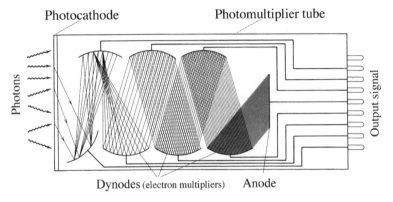

Figure 4.15 Schematic diagram of the main elements of a photomultiplier tube (adapted from Kr88, copyright John Wiley & Sons)

Electrons are emitted from the cathode of the photomultiplier by the photo-electric effect and strike a series of focusing dynodes. These amplify the electrons by secondary emission at each dynode and accelerate the particles to the next stage. The final signal is extracted from the anode at the end of the tube. The electronic pulse can be shorter than 10 ns if the scintillator has a short decay time. The scintillation counter is thus an ideal timing device and it is widely used for 'triggering' other detectors, i.e. its signal is used to decide whether or not to activate other parts of the detector, and whether to record information from the event. Commonly used scintillators are inorganic single crystals (e.g. caesium iodide) or organic liquids and plastics, and a modern complex detector in particle physics may use several tons of detector in combination with thousands of photomultiplier tubes.[13] The robust and simple nature of the scintillation counter

---

[13]For example, the Super Kamiokande experiment mentioned in Chapter 3, which detected neutrino oscillations, although not using scintillation counters, has 13 000 photomultiplier tubes.

has made it a mainstay of experimental nuclear and particle physics since the earliest days of the subject.

Just as direct detection of neutrons is not possible by ionization methods, so the same is true using scintillators. However, the α-particle and the ^{3}H nucleus from the direct nuclear reaction ^{6}Li$(n, \alpha)^3$H mentioned in Section 4.3.2 can produce light in a LiI crystal scintillator and forms the basis for detecting neutrons with energies up to about 20 MeV.

4.4.3   Semiconductor detectors

Solid-state detectors operate through the promotion of electrons from the valence band of a solid to the conduction band as a result of the entry of the incident particle into the solid. The resulting absence of an electron in the valence band (a 'hole') behaves like a positron. Semiconductor detectors are essentially solid-state ionization chambers with the electron–hole pairs playing the role of electron–ion pairs in gas detectors. In the presence of an electric field, the electrons and holes separate and collect at the electrodes, giving a signal proportional to the energy loss of the incident charged particle. Most semiconductor detectors use the principle of the junction diode. Since the band gap in some solids is as small as 1 eV and the energy loss required to produce a pair is only 3–4 eV on average (cf. the 30 eV required in a gas detector), a very large number of electron–hole pairs with only a small statistical fluctuation will be produced by a low-energy particle. Solid-state detectors are therefore very useful in detecting low-energy particles. Semiconductors (principally silicon or germanium) are used as a compromise between materials that have residual conductivity sufficient to enable conduction pulses due to single particles to be distinguished above background and those in which the charges carriers are not rapidly trapped in impurities in the material.

Such detectors have long been used in nuclear physics, where, for example, their excellent energy resolution and linearity, plus their small size and consequent fast response time, make them ideal detectors for γ-ray spectroscopy. Only recently have thin planar detectors become important in particle physics, because of the expense of covering large areas. Nevertheless, several square metres of semiconductor detector are being planned for experiments at the LHC.

One example of a solid-state detector is a *silicon microstrip detector*, where narrow strips of active detector are etched onto a thin slice of silicon, with gaps of the order of 10 μm, to give a tiny analogue of an MWPC. Arrays of such strips can then be used to form detectors with resolutions of the order of 5 μm. These are often placed close to the interaction vertex in a colliding beam experiment, with a view to studying events involving the decay of very short-lived particles. Another example is the *pixel detector*. A single-plane strip detector only gives position information in one dimension (orthogonal to the strip). A pixel detector improves on this by giving information in two dimensions from a single plane.

Solid-state 'vertex detectors' are becoming increasingly important in particle physics and have been incorporated in several of the multi-component detectors designed for use in the new generation of colliders. Their main advantage is their superb spatial resolution; a disadvantage is their limited ability to withstand radiation damage.

4.4.4 Particle identification

Methods of identifying particles are usually based on determining the mass of the particle by a simultaneous measurement of its momentum together with some other quantity. At low values of $\gamma = E/mc^2$, measurements of the rate of energy loss dE/dx can be used, while muons may be characterized by their unique penetrating power in matter, as we have already seen. Here we concentrate on methods based on measuring the velocity or energy, assuming always that the momentum is known. We thus need to start with explaining how momenta are measured.

Measurement of momentum

The momentum of a charged particle is usually determined from the curvature of its track in an applied magnetic field. It is common practice to enclose track chambers in a magnetic field to perform momentum analysis. An apparatus that is dedicated to measuring momentum is called a *spectrometer*. It consists of a magnet and a series of detectors to track the passage of the particles. The precise design depends on the nature of the experiment being undertaken. For example, in a fixed-target experiment at high energies, the reaction products are usually concentrated in a narrow cone around the initial beam direction, whereas in colliding-beam experiments spectrometers must completely surround the interaction region to obtain full angular coverage.

Magnet designs vary. Dipole magnets typically have their field perpendicular to the beam direction. They have their best momentum resolution for particles emitted forward and backward with respect to the beam direction, and are often used in fixed-target experiments at high energies. However, the beam will be deflected, and so at colliders this must be compensated for elsewhere to keep the particles in orbit. Compensating magnets are present in the 'layered detectors' shown in Section 4.6 below. At colliders, the most usual magnet shape is the solenoid, where the field lines are essentially parallel to the beam direction. This device is used in conjunction with cylindrical tracking detectors, like jet chambers, and has its best momentum resolution for particles perpendicular to the beam direction.

We now turn to methods of measuring velocity.

Time-of-flight

The simplest method, in principle, is to measure the time of flight between, for example, two scintillation counters. If the distance between them is L, the time difference for two particles of masses m_1 and m_2 travelling with velocities v_1 and v_2, is

$$\Delta t = t_2 - t_1 = \frac{L}{c} \left(\frac{1}{\beta_1} - \frac{1}{\beta_2} \right), \tag{4.25}$$

where $\beta \equiv v/c$. For a common momentum p, Equation (4.25) may be written, using[14] $E = pc/\beta$,

$$\Delta t = \frac{L}{pc^2} \left[\left(m_2^2 c^4 + p^2 c^2 \right)^{1/2} - \left(m_1^2 c^4 + p^2 c^2 \right)^{1/2} \right]. \tag{4.26}$$

We are interested in the situation where $m_2 \approx m_1 \equiv m$ and $v_2 \approx v_1 \equiv v$. In this case, setting $\Delta m \equiv m_2 - m_1$, the non-relativistic limit of Equation (4.26) is $\Delta t = t \Delta m / m$ and using $v = L/t$, we have

$$\frac{\Delta m}{m} = \Delta t \frac{\beta c}{L}. \tag{4.27}$$

Thus, for example, taking typical values of $\beta = 0.2$, $L = 100 \, \text{cm}$ and $\Delta t = 2 \times 10^{-10} \, \text{s}$ (assuming the timing is done using a scintillation counter), $\Delta m/m$ can be determined to about 1 per cent for low-energy particles. This method is used, for example, in nuclear physics experiments using very low-energy neutron beams.

However, since all high-energy particles have velocities close to the speed of light, the method ceases to be useful for even quite moderate momenta. This can been seen by taking the relativistic limit of Equation (4.26), when we have

$$\Delta t \approx \frac{Lc}{2p^2} \left(m_2^2 - m_1^2 \right), \tag{4.28}$$

which for $m_2 \approx m_1 \equiv m$, $v_2 \approx v_1 \equiv c$, and using $p = \gamma m c$ becomes

$$\frac{\Delta m}{m} = \frac{\gamma^2 c \Delta t}{L}. \tag{4.29}$$

For example, using our previous values for L and Δt, Equation (4.29) shows that the method is not useful for values of γ above about three, which corresponds to a momentum of only about 3 GeV/c for nucleons. Of course this could be extended by taking longer flight paths, but only at greater expense in instrumentation.

---

[14]See Appendix B.

Čerenkov counters

The most important identification method for high-energy particles is based on the Čerenkov effect. When a charged particle with velocity v traverses a dispersive medium of refractive index n, excited atoms in the vicinity of the particle become polarized, and if v is greater than the speed of light in the medium c/n, a part of the excitation energy reappears as coherent radiation emitted at a characteristic angle θ to the direction of motion. The necessary condition $v > c/n$ implies $\beta n > 1$ and by considering how the waveform is produced[15] it can be shown that $\cos \theta = 1/\beta n$ for the angle θ, where $\beta = v/c$ as usual. A determination of θ is thus a direct measurement of the velocity.[16]

Čerenkov radiation appears as a continuous spectrum and may be collected onto a photosensitive detector. Its main limitation from the point of view of particle detection is that very few photons are produced. The number of photons $N(\lambda)\mathrm{d}\lambda$ radiated per unit path length in a wavelength interval $\mathrm{d}\lambda$ can be shown to be

$$N(\lambda)\mathrm{d}\lambda = 2\pi\alpha\left(1 - \frac{1}{\beta^2 n^2}\right)\frac{\mathrm{d}\lambda}{\lambda^2} < 2\pi\alpha\left(1 - \frac{1}{n^2}\right)\frac{\mathrm{d}\lambda}{\lambda^2} \tag{4.30}$$

and so vanishes rapidly as the refractive index approaches unity. The maximum value occurs for $\beta = 1$, which for a particle with unit charge, corresponds to about 200 photons/cm in the visible region in water and glass. These numbers should be compared with the 10^4 photons/cm emitted by a typical scintillator. Because the yield is so small, appreciable lengths are needed to give enough photons, and gas Čerenkov counters in particular can be several metres long.

Čerenkov counters are used in two different modes. The first is as a *threshold counter* to detect the presence of particles whose velocities exceed some minimum value. Suppose that two particles with β values β_1 and β_2 at some given momentum p are to be distinguished. If a medium can be found such that $\beta_1 n > 1 \geq \beta_2 n$, then particle 1 will produce Čerenkov radiation but particle 2 will not. Clearly, to distinguish between highly relativistic particles with $\gamma \gg 1$ also requires $n \approx 1$, so that from Equation (4.30) very few photons are produced. Nevertheless, common charged particles can be distinguished in this way up to at least 30 GeV/c.

Another device is the so-called *ring-image* Čerenkov detector which is a very important device in both fixed-target machines and colliders. If we assume that the particles are not all travelling parallel to a fixed axis, then the radiating medium can be contained within two concentric spherical surfaces of radii R and $2R$ centred on the target or interaction region where the particles are produced, as illustrated in Figure 4.16. The outer surface is lined with a mirror, which focuses

---

[15]This is Huygens' construction in optics.
[16]For the discovery and interpretation of this effect, Pavel Čerenkov, Ilya Frank and Igor Tamm were awarded the 1958 Nobel Prize in Physics.

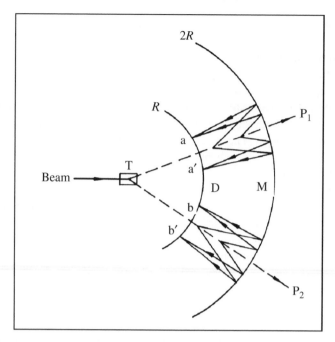

Figure 4.16   Two particles P_1 and P_2, produced from the target T, emit Čerenkov radiation on traversing a medium contained between two spheres of radius R and $2R$. The mirror M on the outer sphere focuses the radiation into ring images at aa′ and bb′ on the inner detector sphere D. The radii of the ring images depend on the angle of emission of the Čerenkov radiation and hence on the velocities of the particles

the Čerenkov radiation into a ring at the inner detector surface. The radius of this ring depends on the angle θ at which the Čerenkov radiation is emitted, and hence on the particle velocity v. It is determined by constructing an image of the ring electronically. This was the technique used in the Super Kamiokande detector discussed in Chapter 3 to detect relativistic electrons and muons produced by neutrino interactions. In that experiment the radiating medium was pure water.

4.4.5   Calorimeters

Calorimeters are an important class of detector used for measuring the energy and position of a particle by its total absorption and are widely used. They differ from most other detectors in that (1) the nature of the particle is changed by the detector, and (2) they can detect neutral as well as charged particles. A calorimeter may be a homogeneous absorber/detector to detect photons and electrons. In early devices this was often a block of lead glass, but is now more likely to be scintillator such as CsI. Alternatively, it can be a sandwich construction with separate layers of absorber (e.g. a metal such as lead) and detector (scintillator, MWPC etc.). The latter are also known as 'sampling calorimeters'. During the absorption process, the particle will interact with the material of the absorber, generating secondary particles which will themselves generate further particles and so on, so that a cascade or shower, develops. For this reason calorimeters are also called 'shower counters'.

The shower is predominantly in the longitudinal direction due to momentum conservation, but will be subject to some transverse spreading due both to multiple Coulomb scattering and the transverse momentum of the produced particles. Eventually all, or almost all, of the primary energy is deposited in the calorimeter, and gives a signal in the detector part of the device.

There are several reasons why calorimeters are important, especially at high energies:

- they can detect neutral particles, by detecting the charged secondaries;

- the absorption process is statistical (and governed by the Poisson distribution), so that the relative precision of energy measurements $\Delta E/E$ varies as $E^{-\frac{1}{2}}$ for large E, which is a great improvement on high-energy spectrometers where $\Delta E/E$ varies as E^2;

- the signal produced can be very fast, of the order of (10–100) ns, and is ideal for making triggering decisions.

Although it is possible to build calorimeters that preferentially detect just one class of particle (electrons and photons, or hadrons) it is also possible to design detectors that serve both purposes. Since the characteristics of electromagnetic and hadronic showers are somewhat different it is convenient to describe each separately. In practice, it is common to have both types in one experiment with the hadron calorimeter stacked behind the electromagnetic one.

Electromagnetic showers

When a high-energy electron or positron interacts with matter we have seen that the dominant energy loss is due to bremsstrahlung, and for the photons produced the dominant absorption process is pair production. Thus the initial electron will, via these two processes, lead to a cascade of $e^\pm$ pairs and photons, and this will continue until the energies of the secondary electrons fall below the critical energy E_C where ionization losses equal those from bremsstrahlung. This energy is roughly given by $E_C \approx 600\,\mathrm{MeV}/Z$.

Most of the correct qualitative features of shower development may be obtained from the following very simple model. We assume:

- each electron with $E > E_C$ travels one radiation length and then gives up half of its energy to a bremsstrahlung photon;

- each photon with $E > E_C$ travels one radiation length and then creates an electron–positron pair with each particle having half the energy of the photon;

- electrons with $E < E_C$ cease to radiate and lose the rest of their energy by collisions;

- ionization losses are negligible for $E > E_C$.

A schematic diagram of the approximate development of a shower in an electromagnetic calorimeter assuming this simple model is shown in Figure 4.17.

If the initial electron has energy $E_0 \gg E_C$, then after t radiation lengths the shower will contain 2^t particles, which consist of approximately equal numbers of electrons, positrons and photons each with an average energy

$$E(t) = E_0/2^t. \tag{4.31}$$

The multiplication process will cease abruptly when $E(t) = E_C$, i.e. at $t = t_{max}$ where

$$t_{max} = t(E_C) \equiv \frac{\ln(E_0/E_C)}{\ln 2} \tag{4.32}$$

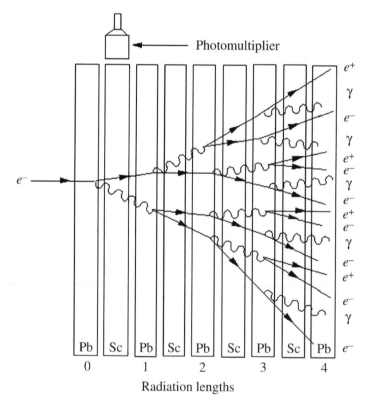

Figure 4.17 Approximate development of an electromagnetic shower in a sampling calorimeter assuming the simple model described in the text; the calorimeter consists of alternate layers of lead (Pb) and a scintillator (Sc), the latter attached to photomultipliers (one only shown)

and the number of particles at this point will be

$$N_{max} = \exp[t_{max}\ln 2] = E_0/E_C. \tag{4.33}$$

The main features of this simple model are observed experimentally, and in particular the maximum shower depth increases only logarithmically with primary energy. Because of this, the physical sizes of calorimeters need increase only slowly with the maximum energies of the particles to be detected. The energy resolution of a calorimeter, however, depends on statistical fluctuations, which are neglected in this simple model, but for an electromagnetic calorimeter it is typically $\Delta E/E \approx 0.05/E^{\frac{1}{2}}$, where E is measured in GeV.

Hadronic showers

Although hadronic showers are qualitatively similar to electromagnetic ones, shower development is far more complex because many different processes contribute to the inelastic production of secondary hadrons. The scale of the shower is determined by the nuclear absorption length defined earlier. Since this absorption length is larger than the radiation length, which controls the scale of electromagnetic showers, hadron calorimeters are thicker devices than electromagnetic ones. Another difference is that some of the contributions to the total absorption may not give rise to an observable signal in the detector. Examples are nuclear excitation and leakage of secondary muons and neutrinos from the calorimeter. The loss of 'visible' or measured energy for hadrons is typically 20–30 per cent greater than for electrons.

The energy resolution of calorimeters is in general much worse for hadrons than for electrons and photons because of the greater fluctuations in the development of the hadron shower. Depending on the proportion of π^0s produced in the early stages of the cascade, the shower may develop predominantly as an electromagnetic one because of the decay $\pi^0 \to \gamma\gamma$. These various features lead to an energy resolution typically a factor of 5–10 poorer than in electromagnetic calorimeters.

4.5 Layered Detectors

As stated earlier, in particle physics it is necessary to combine several detectors in a single experiment to extract the maximum amount of information from it. Typically, working out from the interaction region, there will be a series of wire chambers, followed further out by calorimeters and at the outermost limits, detectors for muons, the most penetrating particles to be detected. The whole device is usually in a strong magnetic field so that momentum measurements may be made. We will illustrate the general features by three examples.

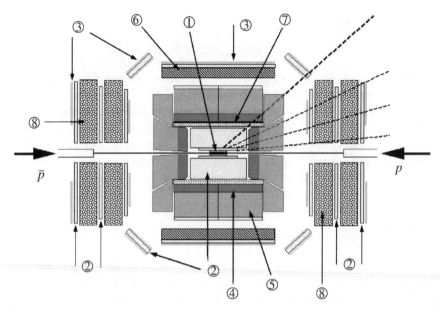

Figure 4.18 The CDF detector at the $p\bar{p}$ collider at Fermilab, USA (Fermilab Graphic, reproduced with permission)

The first is the $p\bar{p}$ Collider Detector at Fermilab (CDF), which is shown schematically in Figure 4.18. The detection of the top quark and the measurement of its mass were first made using this device. The dashed lines indicate some particles produced in the collision. CDF is a large device, being approximately 8 m wide and 26 m in overall length. The beams of protons and antiprotons enter from each end through focusing quadrupole magnets and interact in the central intersection region where there is a silicon vertex detector (1) to detect very short-lived particles. The intersection point is surrounded by a 2000 tonne detector system which, in addition to the vertex detector, consists of inner drift chambers (2), electromagnetic calorimeters (4), hadron calorimeters (5) time-of-flight detectors (not indicated) and further drift chambers (2) on the outside to detect muons. The whole system is in a magnetic field with the solenoid coil shown at (7) and steel shielding at (6). The rest of the detector consists of two symmetrical sets of drift chambers (2) sandwiched between scintillation counters (3) and magnetic toroids (8) to provide momentum measurements, primarily for muons.

The second example, shown schematically in Figure 4.19, is the ATLAS detector currently under construction for use at the Large Hadron Collider (LHC). It is hoped that this and other detectors at the LHC will be able to detect the important Higgs boson, if it exists, and so help solve one the outstanding current problems in particle physics – the origin of mass. The ATLAS detector is even larger than the CDF detector and measures about 22 m high and 44 m long, with an overall weight of approximately 7000 tonnes.

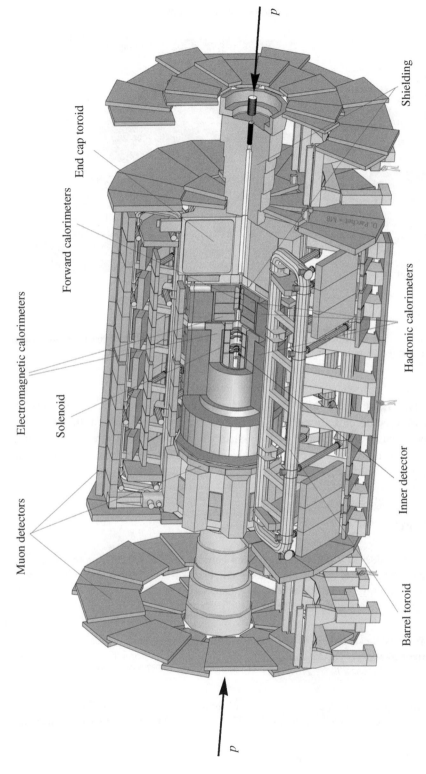

Figure 4.19 The ATLAS detector under construction for use at the *pp* collider LHC (also under construction) at CERN, Geneva, Switzerland (CERN photo, reproduced with permission)

Muon detectors

Electromagnetic calorimeters

Solenoid

Forward calorimeters

End cap toroid

Shielding

Hadronic calorimeters

Inner detector

Barrel toroid

p

p

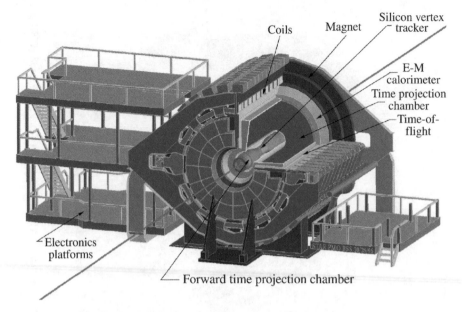

Figure 4.20 The STAR detector at the RHIC accelerator at Brookhaven National Laboratory, USA. (Courtesy of Brookhaven National Laboratory)

Finally, Figure 4.20 shows the STAR detector at the RHIC accelerator at Brookhaven National Laboratory. This detects events resulting from the collisions of heavy ions, typically those of fully-stripped gold nuclei, where the final state may contain many thousands of particles. An example of an event is shown in Figure 9.12.

Problems

4.1 At a collider, a 20 GeV electron beam collides with a 300 GeV proton beam at a crossing angle of $10°$. Evaluate the total centre-of-mass energy and calculate what beam energy would be required in a fixed-target electron machine to achieve the same total centre-of-mass energy.

4.2 What is the length L of the longest drift tube in a linac which operating at a frequency of $f = 20$ MHz is capable of accelerating ^{12}C ions to a maximum energy of $E = 100$ MeV?

4.3 Alpha particles are accelerated in a cyclotron operating with a magnetic field of magnitude $B = 0.8$ T. If the extracted beam has an energy of 12 MeV, calculate the extraction radius and the orbital frequency of the beam (the so-called *cyclotron frequency*).

4.4 Protons with momentum 50 GeV/c are deflected through a collimator slit 2 mm wide by a bending magnet 1.5 m long which produces a field of 1.2 T. How far from the magnet should the slit be placed so that it accepts particles with momenta in the range 49–51 GeV/c?

4.5 Estimate the minimum length of a gas Čerenkov counter that could be used in threshold mode to distinguish between charged pions and charged kaons with momentum 20 GeV/c. Assume that a minimum of 200 photons need to be radiated to ensure a high probability of detection. Assume also that the radiation covers the whole visible spectrum between 400 nm and 700 nm and neglect the variation with wavelength of the refractive index of the gas.

4.6 An e^+e^- collider has a diameter of 8 km and produces beams of energy 45 GeV. Each beam consists of 12 bunches each containing 3×10^{11} particles. The bunches have a cross-sectional area of 0.02 mm^2. What is the luminosity of the machine in units of cm^{-2}s^{-1}?

4.7 What are the experimental signatures and with what detectors would one measure: (a) the decay $Z \rightarrow b\bar{b}$, and (b) $W \rightarrow e\nu$ and $W \rightarrow \mu\nu$.

4.8 The reaction $e^+e^- \rightarrow \tau^+\tau^-$ is studied using a collider with equal beam energies of 5 Gev. The differential cross-section is given by

$$\frac{d\sigma}{d\Omega} = \frac{\alpha^2\hbar^2c^2}{4E_{cm}^2}\left(1 + \cos^2\theta\right)$$

where E_{cm} is the total centre-of-mass energy and θ is the angle between the incoming e^- and the outgoing τ^-. If the detector can only record an event if the $\tau^+\tau^-$ pair makes an angle of at least 30° relative to the beam line, what fraction of events will be recorded? What is the total cross-section for this reaction in nanobarns? If the reaction is recorded for 10^7s at a luminosity of $L = 10^{31}$ cm^{-2}s^{-1}, how many events are expected?

Suppose the detector is of cylindrical construction and at increasing radii from the beam line there is a drift chamber, an electromagnetic calorimeter, a hadronic calorimeter and finally muon chambers. If in a particular event the τ^- decays via $\tau^- \rightarrow \mu^- + \bar{\nu}_\mu + \nu_\tau$ and the τ^+ decays to $\tau^+ \rightarrow e^+ + \bar{\nu}_\tau + \nu_e$, what signals would be observed in the various parts of the detector?

4.9 A charged particle with speed v moves in a medium of refractive index n. By considering the wavefronts emitted at two different times, derive a relation for the angle θ of the emitted Čerenkov radiation relative to the particle's direction in terms of $\beta = v/c$ and n. What is the maximum angle of emission and to what limit does it correspond?

If the momentum p of the particle is known from other detectors, show that the mass squared x of the particle is given by $x = (mc^2)^2 = p^2c^2(n^2\cos^2\theta - 1)$. If the error on the momentum is negligible, show, by taking derivatives of this expression, that for very relativistic particles, the standard error σ_x on x is approximately

$$\sigma_x \approx 2p^2c^2\sqrt{(n^2 - 1)}\sigma_\theta,$$

where σ_θ is the standard error on θ.

4.10   Estimate the thickness of iron through which a beam of neutrinos with energy 300 GeV must travel if 1 in 10^9 of them is to interact. Assume that at high energies the neutrino-nucleon total cross-section is given approximately by $\sigma_\nu \approx 10^{-38} E_\nu \, \text{cm}^2$ where E_ν is given in GeV. The density of iron is $\rho = 7.9 \, \text{g cm}^{-3}$.

4.11   An electron with an initial energy of 2 GeV traverses 10 cm of water with a radiation length of 36.1 cm. Calculate its final energy. How would the energy loss change if the particle were a muon rather than an electron?

4.12   A beam of neutrons with kinetic energy 0.1 eV and intensity $10^6 \, \text{s}^{-1}$ is incident normally on a thin foil of $^{235}_{92}\text{U}$ of effective thickness $10^{-1} \, \text{kg m}^{-2}$. The beam can undergo (1) isotropic elastic scattering, with a cross section $\sigma_{\text{el}} = 3 \times 10^{-2}$ b, (2) radiative capture, with a cross-section $\sigma_{\text{cap}} = 10^2$ b, or (3) it can fission a $^{235}_{92}\text{U}$ nucleus, with a cross-section $\sigma_{\text{f}} = 3 \times 10^2$ b. Calculate the attenuation of the beam and the flux of elastically-scattered particles 5 m from the foil.

4.13   A positron with laboratory energy 50 GeV interacts with the atomic electrons in a lead target to produce $\mu^+ \mu^-$ pairs. If the cross-section for this process is given by $\sigma = 4\pi\alpha^2\hbar^2c^2/3(E_{\text{CM}})^2$, calculate the positron's interaction length. The density of lead is $\rho = 1.14 \times 10^7 \, \text{kg m}^{-3}$.

4.14   A liquid hydrogen target of volume 125 cm^3 and density 0.071 g cm^{-3} is bombarded with a mono-energetic beam of negative pions with a flux $2 \times 10^7 \, \text{m}^{-2} \, \text{s}^{-1}$ and the reaction $\pi^- + p \rightarrow \pi^0 + n$ observed by detecting the photons from the decay of the π^0. Calculate the number of photons emitted from the target per second if the cross-section is 40 mb.

4.15   Assuming the Bethe–Bloch formula is valid for low energies, show that the rate of ionization has a maximum (the Bragg peak) and find the kinetic energy of protons in iron for which this maximum would occur.

4.16   A cylindrical proportional chamber has a central anode wire of radius 0.02 mm and an outer cathode of radius 10 mm with a voltage of 500 V applied between them. What is the electric field at the surface of the anode? If the threshold for ionization by collision is 750 kV m^{-1} and the mean free path of the particles being detected is 4×10^{-6} m, estimate the number of ion pairs produced per primary particle.

5

Quark Dynamics: the Strong Interaction

In Chapter 3 we described the basic properties of quarks and in particular their static properties and how these are used to construct the quark model of hadrons. We now look in more detail at how quarks interact and the role of gluons in the strong interactions. Thus we will be considering dynamical properties and the theoretical framework that describes these interactions.

5.1   Colour

We saw in Chapter 3 that the quark model account of the hadron spectrum is very successful. However, it begs several questions. One is: why are the observed states overwhelmingly of the form $3q$, $3\bar{q}$ and $q\bar{q}$? Another arises from a particular assumption that was implicitly made in Chapter 3. This is: if two quarks of the same flavour uu, dd, ss... are in the same spatial state, they must also be in the same spin state, with their spins parallel. This can be seen very easily by considering the baryon state omega-minus Ω^- that is shown in Table 3.3 and Figure 3.12.[1] From its decay products, it may be deduced that this state has strangeness $S = -3$ and spin $J = \frac{3}{2}$ and thus in the quark model it has the composition $\Omega^- = sss$, where all three quarks have their spins parallel and there is no orbital angular momentum between them. This means that all three like-quarks have the same space and spin states, i.e. the overall wavefunction must be symmetric, which violates the fundamental requirement of the Pauli principle. The latter states that a system of identical fermions has a wavefunction that is overall antisymmetric under the interchange of any two particles, because identical

---

[1]The discovery of the Ω^- was a crucial step in gaining acceptance of the quark model of hadron spectroscopy. The experiment is described in Chapter 15 of Tr75.

Nuclear and Particle Physics   B. R. Martin
© 2006 John Wiley & Sons, Ltd

fermions cannot simultaneously be in the same quantum state. The three s quarks in the Ω^- therefore *cannot* be in the same state. So how do they differ?

The Ω^- is an obvious example of the contradiction, but it turns out that in order for the predictions of the quark model to agree with the observed spectrum of hadron multiplets, it is necessary to assume that overall baryon wavefunctions are symmetric under the interchange of like quarks.[2] In order to resolve this contradiction, it is necessary to assume that a new degree of freedom exists for quarks, but not leptons, which is somewhat whimsically called *colour*. The basic properties of colour are as follows.

1. Any quark u, d, s, ... can exist in three different colour states.[3] We shall see later that there is direct experimental evidence that just three such states exist, which we denote r, g, b for 'red', 'green' and 'blue' respectively.

2. Each of these states is characterized by the values of two conserved *colour charges*, denoted I_3^C and Y^C, which are strong interaction analogues of the electric charge in electromagnetic interactions.[4] These charges depend only on the colour states r, g, b and *not* on the flavours u, d, s, ... The particular values for quarks and antiquarks are given in Table 5.1, and are a consequence of a fundamental symmetry of the strong interaction (called SU(3) colour symmetry), which we will not pursue here. For multiparticle states, the colour charges of the individual states are simply added.

3. Only states with zero values for the colour charges are observable as free particles; these are called *colour singlets*. This is the hypothesis of *colour confinement*. It can be derived, at least approximately, from the theory of strong interactions we shall describe.

Table 5.1 Values of the colour charges I_3^C and Y^C for the colour states of quarks and antiquarks

	(a) Quarks			(b) Antiquarks	
	I_3^C	Y^C		I_3^C	Y^C
r	$\frac{1}{2}$	$\frac{1}{3}$	$\bar{r}$	$-\frac{1}{2}$	$-\frac{1}{3}$
g	$-\frac{1}{2}$	$\frac{1}{3}$	$\bar{g}$	$\frac{1}{2}$	$-\frac{1}{3}$
b	0	$-\frac{2}{3}$	$\bar{b}$	0	$\frac{2}{3}$

---

[2] In Problem 3.4 it was shown explicitly that otherwise the wrong hadron spectrum is predicted.
[3] Needless to say, nothing to do with 'real' colour!
[4] This is one reason we were careful to use the qualifier 'electric' when talking about charge in the context of electromagnetic interactions in earlier chapters.

Returning to the quark model, it can be seen from Table 5.1 that a $3q$ state can only have both $I_3^C = 0$ and $Y^C = 0$ if we have one quark in an r state, one in a g state and one in a b state. Hence in the Ω^-, for example, all three s quarks are necessarily in different colour states, and thus the Pauli principle can be satisfied. Formally, we are assuming that the total wavefunction is the product of a spatial part $\psi_{\text{spatial}}(\mathbf{x})$ and a spin part ψ_{spin}, as usual, but also a colour wavefunction ψ_{colour}, i.e.

$$\Psi = \psi_{\text{spatial}}(\mathbf{x})\, \psi_{\text{spin}}\, \psi_{\text{colour}}. \tag{5.1}$$

The Pauli principle is now interpreted as applying to the total wavefunction including the colour part ψ_{colour}. The combined space and spin wavefunctions can then be symmetric under the interchange of quarks of the same flavour (to agree with experiment) provided the colour wavefunction is antisymmetric. The structure of ψ_{colour} is

$$\psi_{\text{colour}} = \frac{1}{\sqrt{6}}[R_1G_2B_3 + G_1B_2R_3 + B_1R_2G_3 - R_1B_2G_3 - B_1G_2R_3 - G_1R_2B_3], \tag{5.2}$$

where R, G and B represent quarks with colour red, green and blue, respectively.

One can also see from Table 5.1 part of the answer to the first question of this section. Free quarks and fractionally charged combinations like qq and $qq\bar{q}$ are forbidden by colour confinement, in accordance with experimental observation. On the other hand, the combinations $q\bar{q}$ and $3q$ used in the simple quark model are allowed. More unusual combinations like $qq\bar{q}\bar{q}$ and $qqqq\bar{q}$, which could give rise to so-called 'exotic' mesons and baryons, respectively, are not in principle forbidden by colour confinement and, as mentioned in Chapter 3, recent experiments may possibly have provided some evidence for a small number of these, but this has yet to be confirmed.

5.2 Quantum Chromodynamics (QCD)

The theory that describes strong interactions in the standard model is called *quantum chromodynamics*, or QCD for short (chromos means colour in Greek). Although QCD is not tested to the same extent or precision as quantum electrodynamics (QED), the quantum theory of electromagnetic interactions, it is nevertheless in impressive agreement with a large body of experimental data. QCD is similar to QED in that both describe interactions that are mediated by massless spin-1 bosons; gluons in the former case and photons in the latter. Both theories are of the type called *gauge theories* which, as mentioned in Chapter 1, refer to a particular symmetry of the theory. However, there is a very important difference in the two interactions that we now discuss.

Gluons, the force carriers of the strong interaction, have zero electric charge, like photons, but unlike photons, which couple to electric charge, gluons couple to *colour* charges. This leads immediately to the flavour independence of strong interactions discussed in Chapter 3; that is, the different quark flavours $a = u$, d, s, c, b and t have identical strong interactions. We now see that this is because they are postulated to exist in the same three colour states r, g, b, with the same possible values of the colour charges. Flavour independence has its most striking consequences for u and d quarks, which have almost equal masses, where it leads to the phenomenon of isospin symmetry. This results, among other things, in the near equality of the masses of the proton and neutron, and charge states within other multiplets such as pions and kaons, all of which we have seen in Chapter 3 are confirmed by experiment. We will examine the consequence of flavour independence for the bound states of the heavy quarks c and b in Section 5.3.

Although QED and QCD both describe interactions, albeit of very different strengths, that are mediated by massless spin-1 bosons which couple to conserved charges, there is a crucial difference between them that profoundly effects the characters of the resulting forces. While the photons which couple to the electric charge are themselves electrically neutral, gluons have non-zero values of the colour charges to which they couple. This is illustrated in Figure 5.1, which shows a particular example of a quark–quark interaction by gluon exchange.

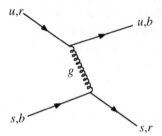

Figure 5.1   Example of quark--quark scattering by gluon exchange; in this diagram, the quark flavours u and s are unchanged, but their colour states can change, as shown

In this diagram, the colour states of the two quarks are interchanged, and the gluon has non-zero values of the colour quantum numbers, whose values follow from colour charge conservation at the vertices, i.e.

$$I_3^C(g) = I_3^C(r) - I_3^C(b) = \frac{1}{2} \tag{5.3}$$

and

$$Y^C(g) = Y^C(r) - Y^C(b) = 1. \tag{5.4}$$

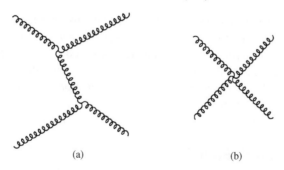

(a)   (b)

Figure 5.2 The two lowest-order contributions to gluon--gluon scattering in QCD: (a) one-gluon exchange, (b) contact interaction

Just as quarks can exist in three colour states, gluons can exist in eight colour states, although we will not need the details of these. The first thing implied by the non-zero values of the colour charges is that gluons, like quarks, are confined and cannot be observed as free particles. The second is that since gluons couple to particles with non-zero colour charges, and since gluons themselves have non-zero colour charges, it follows that gluons couple to other gluons. The two types of gluon self-coupling that occur in QCD are given in Figure 5.2, which shows the two lowest-order contributions to gluon–gluon scattering.

The first is a gluon exchange process in analogy to gluon exchange in quark–quark scattering, which we have encountered previously (see Figure 1.3), while the second involves a so-called 'zero range' or 'contact' interaction. If the forces resulting from these interactions were attractive and sufficiently strong, they could in principle lead to bound states of two or more gluons. These would be a new type of exotic state called *glueballs*. Although some experiments claimed to have detected these, at present there is little compelling evidence that they exist.[5]

The gluon–gluon interactions have no analogue in QED (photons couple to electrically charged particles and hence do not couple directly to other photons) and it can be shown that they lead to properties of the strong interaction that differ markedly from those of the electromagnetic interaction. These properties are *colour confinement*, which we have discussed above, and a new property called *asymptotic freedom*. The latter is the statement that the strong interaction gets weaker at short distances; conversely, as the distance between the quarks increases, the interaction gets stronger.[6] In this strong interaction regime the situation is very complicated, and it has not yet been possible to evaluate the theory precisely. We therefore have to rely on results obtained by numerical simulations of the theory; the approach is called *lattice gauge theory*. In these simulations, the theory is

---

[5]A critical review is given in Ei04.
[6]Asymptotic freedom was postulated in 1973 by David Gross, David Politzer and Frank Wilczek, who were subsequently awarded the 2004 Nobel Prize in Physics.

evaluated at a grid of discrete points on a three-dimensional lattice and by making the lattice spacing small enough it is hoped that the results of the true continuum theory will be approximated. The calculations require very large ultra-fast computers and precise results are difficult to obtain because of the approximations that have to be made. Nevertheless, at present, the demonstration of confinement in QCD rests largely on such simulations.[7]

5.3   Heavy Quark Bound States

Some of the features of QCD discussed above are illustrated by considering the static potential between a heavy quark and an antiquark. Such systems give rise to bound states and because the quarks are so heavy they move slowly enough within the resulting hadrons to be treated non-relativistically to a first approximation. (This is one of the few places in particle physics where a non-relativistic calculation is adequate.) This means that the rest energies of the bound states, and hence their masses, can be calculated from the static potential between the quarks in exactly the same way that the energy levels in the hydrogen atom are calculated, although of course the potential is not Coulombic. In the present case, however, the procedure is reversed, with the aim of determining the form of the static potential from the rather precisely measured energies of the bound states.

The first such state to be discovered, the $J/\psi(3097)$[8], is a bound state of the $c\bar{c}$ system and is part of a family of such states given the name *charmonium,* by analogy with *positronium,* the bound state of an electron and a positron. It is identified with the $n = 1$, $^3S_1$ state of the $c\bar{c}$ system, where n is the principal quantum number and we use the notation $^{2S+1}L_J$, with (L, S) the angular momentum between the quarks and their total spin, respectively. The discovery of the $J/\psi(3097)$ caused considerable excitement because it confirmed the existence of the charm quantum number that had been predicted many years earlier, even though the $J/\psi(3097)$ itself has zero overall charm. It was hence a very important piece of evidence in favour of the standard model.

The interpretation of the $J/\psi(3097)$ as a $c\bar{c}$ bound state follows from its unusually narrow width. For a state decaying predominantly (86 per cent) to hadrons (mostly pions) by the strong interaction one would expect a width measured in MeV, whereas the width of the $J/\psi(3097)$ was only about 90 keV. This meant that there was no possibility of an explanation in terms of just u, d and s quarks. The preferred decay of the $J/\psi(3097)$ would be via the mechanism shown in Figure 5.3(a). However, this is forbidden by energy conservation because

---

[7]Lattice calculations also support the view that gluon–gluon forces are strong enough to give rise to glueballs.

[8]The rather clumsy notation is because it was discovered independently by two groups, led by Burton Richer and Samuel Ting. Richer's group was studying the reactions $e^+e^- \rightarrow$ hadrons and named it the ψ particle. Ting's group discovered it in pBe reactions and called it the J. It is now known as the J/ψ. Richer and Ting shared the 1976 Nobel Prize in Physics for the discovery.

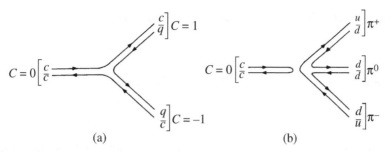

Figure 5.3 Quark diagrams for (a) the decay of a charmonium state to a pair of charmed mesons, and (b) an example of a decay to non-charmed mesons

$M_{J/\psi} < 2M_D$, where M_D is the mass of the lightest meson having non-zero charm, the $D(1870)$. (These latter states had already been seen in neutrino experiments, but not clearly identified.) The mass $2M_D$ is referred to as the *charm threshold*. Since the direct decay to charmed mesons is forbidden, the only hadronic decays allowed must proceed via mechanisms such as that of Figure 5.3(b) and diagrams like this where initial and final quark lines are disconnected are known to be heavily suppressed.[9]

The explanation for this in QCD is that since both the decaying particle and the three pions in the final state are colour singlets, they can only be connected by the exchange of a combination of gluons that is also a colour singlet, i.e. not the exchange of a single gluon. Moreover, the $J/\psi(3097)$ is known to be produced in e^+e^- annihilations via photon exchange, so it must have a charge conjugation $C = -1$. Thus the minimum number of gluons exchanged is three. This is illustrated in Figure 5.4. In contrast, if $M_\psi > 2M_D$ then the decay may proceed via the exchange of low-momentum gluons as usual.

Subsequently, higher-mass charmonium states also with $J^{PC} = 1^{--}$, where $P = (-1)^{L+1}$ and $C = (-1)^{L+S}$, were discovered in e^+e^- reactions and states with other J^{PC} values were identified in their radiative decays. Thus the $n = 1$, $^1S_0$

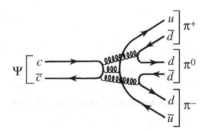

Figure 5.4 OZI-suppressed decay of a charmonium state below the $D\bar{D}$ threshold

---

[9]This is known as the *OZI Rule* after Okubo, Zweig and Iizuka who first formulated it. Another example where it acts is the suppression of the decay $\phi \to \pi^+\pi^-\pi^0$ compared with $\phi \to K\bar{K}$.

ground state $\eta_c(2980)$ has been found from the decays

$$\psi(3686) \rightarrow \eta_c(2980) + \gamma \quad \text{and} \quad J/\psi(3097) \rightarrow \eta_c(2980) + \gamma \tag{5.5}$$

and a series of states $\chi_{ci}(i = 1, 3)$ have been found in the decays

$$\psi(3686) \rightarrow \chi_{ci} + \gamma. \tag{5.6}$$

The latter themselves decay and from an analysis of their decay products they are identified with the $n = 1$ states $^3P_0$, $^3P_1$ and $^3P_2$. Some of these states lie below the charm threshold and like the $J/\psi(3097)$ are forbidden by energy conservation to decay to final states with 'open' charm and thus have widths measured in keV. Others lie above the charm threshold and therefore have 'normal' widths measure in MeV. The present experimental situation for charmonium states with $L \leq 2$ is shown in Table 5.2.

Table 5.2   Predicted $c\bar{c}$ and $b\bar{b}$ states with $L \leq 2$ and masses up to and just above the charm and bottom thresholds (3.74 GeV/c² and 10.56 GeV/c², respectively), compared with experimentally observed states (masses are given in MeV/c²)

$n^{2S+1}L_J$	J^{PC}	$c\bar{c}$ state	$b\bar{b}$ state
1^1S_0	0^{-+}	$\eta_c(2980)$	$\eta_b(9300)$?
1^3S_1	1^{--}	$J/\psi(3097)$	$\Upsilon(9460)$
1^1P_1	1^{+-}	$h_c(3526)$?	
1^3P_0	0^{++}	$\chi_{c0}(3415)$	$\chi_{b0}(9860)$
1^3P_1	1^{++}	$\chi_{c1}(3511)$	$\chi_{b1}(9893)$
1^3P_2	2^{++}	$\chi_{c2}(3556)$	$\chi_{b2}(9913)$
2^1S_0	0^{-+}	$\eta_c(3654)$?	
2^3S_1	1^{--}	$\psi(3686)$	$\Upsilon(10\,023)$
2^3P_0	0^{++}		$\chi_{b0}(10\,232)$
2^3P_1	1^{++}		$\chi_{b1}(10\,255)$
2^3P_2	2^{++}		$\chi_{b3}(10\,269)$
3^3S_1	1^{--}	$\psi(4040)$	$\Upsilon(10\,355)$
4^3S_1	1^{--}	$\psi(4160)$	$\Upsilon(10\,580)$

Later experiments established a spectrum of *bottomium* states, i.e. bound states of the $b\bar{b}$ system. These are also shown in Table 5.2. By analogy with charmonium, those bottomium states below the *bottom threshold* $2M_B = 10.56$ GeV/c², where M_B is the mass of the lightest meson with non-zero beauty quantum number, have widths measured in keV, whereas those above this threshold have 'normal' widths expected of resonances decaying via the strong interaction

The charmonium and bottomium states with $L \leq 2$ are shown in Figure 5.5 as conventional energy level diagrams, where the energies are plotted relative to those

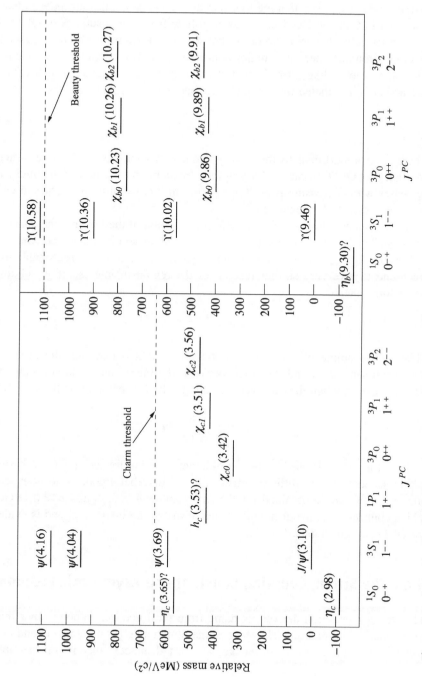

Figure 5.5 Energy levels of the charmonium ($c\bar{c}$) and bottomium ($b\bar{b}$) states for $L \leq 2$; the masses are given in units of GeV/c^2

of the $^3S_1$ ground states. There is a striking similarity in the levels of the two systems, which suggests that the forces in the $c\bar{c}$ and $b\bar{b}$ are flavour independent, as discussed in Chapter 3 and now seen to follow from the postulates of QCD. The level structure is also very similar to that seen in positronium which suggests that, as in positronium, there is a major contribution from a single-particle exchange with the Coulomb-like form. In fact at short interquark distances $r \lesssim 0.1\,\text{fm}$, the interaction is dominated by one-gluon exchange that we can write as

$$V(r) = -\frac{a}{r}, \tag{5.7}$$

where a is proportional to the strong interaction analogue of the fine structure constant α in QED. Because of asymptotic freedom, the strength of the interaction decreases with decreasing r, but for $r < 0.1$ fm this variation is slight and can in many applications be neglected.[10]

In strong interactions we also have to take account of the fact that the quarks are confined. The latter part of the potential cannot at present be calculated from QCD and several forms are used in phenomenological applications. All reasonable forms are found to give very similar results for the region of interest. If we choose a linear form, then

$$V(r) \approx b\,r. \tag{5.8}$$

This is an example of a *confining potential*, in that it does not die away with increasing separation and the force between the quark and antiquark cannot be neglected, even when they are very far apart. The full potential is thus

$$V(r) = -\frac{a}{r} + br. \tag{5.9}$$

If the form (5.9) is used in the Schrödinger equation for the $c\bar{c}$ and $b\bar{b}$ systems, taking account of their different masses, it is found that a good fit to both sets of energy levels can be obtained for the *same* values $a \approx 0.48$ and $b \approx 0.18\,\text{GeV}^2$, which confirms the flavour independence of the strong interaction and is evidence for QCD and the standard model.

5.4   The Strong Coupling Constant and Asymptotic Freedom

The strong interaction derives its name from the force that, among other things, binds quarks into hadrons. However, some remarkable phenomena depend on the fact that the interaction gets weaker at short distances; that is, on asymptotic

---

[10]The equivalent coupling in QED also varies with distance, but the variation is very small and can usually be neglected.

freedom. Such short-distance interactions are associated with large momentum transfers $|\mathbf{q}|$ between the particles, with $|\mathbf{q}| = O(\hbar/r)$, where r is the distance at which the interaction occurs. Hence in discussing scattering from a static potential, like the one above, we can regard the strong coupling α_s as decreasing with increasing momentum transfer, rather than with decreasing r.

In general, the strength of the interaction can be shown to depend on the squared four-momentum transfer

$$Q^2 \equiv E_q^2/c^2 - \mathbf{q}^2, \tag{5.10}$$

which was introduced in Chapter 2. Specifically, it can be shown that the QCD coupling constant α_s is given to a good approximation by

$$\alpha_s = \frac{12\pi}{(33 - 2N_f) \ln(Q^2/\Lambda^2)}, \tag{5.11}$$

where N_f is the number of quark flavours u, d, s,...., with $4m_q^2 c^4 < Q^2$, and $Q^2 \gg \Lambda^2$. The constant Λ is a scale parameter that must be determined from experiment. Thus QCD does not predict the absolute value of α_s, but rather its dependence on Q^2. The value of Λ may be found by measuring the coupling constant in a variety of processes (two of which will be discussed later in this chapter) giving values consistent with

$$\Lambda = 0.2 \pm 0.1 \text{ GeV/c}. \tag{5.12}$$

Because α_s varies with Q^2, it is often referred to as the *running coupling constant*. The values of $\alpha_s(Q^2)$ corresponding to Equation (5.12) are plotted in Figure 5.6. The

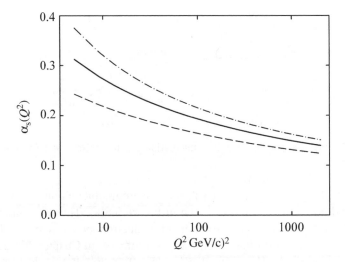

Figure 5.6 The running coupling constant α_s corresponding to four flavours and a scale parameter $\Lambda = 0.2 \pm 0.1$ GeV/c; the dashed, solid and dot-dashed curves correspond to $\Lambda = 0.1$, 0.2 and 0.3, respectively

variation with Q^2 is small at large Q^2 and over limited Q^2 regions it can often be neglected. In this large Q^2 region, the coupling is sufficiently weak that calculations can be performed with reasonable accuracy by retaining only diagrams of lowest and next-to-lowest order; and sometimes the short-range strong interaction can be neglected to a first approximation, as we shall see.

Although there are other forces that increase with increasing separation (for example, the force between two particles connected by a spring or elastic string), the difference between those and the present case is that in the former cases eventually something happens (for example, the string breaks) so that the particles (or the ends of the string) become free. This does not happen with the strong force. Instead, the energy stored in the colour field increases until it becomes sufficiently large to create $q\bar{q}$ pairs and eventually combinations of these will appear as physical hadrons. This latter process is called *fragmentation* and is rather poorly understood. The behaviour of the strong interaction as a function of distance (or equivalently momentum transfer) is so unlike the behaviour of other forces we are familiar with (e.g. gravity and electromagnetism) that it is worth looking at why this is.

In QED, single electrons are considered to emit and reabsorb photons continually, as shown in Figure 5.7(a). Such a process is an example of a so-called *quantum fluctuation*, i.e. one particle converting to two or more particles for a finite time. This is allowed provided the time and the implied violation of energy conservation are compatible with the uncertainty principle. Of course if another electron is nearby, then it may absorb the photon and we have the usual one-photon exchange scattering process of Figure 5.7(b).

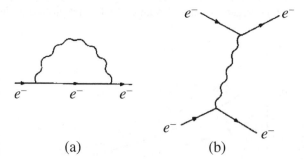

(a)   (b)

Figure 5.7 (a) The simplest quantum fluctuation of an electron, and (b) the associated exchange process

The emitted photon may itself be subject to quantum fluctuations, leading to more complicated diagrams like those shown in Figure 5.8(a). Thus the initial electron emits not only photons, but also indirectly electron–positron pairs. These are referred to as a 'sea' of virtual electrons (cf. comments in Chapter 3 in the context of the quark model). The equivalent contribution to elastic electron–electron scattering is shown in Figure 5.8(b).

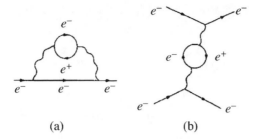

(a)   (b)

Figure 5.8 (a) A more complicated quantum fluctuation of the electron, and (b) the associated exchange process

These virtual processes are collectively referred to as *vacuum polarization effects*.[11] The production of virtual e^+e^- pairs produces a shielding effect, so that the charge and the strength of the interaction α, as seen from a distance, will appear altered. Detailed calculations show that if we write the Coulomb potential as

$$\phi_{\text{eff}}(r) = \frac{\alpha_{\text{eff}}(r)\hbar c}{r}, \qquad (5.13)$$

then

$$\alpha_{\text{eff}} = \alpha \approx 1/137 \qquad (5.14)$$

for

$$r \gg r_C \equiv \hbar/mc = 3.9 \times 10^{-13}\,\text{m}, \qquad (5.15)$$

but for $r \le r_C$, the value of α is somewhat larger and increases as r becomes smaller. In other words, the strength of the interaction increases at very short distances. Formally, without proof, the QED coupling $\alpha_{\text{em}}(Q^2)$ is given by

$$\alpha_{\text{em}}(Q^2) = \alpha(\mu^2)\left[1 - \frac{1}{3\pi}\alpha(\mu^2)\ln\left(\frac{Q^2}{\mu^2}\right)\right]^{-1}, \qquad (5.16)$$

where μ^2 is a low-energy value of Q^2 at which the value of α is known. If, for example, we take $\mu = 1\,\text{MeV}/c$ and $\alpha = 1/137$, i.e. the value of the fine structure constant as found from low-energy interactions, then at the mass of the Z^0 boson, $\alpha \approx 1/135$. Thus the electromagnetic coupling increases with energy-transfer, but only very slowly.

---

[11]The name arises from the analogy of placing a charge in a dielectric medium. This aligns the particles of the medium and produces a net polarization.

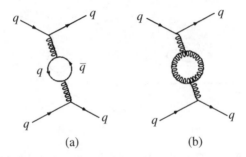

Figure 5.9 The two lowest-order vacuum polarization corrections to one-gluon exchange in quark--quark scattering

Vacuum polarization effects have measurable consequences. For example, the 2S state in hydrogen is predicted to be more tightly bound than it would be in a pure Coulomb potential. The increased binding is only 2.2×10^{-7} eV, but nevertheless it is confirmed by extremely accurate measurements on the hydrogen spectrum. There are also very small corrections to the magnetic moment of the electron that have been verified experimentally to extraordinary precision.

Quantum fluctuations also exist in QCD and also give rise to a variation of the interaction strength with distance. If, by analogy with QED, we consider quark–quark scattering, then the two lowest-order vacuum polarization corrections are shown in Figure 5.9. The first of these (Figure 5.9(a)) is analogous to virtual e^+e^- production in QED and also leads to a screening effect. However, the second diagram (Figure 5.9(b)) has no counterpart in QED, because there are no direct photon self-couplings. Calculations show that this diagram leads to an *antiscreening* effect that is larger than the screening effect from Figure 5.9(a) and so the net effect is that the interaction grows *weaker* at short distances, i.e. asymptotic freedom. Formally, the strong interaction coupling α_s is given by a formula analogous to that for α_{em} above, except the coefficient of the logarithmic term is different and, crucially, its sign is positive:

$$\alpha_s(Q^2) = \alpha_s(\mu^2)\left[1 + \frac{\alpha_s(\mu^2)}{12\pi}(33 - 2N_f)\ln(Q^2/\mu^2)\right]^{-1}, \qquad (5.17)$$

where again μ^2 is a low-energy value of Q^2 at which the value of α_s is known and N_f is the number of quark flavours that take part in the interaction.

5.5 Jets and Gluons

A striking feature of many high-energy particle collisions is the occurrence of jets of hadrons in the final state. We have already mentioned these in Section 3.2.1 when we discussed the experimental evidence for quarks and again when we discussed basic properties of quarks and gluons interactions earlier in this chapter. They have been extensively studied in the reaction

$$e^+ + e^- \rightarrow \text{hadrons} \tag{5.18}$$

at high energies using colliding beam experiments, which were discussed in Chapter 4. High-energy electrons and positrons collide head-on, with equal and opposite momenta, so that the total momentum of the hadrons produced cancels out to zero in order to conserve momentum. This is a particularly 'clean' reaction, because the initial particles are elementary, without internal structure.

In the centre-of-mass energy range 15–40 GeV, electron–positron annihilation into hadrons is dominated by the production of jets. These can be regarded as occurring in two stages: (1) a primary electromagnetic process $e^+ + e^- \rightarrow q + \bar{q}$ (due to photon exchange) leading to the production of a quark–antiquark pair, followed by (2) fragmentation (the concept we met in discussing asymptotic freedom) which converts the high-energy $q\bar{q}$ pair into two jets of hadrons. This is illustrated in Figure 5.10.

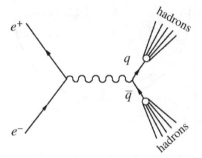

Figure 5.10 Basic mechanism of two-jet production in electron--positron annihilation

The fragmentation process that converts the quarks into hadrons is very complicated, and the composition of the jets – i.e. the numbers and types of particles in the jet and their momenta – varies from event to event. However, the direction of a jet, defined by the total momentum vector

$$\mathbf{P} = \sum_i \mathbf{p}_i, \tag{5.19}$$

where the sum extends over all the particles within the jet, closely reflects the parent quark or antiquark direction. This is because the QCD interaction is relatively weak at very short distances (asymptotic freedom), and the quark and antiquark do not interact strongly until they are separated by a distance r of order 1 fm. At these relatively large distances, only comparatively small momenta can be transferred, and hence the jets that subsequently develop point almost exactly in the initial quark and antiquark directions. That is, the jet angular distribution relative to the electron beam direction reflects the angular distributions of the quark and antiquark in the basic reaction $e^+ + e^- \rightarrow q + \bar{q}$. The latter can easily be calculated in QED as it is a purely electromagnetic process, and is in excellent agreement with the observed angular distribution of the jets. This is one of the pieces of evidence for the existence of quarks that was cited in Chapter 3 and again at the start of the present chapter.

Although the dominant process in electron–positron annihilation into hadrons is the formation of two 'back-to-back' jets, occasionally we would expect a high-momentum gluon to be emitted by the quark or anti-quark before fragmentation occurs, in much the same way as a high-energy electron sometimes emits a photon (i.e. bremssrahlung). The quark, antiquark and gluon then all fragment into hadrons, leading to a three-jet event. A computer reconstruction of such an event in a jet chamber is shown in Figure 5.11.

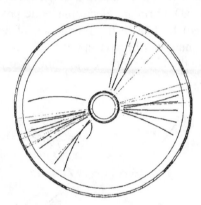

Figure 5.11   Computer reconstruction of a three-jet event in electron--positron annihilation

Events like these provided the first unambiguous evidence for gluons, because the angular distributions of the jets are found to be in good agreement with the theoretical expectation for spin-1 gluons, but are inconsistent with what would be expected if, for example, the third jet originated from a particle of spin-0. The ratio of three-jet to two-jet events can also be calculated, assuming that the third jet is a gluon, because the probability that a quark or antiquark will emit a gluon is determined by the strong coupling α_s, in the same way that the probability that an electron or positron will emit a photon is determined by the fine structure constant α. This leads to a value of α_s and hence Λ, the QCD scale parameter. The value obtained is consistent with Equation (5.12) found from other determinations and lends further support for the whole picture of quarks interacting via the exchange of gluons.

5.6   Colour Counting

What evidence is there that quarks exist in just three colour states? This question can be settled by using data from electron–positron annihilation. The cross-sections for electron–positron annihilation to hadrons and for electron–positron annihilation to muons[12] both decrease rapidly with energy, but their ratio

$$R \equiv \frac{\sigma(e^+e^- \rightarrow \text{hadrons})}{\sigma(e^+e^- \rightarrow \mu^+\mu^-)} \tag{5.20}$$

is almost energy independent. The near constancy of this ratio follows from the dominance of the two-step mechanism of Figure 5.10, with the total annihilation rate being determined by that of the initial reaction $e^+e^- \rightarrow q + \bar{q}$. The value of the ratio R then directly confirms the existence of three colour states, each with the same electric charge, for each quark flavour.

To understand this, let us suppose that each quark flavour $f = u, d, s \ldots$ exists in N_C colour states, so that $N_C = 3$ according to QCD, while $N_C = 1$ if the colour degree of freedom does not exist. Since the different colour states all have the same electric charge, they will all be produced equally readily by the mechanism of Figure 5.10, and the rate for producing quark pairs of any given flavour $f = u, d, s, \ldots$ will be proportional to the number of colours N_C. The cross-section is also proportional to the squared charge of the produced pair (because this is a first-order electromagnetic process), and since muon pairs are produced by an identical mechanism, we obtain

$$\sigma(e^+e^- \rightarrow q\bar{q}) = N_C\, e_f^2 \sigma(e^+e^- \rightarrow \mu^+\mu^-), \tag{5.21}$$

where e_f is the electric charge, in units of e, on a quark of flavour f.

The cross-section for $e^+ + e^- \rightarrow$ hadrons will receive an additional contribution of the form of Equation (5.21) when the energy passes a threshold for a new quark flavour to be produced. Thus R at low energies will have a series of 'steps' corresponding to the production of pairs of new quarks and this is what is observed experimentally. At high energies above the threshold for the production of $b\bar{b}$ pairs and assuming that hadron production is completely dominated by the two-step process of Figure 5.10, we would have[13]

$$R = R_0 \equiv N_C(e_u^2 + e_d^2 + e_s^2 + e_c^2 + e_b^2) = 11N_C/9. \tag{5.22}$$

When the small contribution from the three-jet events and other corrections of order α_s are taken into account, this expression for R is modified to

$$R = R_0(1 + \alpha_s/\pi), \tag{5.23}$$

---

[12]The cross-section for the production of muon pairs is essentially a purely electromagnetic one, except at very high energies where the effect of the weak interaction may be seen. This will be discussed in Chapter 6.
[13]There is no contribution from the top quark because it is too heavy to be produced, even at the high energies we are considering.

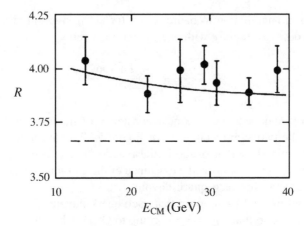

Figure 5.12 Measured values of the cross-section ratio R and the theoretical prediction from QCD for $N_C = 3$ colours; the dashed line shows the prediction without QCD corrections

giving rise to a weak energy dependence of R from the energy dependence of α_s discussed earlier (Equation (5.17)). Although these corrections of order α_s are small compared to the dominant contribution, they must be included if the most precise experimental data on R are to be accounted for. The data are in excellent agreement with the theoretical prediction for the value $N_C = 3$ (see Figure 5.12) and hence prove that quarks exist in just three colour states.

5.7 Deep Inelastic Scattering and Nucleon Structure

In Chapter 2 we discussed the scattering of electrons from nuclei to determine their radial charge distributions. This was done by assuming a form for the charge distribution, calculating the resulting form factor (i.e. the Fourier transform of the charge distribution) and using it to fit experimental cross-sections. In a somewhat similar way we can use high-energy inelastic scattering to investigate the charge distribution within nucleons. This is referred to as *deep inelastic scattering*, because the projectiles probe deep into the internal structure of the nucleon. This type of interaction was mentioned in Section 2.9 in the context of classifying nuclear reaction mechanisms. The original experiments of this type in particle physics were done in the 1960s and provided the first definitive evidence for the existence of quarks. We will deduce that nucleons have a sub-structure of point-like charged constituents.[14]

---

[14]The pioneering work on deep inelastic scattering done by Jerome Friedman, Henry Kendall and Richard Taylor resulted in their receiving the 1990 Nobel Prize in Physics.

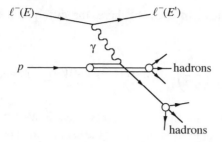

Figure 5.13 Dominant one-photon exchange mechanism for inelastic lepton–proton scattering where $\ell = e \ or \ \mu$

The dominant one-photon contribution to the inelastic scattering of a charged lepton from a proton in the spectator quark model is shown in Figure 5.13. Unlike elastic scattering, where at a given lepton energy E there is only one free variable (e.g. the scattering angle), in inelastic scattering the excitation energy of the nucleon adds a further degree of freedom, so we can define two independent variables. These are usually taken to be ν, defined by

$$2M\nu \equiv W^2c^2 + Q^2 - M^2c^2 \qquad (5.24)$$

and a dimensionless quantity (called the *scaling variable*) given by

$$x \equiv Q^2/2M\nu. \qquad (5.25)$$

Here, M is the proton mass, W is the invariant mass of the final-state hadrons and Q^2 is the squared energy–momentum transfer

$$Q^2 = (E - E')^2/c^2 - (\mathbf{p} - \mathbf{p}')^2. \qquad (5.26)$$

The physical interpretation of x will be discussed below. In the rest frame of the initial proton, ν reduces to

$$\nu = E - E' \qquad (5.27)$$

and so is the Lorentz-invariant generalization for the energy transferred from the lepton to the proton.

In Chapter 2 we discussed several modifications to the formalism for describing the structure of nuclei obtained from scattering experiments. Here we are dealing with high-energy projectiles and so we will need to take all those corrections into account. In particular, the magnetic interaction introduces a second form factor. (cf Equation (2.14)). The two form factors, denoted W_1 and W_2, are called *structure functions* in this context. In terms of these, the differential cross-section may be written

$$\frac{d^2\sigma}{d\Omega dE'} = \left(\frac{d\sigma}{d\Omega}\right)_{\text{Mott}} [W_2(Q^2, \nu) + 2W_1(Q^2, \nu)\tan^2(\theta/2)], \qquad (5.28)$$

where θ is the lepton scattering angle. For values of $W \leq 2.5\,\text{GeV}/c^2$, the cross-sections show considerable structure due to the excitation of nucleon resonances, but above this mass they are smoothly varying. In the latter region, the values of the structure functions can be extracted from the data by choosing suitable parameter-izations and fitting the available data in an analogous way to the way charge distributions of nuclei were deduced in Chapter 2.

Rather than W_1 and W_2, it is usual to work with two related dimensionless structure functions defined by

$$F_1(x, Q^2) \equiv Mc^2 W_1(Q^2, \nu) \quad \text{and} \quad F_2(x, Q^2) \equiv \nu W_2(Q^2, \nu). \qquad (5.29)$$

It is a remarkable fact that at fixed values of x the structure functions have only a very weak dependence on Q^2. This behaviour is referred to as *scaling* and is illustrated in Figure 5.14. As the Fourier transform of a spherically symmetric point-like distribution is a constant, we conclude that the proton has a sub-structure of point-like charge constituents.

The interpretation of scaling is simplest in a reference frame where the target nucleon is moving with a very high velocity, so that the transverse momenta and

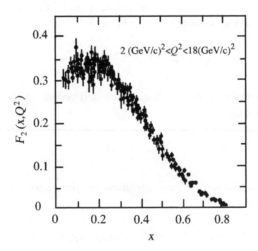

Figure 5.14 The structure function F_2 of the proton as a function of x, for Q^2 between 2 and 18 $(\text{GeV}/c)^2$ (reproduced from At82 with kind permission of Springer Science and Business Media)

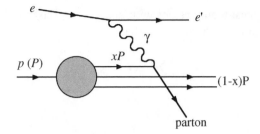

Figure 5.15 The parton model of deep inelastic scattering

rest masses of its constituents may be neglected. The structure of the nucleon is then given by the longitudinal momentum of its constituents. This approach was first adopted by Feynman and Bjorken, who called the constituents *partons*. (We now identify charged partons with quarks and neutral partons with gluons.) In the *parton model*, deep inelastic scattering is visualized as shown in Figure 5.15. The target nucleon is a stream of partons each with four-momentum xP, where $P = (p, \mathbf{p})$ is the four-momentum of the nucleon and $p = |\mathbf{p}|$, is its (very large) three-momentum, so that the nucleon mass may be neglected.

Suppose now that one parton of mass m is scattered elastically by the exchanged photon of four-momentum Q. Then

$$(xP + Q)^2 = (x^2P^2 + 2xP \cdot Q + Q^2) = m^2c^4 \approx 0. \tag{5.30}$$

If $\left|x^2P^2\right| = x^2M^2c^4 \ll Q^2$, then

$$x = -\frac{Q^2}{2P \cdot Q} = \frac{Q^2}{2M\nu}, \tag{5.31}$$

where the invariant scalar product has been evaluated in the laboratory frame in which the energy transfer is ν and the nucleon is at rest. This is our previous definition Equation (5.25). Thus, the physical interpretation of x is the fractional three-momentum of the parton in the reference frame where the nucleon has a very high velocity. This is equivalent to having a parton of mass m stationary in the laboratory system, with the elastic relation $Q^2 = 2m\nu$. So provided $Q^2 \gg M^2$,

$$x = \frac{Q^2}{2M\nu} = \frac{m}{M}, \tag{5.32}$$

i.e. x may also be interpreted as the fraction of the nucleon mass carried by the struck parton.

To identify the constituent partons with quarks we need to know their spins and charges. For the spin, it can be shown that

$$F_1(x, Q^2) = 0 \quad \text{(spin 0)} \tag{5.33}$$

and

$$2xF_1(x, Q^2) = F_2(x, Q^2) \quad \text{(spin 1/2)}. \tag{5.34}$$

The latter relation, known as the *Callan–Gross relation*, follows by comparing the coefficients in the equation for the double differential cross-section Equation (5.28) with that in Chapter 2 (Equation (2.14)). This gives

$$2W_1/W_2 = 2\tau, \tag{5.35}$$

where $\tau = Q^2/4m^2 c^2$ and m is the mass of the target, in this case the mass of the struck parton. Replacing W_1 by F_1/Mc^2 and W_2 by F_2/ν, gives

$$\frac{\nu}{Mc^2} \frac{F_1}{F_2} = \frac{Q^2}{4m^2c^2} \tag{5.36}$$

and since now $Q^2 = 2m\nu$, we have $m = Q^2/2\nu = xM$. Finally, using this mass in Equation(5.36) yields the Callan–Gross relation. Figure 5.16 shows some results for the ratio $2xF_1/F_2$. It is clear that spin-$\frac{1}{2}$ is strongly favoured.

To deduce the parton charges is more complicated. We will assume that the constituent partons are quarks and show that this is consistent with experimental data. We start by defining $q_f(x)$ to be the momentum distribution of a quark of flavour f, i.e. $q_f(x)dx$ is the probability of finding in a nucleon a quark of flavour f, with momentum fraction in the interval x to $x + dx$. A given nucleon will consist of a combination of valence quarks (i.e. those that give rise to the observed quantum numbers in the quark model) and additional quark–antiquark pairs that are continually produced and annihilated by the radiation of virtual gluons by the quarks.[15] (Recall the discussion of quantum fluctuations in electrodynamics in Section 5.4.) Thus, in general, a structure function can be written as the sum of contributions from quarks and antiquarks of all flavours. Also, from the cross-section formula Equation (5.28), we would expect the structure functions to involve the quark distributions weighted by the squares of the quark charges z_f (in units of e) for a given quark flavour f.

---

[15]These are the 'sea' quarks referred to in the discussion of the static quark model in Chapter 3.

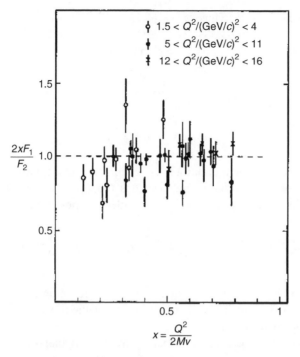

Figure 5.16 The ratio $2xF_1/F_2$ at fixed x

Thus, for example, F_2 is

$$F_2(x) = x \sum_f z_f^2 \left[q_f(x) + \bar{q}_f(x) \right]. \tag{5.37}$$

If we concentrate on the scattering of charged leptons, i.e. electrons or muons, and consider just the possibility of light quarks u, d and s within nucleons, then we have (for $\ell = e, \mu$)

$$F_2^{\ell p}(x) = x \left[\frac{1}{9}(d^p + \bar{d}^p) + \frac{4}{9}(u^p + \bar{u}^p) + \frac{1}{9}(s^p + \bar{s}^p) \right] \tag{5.38a}$$

and

$$F_2^{\ell n}(x) = x \left[\frac{1}{9}(d^n + \bar{d}^n) + \frac{4}{9}(u^n + \bar{u}^n) + \frac{1}{9}(s^n + \bar{s}^n) \right], \tag{5.38b}$$

where, for example, $u^{n,p}$ is the distribution of u quarks in the neutron and proton. Using isospin symmetry, interchanging u and d quarks changes neutron to proton, i.e. $u \leftrightarrow d$ implies $n \leftrightarrow p$. Thus,

$$u^p(x) = d^n(x) \equiv u(x), \qquad (5.39a)$$
$$d^p(x) = u^n(x) \equiv d(x), \qquad (5.39b)$$

and

$$s^p(x) = s^n(x) \equiv s(x), \qquad (5.39c)$$

with similar relations for the antiquarks. Then if we work with a target nucleus with equal numbers of protons and neutrons (an *isoscalar* target), its structure function will have the approximate form (neglecting purely nuclear effects)

$$F_2^{\ell N}(x) = \frac{1}{2}[F_2^{\ell p}(x) + F_2^{\ell n}(x)] = \frac{5}{18}x\sum_{q=d,u}[q(x) + \bar{q}(x)] + \frac{1}{9}x[s(x) + \bar{s}(x)]. \quad (5.40)$$

The second term is small because s quarks are only present in the sea component at the level of a few percent. Thus the mean squared value of the charges of the u and d quarks is approximately $\frac{5}{18}$.

The final step is to extract information from deep inelastic scattering using neutrinos and antineutrinos as projectiles. This is more complicated because, as we shall see in Chapter 6, neutrinos and antineutrinos couple differently to the different quarks and antiquarks and there is also a third form factor involved. Without proof, we shall just quote the result:

$$F_2^{\nu N}(x) = x\sum_{q=d,u}[q(x) + \bar{q}(x)]. \qquad (5.41)$$

There is no electric charge factor outside the summation because, just as quarks form strong interaction isospin multiplets with different electric charges, the leptons also form weak isospin multiplets, but in this case the resulting weak charge is the same for all quarks.[16]

From Equation (5.40) and (5.41), we expect

$$F_2^{\nu N}(x) \leq \frac{18}{5}F_2^{\ell N}(x). \qquad (5.42)$$

The experimental data illustrated in Figure 5.17 show that $F_2^{\ell N}(x)$ and $F_2^{\nu N}(x)$ are equal within errors except possibly at small values of x where antiquarks are more important. Thus one can conclude that the partons do have charges $\frac{2}{3}$ and $-\frac{1}{3}$, which completes the evidence for identifying partons with quarks.

---

[16]Weak isospin will be discussed briefly in Chapter 6.

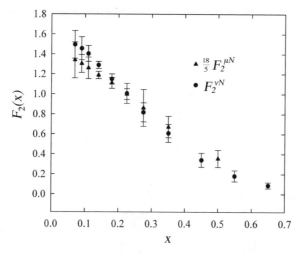

Figure 5.17 Comparison of $F_2(x)$ from deep inelastic muon (data from Ar97) and neutrino (data from Se97) scattering experiments; the data points are the average over a range of $Q^2 > 2\,(\mathrm{GeV/c})^2$ and the error bars express the range of data values within the Q^2 ranges

Combining data from different experiments, with electrons, muons, neutrinos and antineutrinos as projectiles, enables individual quark/parton momentum distributions to be extracted from combinations of cross-sections. Some typical results at $Q^2 = 10\,(\mathrm{GeV/c})^2$ are shown in Figure 5.18 for the combinations

$$Q(x) = d(x) + u(x) \tag{5.43a}$$

and

$$\bar{Q}(x) = \bar{d}(x) + \bar{u}(x). \tag{5.43b}$$

The difference

$$Q_v(x) \equiv Q(x) - \bar{Q}(x) \tag{5.44}$$

can be identified as the distribution of the valence quarks of the quark model. It can be seen that Q_v is concentrated around $x \approx 0.2$ and dominates except at small values of x where the antiquarks $\bar{q}$ in the sea distribution are important.

The results of Figure 5.18 reveal an interesting and unexpected result concerning gluons within the nucleon. If we integrate the momentum distributions for quarks and antiquarks over all x we might expect to recover the total momentum of the nucleon, whereas the curves of Figure 5.18 yield a value of approximately 0.5. Thus it follows that about 50 per cent of the momentum is carried by gluons.

Although scaling is approximately correct, it is certainly not exact. In Figure 5.19 we show some deep inelastic scattering data plotted in more detail. The

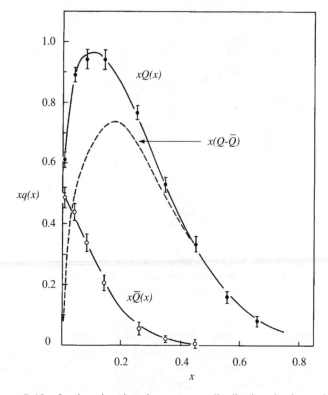

Figure 5.18   Quark and antiquark momentum distributions in the nucleon

deviations from scaling are due to QCD corrections to the simple quark model, i.e. the quark in the proton that is struck by the exchanged photon can itself radiate gluons. Again, without further details, the scaling violations are explained by QCD using a value of the strong interaction parameter Λ that is consistent with that obtained from other sources (e.g. the three-jet events that we have discussed above).[17]

Finally, it is worth noting that the nucleon structure functions and hence the quark densities are found from lepton scattering experiments using a range of different nuclear targets. We have seen that the average binding energy of nucleons in heavy nuclei is of the order of 7–8 MeV per nucleon. As this energy is much smaller than those used in deep inelastic scattering experiments, it might be thought safe to ignore nuclear effects (except those due to the internal motion of the nucleons – the Fermi momentum – that are typically about 200 MeV/c). However, experiments have shown that the structure functions do in fact depend

---

[17]Scaling violations are discussed in detail, but at a more advanced level than here in, for example Ha84.

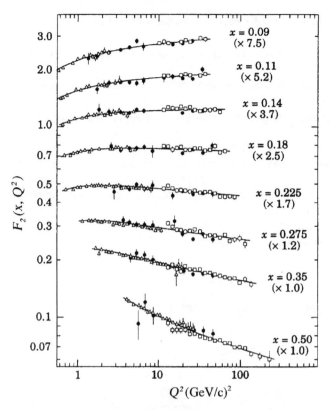

Figure 5.19 A compilation of values of F_2 measured in deep inelastic electron and muon scattering from a deuterium target -- different symbols denote different experiments; for clarity, the data at different values of x have been multiplied by the factors shown in brackets and the solid line is a QCD fit with $\Lambda = 0.2$ GeV (adapted from Mo94, copyright the American Physical Society)

slightly on the nuclear medium. Although the effects are very small and not enough to alter the conclusions of this chapter, it is a reminder that there are still things to be learnt about the role of nuclear matter and that this may hold information on the nuclear force in terms of the fundamental quark–gluon interaction.

Problems

5.1 The general combination of m quarks and n antiquarks $q^m \bar{q}^n$, with baryon number $B > 0$ has a colour wavefunction that may be written $r^\alpha g^\beta b^\gamma \bar{r}^{\bar\alpha} \bar{g}^{\bar\beta} \bar{b}^{\bar\gamma}$, where r^α means that there are α quarks in the r colour state, etc.. By imposing the condition of colour confinement, show that $m - n = 3p$, where p is a non-negative integer and hence show that states with the structure qq are not allowed.

5.2 Draw the lowest-order Feynman diagrams for the following processes:

  (a) the interaction of a quark and a gluon to produce a quark and a photon;

  (b) the production of a single Z^0-boson in a collision of protons and antiprotons;

  (c) the annihilation of an electron and a positron to produce a pair of W-bosons.

5.3 A $p\bar{p}$ collider with equal beam energies is used to produce a pair of top quarks. Draw a Feynman diagram for this process that involves a single gluon. If the three quarks of the proton (or antiproton) carry between them 50 per cent of the hadron total energy–momentum, calculate the minimum beam momentum required to produce the $t\bar{t}$ pair.

5.4 The lowest Feynman diagram for inelastic electron–proton scattering at high energies

$$e^-(E,\mathbf{p}c) + p(E_P,\mathbf{P}_pc) \rightarrow e^-(E',\mathbf{p}'c) + X(\text{hadrons})$$

is shown in Figure 5.20.

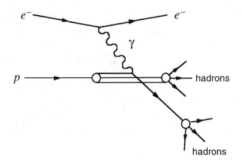

Figure 5.20   Kinematics of inelastic electron--proton scattering

Use energy–momentum conservation to show that the variable ν defined in Equation (5.24) becomes $\nu = E - E'$ in the rest frame of the proton. Hence show that the variable x defined in Equation (5.25) lies in the range $0 \le x \le 1$ if the mass of the electron is neglected.

5.5 If hadron–hadron total cross-sections are assumed to be the sum of the cross-sections between their constituent quarks, show that the quark model predicts the relationship:

$$\sigma(\Lambda p) = \sigma(pp) + \sigma(K^- n) - \sigma(\pi^+ p).$$

5.6 The 3γ decay of positronium (the bound state of e^+e^-) has a width that in QED is predicted to be $\Gamma(3\gamma) = 2(\pi^2 - 9)\alpha^6 m_e c^2/9\pi$, where α is the fine structure constant. If the hadronic decay of the $c\bar{c}$ bound state $J/\Psi(3100)$ proceeds via an analogous mechanism, but involving three gluons, use the experimental hadronic width $\Gamma(3g) = 80\,\text{keV}$ to estimate the strong interaction coupling constant α_s. Use an analogous assumption to estimate α_s from the radiative width $\Gamma(gg\gamma) = 0.16\,\text{keV}$ of the $b\bar{b}$ bound state $\Upsilon(9460)$.

5.7 Use Equations (5.38) and (5.39) to derive the *Gottfried sum rule*,

$$\int\limits_0^1 [F_2^{ep}(x) - F_2^{en}(x)]\,\frac{dx}{x} = \frac{1}{3} + \frac{2}{3}\int\limits_0^1 [\bar{u}(x) - \bar{d}(x)]dx,$$

where the quark distributions refer to the proton.

5.8 Estimate the cross-section ratio R defined in Equation (5.20) at centre-of-mass energies $E_{CM} = 2.8\,\text{GeV}$ and 15 GeV. How would R change if the energy were increased so that top quark pairs could be produced?

5.9 Common forms assumed for the momentum distributions of valence quarks in the proton are:

$$F_u(x) = xu(x) = a(1 - x)^3; \quad F_d(x) = xd(x) = b(1 - x)^3.$$

If the valence quarks account for half the proton's momentum, find the values of a and b.

5.10 The cross-section $\sigma(u\bar{d} \to W^+)$ near the mass of the W^+ is given by the Breit–Wigner form

$$\sigma = \frac{\pi(\hbar c)^2 \lambdabar^2 \Gamma\Gamma_{u\bar{d}}}{3[4(E - M_W c^2)^2 + \Gamma^2]},$$

where (M_W, Γ) are the mass and total width of the W^+, $\Gamma_{u\bar{d}}$ is the partial width for $W^+ \to u\bar{d}$, E is the total centre-of-mass energy of the $u\bar{d}$ pair and $\lambdabar = 2/E$. Find the maximum value of σ, i.e. $\sigma_{\max}$, given that the branching ratio for $W^+ \to u\bar{d}$ is $1/3$. Use this result and the quark distributions of Question 5.9 to find an expression for the cross-section $\sigma(p\bar{p} \to W^+ + \cdots)$ in terms of the $p\bar{p}$ total centre-of-mass energy $\sqrt{s}$ and $\sigma_{\max}$ and evaluate your result for $\sqrt{s} = 1\,\text{TeV}$. (Use the narrow width, i.e. delta function, approximation

$$\sigma_{u\bar{d}}(E) = \pi \frac{\Gamma_W}{M_W c^2} \sigma_{\max} \delta\left(1 - \frac{E^2}{(M_W c^2)^2}\right)$$

in integrals.)

6

Electroweak Interactions

We have already discussed some aspects of weak and electromagnetic interactions when we discussed nuclear stability in Chapter 2 and again when we introduced the basic properties of leptons in Chapter 3. In this chapter we will consider wider aspects of the weak interaction and also its unification with electromagnetism to produce the spectacularly successful *electroweak theory*.

6.1 Charged and Neutral Currents

Like the strong and electromagnetic interactions, the weak interaction is also associated with elementary spin-1 bosons, which act as 'force carriers' between quarks and/or leptons. Until 1973 all observed weak interactions were consistent with the hypothesis that they were mediated by the exchange of the charged bosons $W^\pm$ only. However, in the 1960s, a theory was developed which unified electromagnetic and weak interactions in a way that is often compared with the unification of electric and magnetic interactions by Maxwell a century earlier. This new theory made several remarkable predictions, including the existence of the heavy neutral vector boson Z^0 and of weak reactions arising from its exchange. The latter processes are called *neutral current* reactions (the word neutral referring to the charge of the exchanged particle) to distinguish them from the so-called *charged current* reactions arising from charged $W^\pm$ boson exchange. In particular, neutral current reactions of the type $\nu_\mu + N \to \nu_\mu + X$ were predicted to occur via the mechanism of Figure 6.1, where N is a nucleon and X is any set of hadrons allowed by the conservation laws. Although difficult to detect, such reactions were first observed in a bubble chamber experiment in 1973.

The prediction of the existence and properties of neutral currents, prior to their discovery, is only one of many remarkable successes of the unified theory of electromagnetic and weak interactions. Others include the prediction of the existence of the charmed quark, prior to its discovery in 1974 and the prediction

Nuclear and Particle Physics   B. R. Martin
© 2006 John Wiley & Sons, Ltd

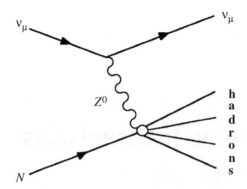

Figure 6.1   Feynman diagram for the weak neutral current reaction $\nu_\mu + N \rightarrow \nu_\mu + X$

of the masses of the $W^\pm$ and Z^0 bosons prior to the long-awaited detection of these particles in 1983. In general, the theory is in agreement with all data on both weak and electromagnetic interactions, which are now referred to collectively as the *electroweak interaction*, in the same way that electric and magnetic interactions are referred to collectively as electromagnetic interactions. Furthermore, the theory predicts the existence of a new spin-0 boson, the so-called *Higgs boson*, which is associated with the origin of particle masses within the model. This was mentioned in passing in earlier chapters. Although a detailed discussion of the Higgs boson is beyond the scope of this book, there is a brief discussion of the role of this very important particle in Chapter 9.

The new unification only becomes manifest at high energies, and at low energies weak and electromagnetic interactions can still be clearly separated. This follows from the general form of the amplitude Equation (1.41):

$$F(\mathbf{q}^2) = \frac{-g^2 \hbar^2}{|\mathbf{q}|^2 + M_X^2 c^2}, \tag{6.1}$$

where M_X^2 is the mass of the exchanged particle and g is the appropriate coupling. For weak interactions, $M_X = M_{W,Z} \approx 80\,\text{GeV}/c^2$ and for the electromagnetic interaction $M_X = M_\gamma = 0$. Thus, even with $g_{\text{weak}} \sim g_{\text{em}}$, the amplitudes for the two interactions will only become of comparable size for $|\mathbf{q}|^2 \sim M_X^2 c^2$, i.e. at high energies. We therefore start by considering the weak interaction at low energies and deduce some of its general properties that are valid at all energies. Later we will consider how unification arises and some of its consequences.

6.2   Symmetries of the Weak Interaction

In this section we will discuss the parity (P) and charge conjugation (C) operators, which were introduced in Chapter 1. These are conserved in the strong and electromagnetic interactions. The first indication that parity might be violated in weak interactions came from observations on the pionic decays of K-mesons, i.e. $K \rightarrow \pi\pi$ and $K \rightarrow \pi\pi\pi$,[1] and these led Lee and Yang in 1956 to make a thorough study of all previous experiments in which parity conservation had been assumed or apparently proved. They came to the startling conclusion that there was in fact no firm evidence for parity conservation in weak interactions; and they suggested experiments where the assumption could be tested.[2] This led directly to the classic demonstration of parity violation from a study of the β-decay of polarized ^{60}Co nuclei. We shall just describe the principles of this experiment.[3]

The experiment was done in 1957 by Wu and co-workers, who placed a sample of ^{60}Co inside a magnetic solenoid and cooled it to a temperature of 0.01 K. At such temperatures, the interaction of the magnetic moments of the nuclei with the magnetic field overcomes the tendency to thermal disorder, and the nuclear spins tend to align parallel to the field direction. The polarized ^{60}Co nuclei produced in this way decay to an excited state of ^{60}Ni by the β-decay

$$^{60}\text{Co} \rightarrow {}^{60}\text{Ni}^* + e^- + \bar{\nu}_e. \tag{6.2}$$

Parity violation was established by the observation of a 'forward–backward decay asymmetry', i.e. the fact that fewer electrons were emitted in the forward hemisphere than in the backward hemisphere with respect to the spins of the decaying nuclei.

We can show that this implies parity violation as follows. The parity transformation reverses all particle momenta **p** while leaving their orbital angular momenta **r** × **p**, and by analogy their spin angular momenta, unchanged. Hence in the rest frame of the decaying nuclei its effect is to reverse the electron velocity while leaving the nuclear spins unchanged, as shown in Figure 6.2. Parity invariance would then require that the rates for the two processes Figure 6.2(a) and Figure 6.2(b) were equal, so that equal numbers of electrons would be emitted in the forward and backward hemispheres with respect to the nuclear spins, in contradiction to what was observed. The discovery of parity violation was a watershed in the history of weak interactions because the effect is large, and an understanding of weak interactions is impossible if it is neglected.

The charge conjugation operator C changes all particles to antiparticles and as we will see presently is also not conserved in weak interactions. In examining these operators, two interconnected themes will emerge. The first is that these effects have their origin in the spin dependence of weak interactions; the second is

---

[1]Two particles, called at that time τ and θ, were observed to decay via the weak interaction to $\pi\pi$ and $\pi\pi\pi$ final states, respectively, which necessarily had different final-state parities. However, the τ and θ had properties, including the near equality of their masses, which strongly suggested that they were in fact the same particle. Analysis of the 'τ–θ puzzle' suggested that parity was not conserved in the decays.

[2]For their work on parity non-conservation, Chen Yang and Tsung-Dao Lee were awarded the 1957 Nobel Prize in Physics.

[3]This classic experiment is described in readable detail in Chapter 10 of Tr75.

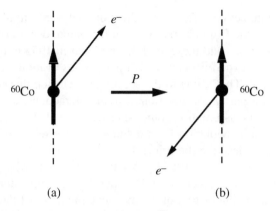

Figure 6.2 Effect of a parity transformation on ^{60}Co decay: the thick arrows indicate the direction of the spin of the ^{60}Co nucleus, while the thin arrows show the direction of the electron's momentum

that while P-violation and C-violation are large effects, there is a weaker combined symmetry, called CP-invariance, which is almost exactly conserved. This has its most striking consequences for the decays of neutral mesons, which are also discussed below. We start by considering the P and C operators in more detail.

C-violation and P-violation are both conveniently illustrated by considering the angular distributions of the electrons and positrons emitted in the decays

$$\mu^- \to e^- + \bar{\nu}_e + \nu_\mu \tag{6.3a}$$

and

$$\mu^+ \to e^+ + \nu_e + \bar{\nu}_\mu \tag{6.3b}$$

of polarized muons. In the rest frame of the decaying particle these were found to be of the form

$$\Gamma_{\mu^\pm}(\cos\theta) = \frac{1}{2}\Gamma_\pm \left[1 - \frac{\xi_\pm}{3}\cos\theta\right], \tag{6.4}$$

where θ is the angle between the muon spin direction and the direction of the outgoing electron or positron, as shown in Figure 6.3(a). The quantities $\xi_\pm$ are called the asymmetry parameters, and $\Gamma_\pm$ are the total decay rates, or equivalently the inverse lifetimes, i.e.

$$\tau_\pm^{-1} \equiv \int_{-1}^{+1} d\cos\theta\, \Gamma_{\mu^\pm}(\cos\theta) = \Gamma_\pm, \tag{6.5}$$

as may easily be checked by direct substitution.

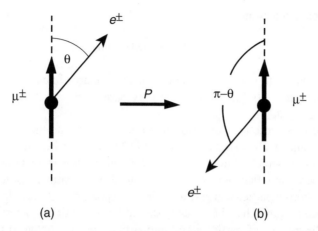

Figure 6.3 Effect of a parity transformation on muon decays: the thick arrows indicate the direction of the muon spin, while the thin arrows indicate the direction of the electron's momentum

We consider now the consequences of assuming parity and charge conjugation for these decays, starting with the latter as it is the simpler. Under charge conjugation, μ^- decay converts to μ^+ decay. C-invariance then implies that the rates and angular distributions for these decays should be the same, i.e.

$$\Gamma_+ = \Gamma_- \quad (C\text{-invariance}) \tag{6.6}$$

and

$$\xi_+ = \xi_- \quad (C\text{-invariance}). \tag{6.7}$$

The parity transformation preserves the identity of the particles, but reverses their momenta while leaving their spins unchanged. Its effect on muon decay is shown in Figure 6.3, where we see that it changes the angle θ to $\pi - \theta$, so that $\cos \theta$ changes sign. Hence P-invariance implies

$$\Gamma_{\mu^\pm}(\cos \theta) = \Gamma_{\mu^\pm}(-\cos \theta) \quad (P\text{-invariance}). \tag{6.8}$$

Substituting Equation (6.4), leads to the prediction that the asymmetry parameters vanish,

$$\xi_\pm = 0 \quad (P\text{-invariance}). \tag{6.9}$$

Experimentally, the $\mu^\pm$ lifetimes are equal to a very high level of precision, so that the prediction for the lifetimes is satisfied; but the measured values of the asymmetry parameters are

$$\xi_- = -\xi_+ = 1.00 \pm 0.04, \tag{6.10}$$

which shows that both C-invariance and P-invariance are violated. The violation is said to be 'maximal', because the asymmetry parameters are defined to lie in the range $-1 \leq \xi_\pm \leq 1$.

In view of these results, a question that arises is: why do the μ^+ and μ^- have the same lifetime if C-invariance is violated? The answer lies in the principle of *CP-conservation*, which states that the weak interaction is invariant under the combined operation CP even though both C and P are separately violated. The CP operator transforms particles at rest to their corresponding antiparticles at rest, and CP-invariance requires that these states should have identical properties. Thus, in particular, the masses of particles and antiparticles are predicted to be the same. Specifically, if we apply the CP operator to muon decays, the parity operator changes θ to $\pi - \theta$ as before, while the C operator changes particles to antiparticles. Hence CP-invariance alone implies that the condition obtained from P-invariance is replaced by the weaker condition

$$\Gamma_{\mu^+} (\cos \theta) = \Gamma_{\mu^-}(- \cos \theta). \tag{6.11}$$

Again, substituting Equation (6.4) into Equation (6.11), gives

$$\Gamma_+ = \Gamma_- \quad (CP\text{-invariance}), \tag{6.12}$$

implying equal lifetimes and also

$$\xi_+ = -\xi_- \quad (CP\text{-invariance}), \tag{6.13}$$

in agreement with the experimental results. Thus CP-invariance retains the symmetry between particles and antiparticles as observed by experiment, at least for μ-decays. In fact CP-invariance has been verified in a wide variety of experiments involving weak interactions, and it is believed to be exact for purely leptonic processes (i.e. ones involving only leptons) and a very good approximation for those involving hadrons. (The only known violations will be discussed in Section 6.6.1.) Particles and antiparticles have the same masses and lifetimes even if CP is not conserved.

6.3   Spin Structure of the Weak Interactions

We turn now to the spin structure of the weak interactions, which is closely related to the symmetry properties discussed above. As this spin structure takes its simplest form for zero-mass particles, we will discuss the case of neutrinos and antineutrinos first, assuming that they have zero mass for the purpose of this discussion.

6.3.1 Neutrinos

In discussing neutrinos, it is convenient to use the so-called *helicity states*, in which the spin is quantized along the direction of motion of the particle, rather than along some arbitrarily chosen 'z-direction'. For a spin-$\frac{1}{2}$ particle, the spin component along the direction of its motion can be either $+\frac{1}{2}$ or $-\frac{1}{2}$ (in units of $\hbar$), as illustrated in Figure 6.4, corresponding to positive or negative helicity respectively. These states are called *right-handed* or *left-handed*, respectively, since the spin direction corresponds to rotational motion in a right-handed or left-handed sense when viewed along the momentum direction.

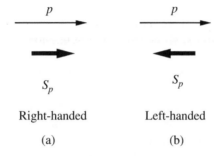

Figure 6.4 Helicity states of a spin-$\frac{1}{2}$ particle: the long thin arrows represent the momenta of the particles and the shorter thick arrows represent their spins

We will denote these states by a subscript R or L, so that, for example, ν_L means a left-handed neutrino. The remarkable fact about neutrinos and antineutrinos, which only interact via the weak interaction, is that *only left-handed neutrinos ν_L and right-handed antineutrinos $\bar{\nu}_R$ are observed in nature*. This obviously violates *C*-invariance, which requires neutrinos and antineutrinos to have identical weak interactions. It also violates *P*-invariance, which requires the states ν_L and ν_R to also have identical weak interactions, since the parity operator reverses the momentum while leaving the spin unchanged and so converts a left-handed neutrino into a right-handed neutrino. It is, however, compatible with *CP*-invariance, since the *CP* operator converts a left-handed neutrino to a right-handed antineutrino, as illustrated in Figure 6.5.

The helicity of the neutrino was first measured in an ingenious experiment by Goldhaber and co-workers in 1958. Again, we will only discuss the principles of the experiment. They studied electron capture in ^{152}Eu, i.e.

$$e^- + {}^{152}\text{Eu}(J = 0) \rightarrow {}^{152}\text{Sm}^*(J = 1) + \nu_e, \tag{6.14}$$

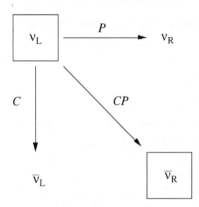

Figure 6.5   Effect of C, P and CP transformations; only the states shown in boxes are observed in nature

where the spins of the nuclei are shown in brackets. The excited state of samarium that is formed decays to the ground state by γ-emission

$$^{152}\text{Sm}^*(J=1) \rightarrow {}^{152}\text{Sm}(J=0) + \gamma \qquad (6.15)$$

and it is these γ-rays which were detected in the experiment. In the first reaction (Equation (6.14)), the electrons are captured from the K-shell and the initial state has zero momentum, so that the neutrino and the $^{152}\text{Sm}^*$ nucleus recoil in opposite directions. The experiment selected events in which the photon was emitted in the direction of motion of the decaying $^{152}\text{Sm}^*$ nucleus, so that overall the observed reaction was

$$e^- + {}^{152}\text{Eu}(J=0) \rightarrow {}^{152}\text{Sm}(J=0) + \nu_e + \gamma, \qquad (6.16)$$

where the three final-state particles were co-linear, and the neutrino and photon emerged in opposite directions, as shown in Figure 6.6.

The helicity of the neutrino can then be deduced from the measured helicity of the photon by applying angular momentum conservation about the event axis to the overall reaction. In doing this, no orbital angular momentum is involved, because the initial electron is captured from the atomic K-shell and the final-state particles all move along the event axis. Hence the spin components of the neutrino and photon, which can be $\pm\frac{1}{2}$ and ± 1 respectively, must add to give the spin component of the initial electron, which can be $\pm\frac{1}{2}$. This gives two possible spin configurations, as shown in Figures 6.6(a) and 6.6(b). In each case the photon and neutrino have the same helicities. In the actual experiment, the polarization of the photons was determined by studying their absorption in magnetized iron (which depends on the polarization of the photon) and the results obtained were consistent with the occurrence of left-handed neutrinos only, corresponding to Figure 6.6(a).

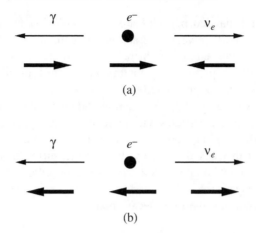

Figure 6.6 Possible helicities of the photon and neutrinos emitted in the reaction $e^- + {}^{152}\text{Eu}(J=0) \rightarrow {}^{152}\text{Sm}(J=0) + \nu_e + \gamma$ for those events in which they are emitted in opposite directions. Experiment selects configuration (a)

Later experiments have shown that only right-handed antineutrinos take part in weak interactions.

6.3.2 Particles with mass: chirality

To see the effect of the spin dependence in weak interactions involving particles with mass, we will look at the decays of the pion and muon which are, of course, examples of charged current reactions. The spin dependence is of a special form, called a *V–A interaction*. This name is derived from the behaviour under a parity transformation of the weak interaction analogue of the electromagnetic current. The letter **V** denotes a *proper vector*, which is one whose direction is reversed by a parity transformation (an example is momentum **p**). The familiar electric current, to which photons couple, transforms as a proper vector under parity. Because parity is not conserved in weak interactions, the corresponding weak current, to which $W^{\pm}$-bosons couple, has in addition to a vector (**V**) component another component whose direction is unchanged by a parity transformation. Such a quantity is called an *axial-vector* (**A**) (an example of an axial-vector is orbital angular momentum $\mathbf{L} = \mathbf{r} \times \mathbf{p}$). Since observables are related to the modulus squared of amplitudes, either term would lead by itself to parity conservation. Parity non-conservation is an interference effect between the two components.

Here we shall consider only the most important characteristic of this spin dependence, which is that the results discussed above for neutrinos, hold for all fermions in the ultra-relativistic limit. That is, in the limit that their velocities approach that of light, only left-handed fermions ν_{L}, e_{L}^- etc. and right-handed antifermions $\bar{\nu}_{\text{R}}$, e_{R}^+ etc. are emitted in charged current interactions. These

right-handed and left-handed particles are called *chiral* states and these are the
eigenstates that take part in weak interactions. In general, chiral states are linear
combinations of helicity states,[4] with the contributions of the 'forbidden' helicity states e_R^-, e_L^+ etc. suppressed by factors which are typically of the order of
$(mc^2/E)^2$, where m is the appropriate fermion mass and E its energy. For massless
neutrinos this is always a good approximation and chiral states and helicity states
are identical. However, for particles with mass, it is only a good approximation for
large energies E. These spin properties can be verified most easily for the electrons
and muons emitted in weak decays, by directly measuring their spins. Here we
shall assume them to hold and use them to understand some interesting features of
pion and muon decays.

We start by considering the pion decay mode

$$\pi^+ \rightarrow \ell^+ + \nu_\ell . \quad (\ell = e, \mu) \tag{6.17}$$

In the rest frame of the decaying pion, the charged lepton and the neutrino recoil in
opposite directions, and because the pion has zero spin, their spins must be
opposed to satisfy angular momentum conservation about the decay axis. Since the
neutrino (assumed to be zero mass) is left-handed, it follows that the charged
lepton must also be left-handed, as shown in Figure 6.7, in contradiction to the
expectations for a relativistic antilepton.

Figure 6.7   Helicities of the charged leptons in pion decays: the short arrows denote spin
vectors and the longer arrows denote momentum vectors

For the case of a positive muon this is unimportant, since it is easy to check
that it recoils non-relativistically and so both chirality states are allowed. However,
if a positron is emitted it recoils *relativistically*, implying that this mode is
suppressed by a factor that we can estimate from the above to be of the order
of $(m_e/m_\pi)^2 \approx 10^{-5}$. Thus the positron decay mode is predicted to be much rarer
than the muonic mode. This is indeed the case, and the measured ratio

$$\frac{\Gamma(\pi^+ \rightarrow e^+ + \nu_e)}{\Gamma(\pi^+ \rightarrow \mu^+ + \nu_\mu)} = (1.218 \pm 0.014) \times 10^{-4} \tag{6.18}$$

is in excellent agreement with a full calculation that takes into account both the
above suppression and the difference in the density-of-final states (i.e. the
difference in the Q-values) for the two reactions.

---

[4]This is another example where linear combinations of states are the ones of physical interest; compare
neutrino mixing (Section 3.1.3).

A second consequence of the chirality argument is that the muons emitted in pion decays are 100 per cent polarized (see Figure 6.7).[5] We have mentioned this earlier in connection with measuring the muon decay asymmetries. These have their origins in the spin structure of the interaction, as we shall illustrate for the highest-energy electrons emitted in the decay of the muon. These have energy

$$E = \frac{m_\mu c^2}{2} \left(1 + \frac{m_e^2}{m_\mu^2} \right) \gg m_e c^2 \tag{6.19}$$

and correspond to decays in which the neutrino and antineutrino are both emitted in the direction opposite to the electron. This is illustrated in Figure 6.8 for the two simplest cases in which the electron is emitted in the muon spin direction (Figure 6.8(a)) and opposite to it (Figure 6.8(b)).

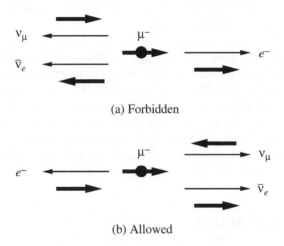

Figure 6.8   Muon decays in which electrons of the highest possible energy are emitted: (a) in the muon spin direction, and (b) opposite to the muon spin direction

Since the neutrino and antineutrino have opposite helicities, the muon and electron must have the same spin component along the event axis in order to conserve angular momentum, implying the electron helicities shown in Figure 6.8. When combined with the fact that the relativistic electrons emitted must be left-handed, this implies that electrons cannot be emitted in the muon spin direction. We thus see that the spin structure of the interaction automatically gives rise to a forward–backward asymmetry in polarized muon decays. Of course not all the electrons have the maximum energy and the actual asymmetry, averaged over all electron energies, can only be calculated by using the full form of the **V–A**

---

[5]This is in the rest frame of the decaying pion and assumes that the neutrino has zero mass. The degree of polarization in the laboratory frame is a function of the muon momentum.

interaction.[6] The resulting prediction is in excellent agreement with the measured values.

Finally, we have seen in earlier chapters that there is increasing evidence that neutrinos are not strictly massless. How then can we ensure that the weak interactions only couple to ν_L and $\bar{\nu}_R$? To understand this we return to the Dirac equation, which was mentioned in Chapter 1. As was stated there, the solution of this equation for a massive spin-$\frac{1}{2}$ particle is in the form of a four-component spinor, whose components are interpreted as the two possible spin projections for the particle and its antiparticle of a given energy (see Section 1.2 and Equation (1.4)). However, in the case of a massless fermion the Hamiltonian of Equation (1.2) consists only of a spin projection term and there is a simpler solution of the Dirac equation consisting of two independent two-component wavefunctions. If we assume for definiteness the case of neutrinos (assumed to be massless), then these would correspond to the pairs $(\nu_L, \bar{\nu}_R)$ and $(\nu_R, \bar{\nu}_L)$. This observation was first made by Weyl in 1929, but was rejected as unphysical because under a parity transformation $\nu_L \to \nu_R$ (see Figure 6.5) and hence the interaction would not be invariant under parity. However, we now know that parity is not conserved in the weak interactions, so this objection is no longer valid. A possible solution is therefore to make the neutrino its own antiparticle. In this case $(\nu_L, \bar{\nu}_R)$ are identified as two helicity components of a four-component spinor and the other two components $(\nu_R, \bar{\nu}_L)$, if they exist, can then be a fermion of a different mass. This scheme is due to Majorana and is very different to the structure of a spinor describing a massive spin-$\frac{1}{2}$ fermion such as an electron. A test of this idea would be the observation of neutrinoless double β-decay, such as that given in Equation (3.37), which is only possible if $\nu_e \equiv \bar{\nu}_e$.

6.4 $W^{\pm}$ and Z^0 Bosons

The three intermediate vector bosons mediating weak interactions, the two charged bosons W^+ and W^- and the neutral Z^0, were all discovered at CERN in 1983 in the reactions

$$\bar{p} + p \to W^+ + X^-, \quad \bar{p} + p \to W^- + X^+, \quad \text{and} \quad \bar{p} + p \to Z^0 + X^0, \quad (6.20)$$

where $X^{\pm}$ and X^0 are arbitrary hadronic states allowed by the conservation laws. The beams of protons and antiprotons were supplied by a proton–antiproton collider, which was specifically built for this purpose. At the time it had proton and antiproton beams with maximum energies of 270 GeV each, giving a total centre-of-mass energy of 540 GeV. Two independent experiments were mounted (called

---

[6]See, for example, Chapter 12 of Ha84.

UA1 and UA2), both of which were examples of the 'layered' detector systems that were discussed in Chapter 4.[7] One of the main problems facing the experimenters was that for each event in which a $W^{\pm}$ or Z^0 is produced and decays to leptons, there were more than 10^7 events in which hadrons alone are produced and so the extraction of the signal required considerable care.

In contrast to the zero mass photons and gluons, the $W^{\pm}$ and Z^0 bosons are both very massive particles, with measured masses

$$M_W = 80.6\,\mathrm{GeV}/c^2, \quad M_Z = 91.2\,\mathrm{GeV}/c^2, \tag{6.21}$$

while their lifetimes are about 3×10^{-25} s. Their dominant decays lead to jets of hadrons, but the leptonic decays

$$W^+ \to \ell^+ + \nu_\ell, \quad W^- \to \ell^- + \bar{\nu}_\ell \tag{6.22}$$

and

$$Z^0 \to \ell^+ + \ell^-, \quad Z^0 \to \nu_\ell + \bar{\nu}_l, \tag{6.23}$$

where $\ell = e$, μ or τ as usual, are also important. The particles are detected as resonance-like enhancements in plots of the invariant mass of suitable final states seen in reactions such as Equation (6.20).[8]

We have seen that an important feature of an exchange interaction is its strength. As in the case of electromagnetism, Feynman diagrams for weak interactions are constructed from fundamental three-line vertices. Those for lepton–$W^{\pm}$ interactions are shown in Figure 6.9.

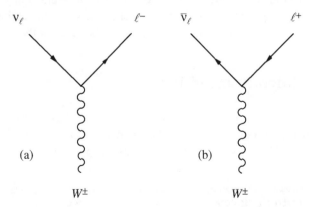

Figure 6.9　The two basic vertices for $W^{\pm}$-lepton interactions

[7]Simon van der Meer lead the team that built the accelerator and Carlo Rubbia lead the UA1 experimental team that subsequently discovered the bosons. They shared the 1984 Nobel Prize in Physics for their work.
[8]A more detailed description of the UA1 experiment is given in, for example, Section 8.1 of Ma97.

At each vertex a boson is emitted or absorbed; while both fermion lines belong to the same generation $\ell = e$, μ or τ, with one arrow pointing inwards and one outwards to guarantee conservation of each lepton number N_e, N_μ and N_τ. Finally, associated with each vertex is a dimensionless parameter with the same value

$$\alpha_W = g_W^2/4\pi\hbar c \approx 1/400 \tag{6.24}$$

at high energies for *all three generations* (because of lepton universality). This constant is the weak analogue of the fine structure constant $\alpha \approx 1/137$ in electromagnetic interactions, with g_W the weak analogue of the electronic charge e in appropriate units.

We see from the above that, despite its name, the weak interaction has a similar intrinsic strength to the electromagnetic interaction. Its apparent weakness in many low-energy reactions, is solely a consequence of its short range, which arises because the exchange bosons are heavy. From Equation (6.1) we see that the scattering amplitude has a denominator that contains the squared mass of the exchanged particle and so at energies where the de Broglie wavelengths $\lambda = h/p$ of the particles are large compared with the range of the weak interaction, which is an excellent approximation for all lepton and hadron decays, the range can be neglected altogether. In this approximation the weak interaction becomes a *point* or *zero range* interaction, whose effective interaction strength can be shown to be

$$\alpha_{\text{eff}} = \alpha_W \left(\bar{E}/M_W c^2\right)^2, \quad \bar{E} \ll M_W c^2, \tag{6.25}$$

where $\bar{E}$ is a typical energy scale for the process in question. (For example in muon decay it would be the mass of the muon.) Thus we see that the interaction is both weak and very energy dependent at 'low energies', but becomes comparable in strength with the electromagnetic interaction at energies on the scale of the W-boson mass.

6.5 Weak Interactions of Hadrons

The weak decays of hadrons are understood in terms of basic processes in which $W^\pm$ bosons are emitted or absorbed by their constituent quarks. In this section we will consider both decays and scattering processes, starting with the former.

6.5.1 Semileptonic decays

A typical semileptonic decay (i.e. one that involves both hadrons and leptons) is that of the neutron, which at the quark level is

$$d \rightarrow u + e^- + \bar{\nu}_e, \tag{6.26}$$

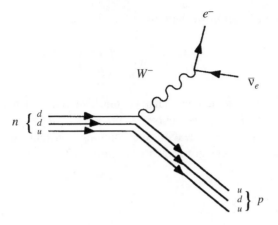

Figure 6.10 Quark diagram for the decay $n \to pe^- \bar{\nu}_e$

as illustrated in Figure 6.10, where the other two quarks play the role of spectators. Similarly, in the pion decay process

$$\pi^- (d\bar{u}) \to \mu^- + \bar{\nu}_\mu \tag{6.27}$$

the initial quarks annihilate to produce a W boson as shown in Figure 6.11. However, the weak interactions of quarks are more complicated than those of leptons, and are best understood in terms of two ideas: *lepton–quark symmetry*, and *quark mixing*.

For simplicity, we will look firstly at the case of just two generations of quarks and leptons. In this case, lepton–quark symmetry asserts that the first two generations of quarks

$$\begin{pmatrix} u \\ d \end{pmatrix} \quad \text{and} \quad \begin{pmatrix} c \\ s \end{pmatrix} \tag{6.28}$$

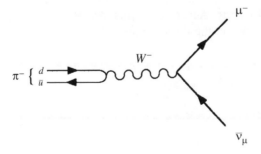

Figure 6.11 Quark diagram for the process $\pi^- \to \mu^- + \bar{\nu}_\mu$

and the first two generations of leptons

$$\begin{pmatrix} \nu_e \\ e^- \end{pmatrix} \quad \text{and} \quad \begin{pmatrix} \nu_\mu \\ \mu^- \end{pmatrix} \tag{6.29}$$

have *identical* weak interactions. That is, one can obtain the basic $W^\pm$ quark vertices by making the replacements $\nu_e \to u$, $e^- \to d$, $\nu_\mu \to c$, $\mu^- \to s$ in the basic $W^\pm$ lepton vertices, leaving the coupling constant g_W unchanged. The resulting $W^\pm$ quark vertices are shown in Figure 6.12.

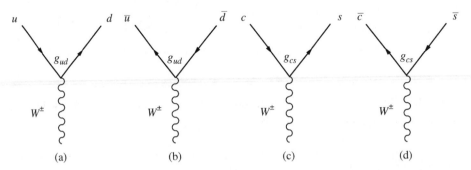

Figure 6.12 The $W^\pm$ quark vertices obtained from quark--lepton symmetry, without quark mixing

Quark symmetry in the simple form stated above then implies that the fundamental processes $d + \bar{u} \to W^-$ and $s + \bar{c} \to W^-$ occur with the *same* couplings g_W as the corresponding leptonic processes, i.e. in Figure (6.12) we have $g_{cs} = g_{ud} = g_W$, while the processes $s + \bar{u} \to W^-$ and $d + \bar{c} \to W^-$ are forbidden. This works quite well for many reactions, like the pion decay $\pi^- \to \mu^- + \bar{\nu}_\mu$, but many decays that are forbidden in this simple scheme are observed to occur, albeit at a rate which is suppressed relative to the 'allowed' decays. An example of this is the kaon decay $K^- \to \mu^- + \bar{\nu}_\mu$, which requires a $s + \bar{u} \to W^-$ vertex, which is not present in the above scheme.

All these suppressed decays can be successfully incorporated into the theory by introducing *quark mixing*. According to this idea, the d and s quarks participate in the weak interactions via the linear combinations

$$d' = d \cos \theta_C + s \sin \theta_C \tag{6.30a}$$

and

$$s' = -d \sin \theta_C + s \cos \theta_C, \tag{6.30b}$$

where the parameter θ_C is called the *Cabibbo angle.*[9] That is, lepton–quark symmetry is assumed to apply to the doublets

$$\begin{pmatrix} u \\ d' \end{pmatrix} \quad \text{and} \quad \begin{pmatrix} c \\ s' \end{pmatrix}. \tag{6.31}$$

This then generates new vertices previously forbidden. For example, the usW vertex required for the decay $K^- \rightarrow \mu^- + \bar{\nu}_\mu$ arises from the interpretation of the $ud'W$ vertex shown in Figure 6.13. In a similar way a new cdW vertex is also generated.

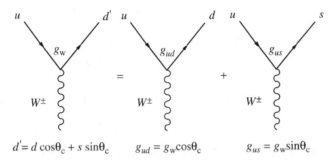

$$d' = d\cos\theta_c + s\sin\theta_c \qquad g_{ud} = g_w\cos\theta_c \qquad g_{us} = g_w\sin\theta_c$$

Figure 6.13   The $ud'W$ vertex and its interpretation in terms of udW and usW vertices

Quark mixing enables theory and experiment to be brought into good agreement by choosing a value $\theta_C \approx 13°$ for the Cabibbo angle. One then finds that the rates for the previously 'allowed' decays occur at rates which are suppressed by a factor $\cos^2\theta_C \approx 0.95$, while the previously 'forbidden' decays are now allowed, but with rates which are suppressed by a factor $\sin^2\theta_C \approx 0.05$.

Historically, the most remarkable thing about these ideas is that they were formulated before the discovery of the charmed quark. In 1971 seven fundamental fermions were known: the four leptons ν_e, e^-, ν_μ and μ^-, and the three quarks u, d and s. This led Glashow, Iliopolous and Maiani to propose the existence of a fourth quark c to complete the lepton–quark symmetry and to solve problems associated with neutral currents that we will discuss in Section 6.7. The existence of the charmed quark was subsequently confirmed in 1974 with the discovery of the first charmonium states (this is why their discovery was so important – see the discussion in Section 5.3) and its measured weak couplings are consistent with the predictions of lepton–quark symmetry and quark mixing.

We now know that there are six leptons

$$\begin{pmatrix} \nu_e \\ e^- \end{pmatrix} \quad \begin{pmatrix} \nu_\mu \\ \mu^- \end{pmatrix} \quad \begin{pmatrix} \nu_\tau \\ \tau^- \end{pmatrix} \tag{6.32}$$

[9]This is yet another example of physical states being mixtures of other states.

and six known quarks

$$\begin{pmatrix} u \\ d \end{pmatrix} \quad \begin{pmatrix} c \\ s \end{pmatrix} \quad \begin{pmatrix} t \\ b \end{pmatrix}. \tag{6.33}$$

When the third generation is taken into account, the mixing scheme becomes more complicated, as we must allow for the possibility of mixing between all three 'lower' quarks d, s and b instead of just the first two and more parameters are involved. In general the mixing can be written in the form

$$\begin{pmatrix} d' \\ s' \\ b' \end{pmatrix} = \begin{pmatrix} V_{ud} & V_{us} & V_{ub} \\ V_{cd} & V_{cs} & V_{cb} \\ V_{td} & V_{ts} & V_{tb} \end{pmatrix} \begin{pmatrix} d \\ s \\ b \end{pmatrix}, \tag{6.34}$$

where $V_{ij}(i = u, c, t; j = d, s, b)$ the so-called *CKM matrix*,[10] is unitary to ensure the orthonormality of the new states generated by the transformation. The matrix elements V_{ij} are all obtainable from various decay processes and values exist for them, although the smaller off-diagonal terms are not very well measured.[11] For the first two generations, the changes introduced by this more complex mixing are very small. However, a new feature that emerges is the possibility of *CP* violation. We shall see in the Section 6.6.1 that *CP* violation does actually occur in the decays of neutral *K*-mesons and neutral *B*-mesons and it is of considerable interest to see if the size of the violation is consistent with the CKM mixing formalism and the standard model.

6.5.2 Neutrino scattering

Consider the elastic scattering process $\nu_e + e^- \rightarrow \nu_e + e^-$ at high energies, proceeding via the exchange of a *W*-meson, i.e. a charged current weak interaction. We know the *W*-meson couples only to left-handed fermions and from the discussion of Section 6.3.1 that neutrinos have negative helicity, i.e. they are polarized along the direction of their motion (which we will take to be the *z*-axis). We also know from the work of Section 6.3.2 that in the relativistic limit, the same is true of electrons. We are therefore led to the centre-of-mass spin/momentum configurations before the collision shown in Figure 6.14(a). If the interaction scatters the particles through an angle of 180°, then the centre-of-mass spin/ momentum configurations after the collision are those shown in Figure 6.14(b). In

---

[10]The initials stand for Cabibbo, Kobayashi and Maskawa, the last two of whom extended the original Cabibbo scheme to three generations of quarks.
[11]A review is given Ei04.

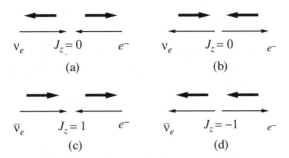

Figure 6.14   Spin (thick arrows) and momentum (thin arrows) configurations for $\nu_e e^-$ and $\bar{\nu}_e e^-$ interactions: (a) $\nu_e e^-$ before collision; (b) $\nu_e e^-$ after scattering through 180°; (c) $\bar{\nu}_e e^-$ before collision; (d) $\bar{\nu}_e e^-$ after scattering through 180°

both cases the total spin component along the z-axis is zero. This result is true for all angles and the scattering is isotropic.

From this we can calculate the differential cross-section using the formulae of Chapter 1. We will assume that the squared momentum transfer Q^2 is such that $Q^2_{max} \ll M_W^2 c^2$, so that the matrix element may be written [cf. Equation (6.1)]

$$f(\nu_e + e^- \rightarrow \nu_e + e^-) = -G_F, \tag{6.35}$$

where G_F is the Fermi coupling constant of Equation (1.42), i.e.

$$G_F = \frac{4\pi(\hbar c)^3 \alpha_W}{(M_W c^2)^2} \tag{6.36}$$

and $\alpha_W = g^2/4\pi\hbar c$ is the equivalent of the fine structure constant for charged current weak interactions. Hence, using Equation (1.57) and recalling that the velocities of both the neutrino and electron are equal to c,

$$\frac{d\sigma}{d\Omega}(\nu_e e^-) = \frac{1}{4\pi^2} \frac{G_F^2}{(\hbar c)^4} E_{CM}^2. \tag{6.37}$$

At high energies E_{CM}^2 is given by

$$E_{CM}^2 \approx 2m_e c^2 E_\nu, \tag{6.38}$$

where E_ν is the energy of the neutrino. So finally the total cross-section is

$$\sigma_{tot}(\nu_e e^-) = \frac{2m_e c^2 G_F^2}{\pi(\hbar c)^4} E_\nu \tag{6.39}$$

and increases linearly with E_ν.[12]

If we apply the same argument to the scattering of antineutrinos, we are lead to the configurations shown in Figures 6.14(c) and 6.14(d). The initial state has $J_z = 1$, but the final state has $J_z = -1$. Thus scattering through $180°$ is forbidden by angular momentum conservation and the amplitude must contain a factor $(1 + \cos\theta)$. This is borne out by a full calculation using the **V–A** formalism which gives, in the same approximation,

$$\frac{d\sigma}{d\Omega}(\bar{\nu}_e e^-) = \frac{1}{16\pi^2}\frac{G_F^2}{(\hbar c)^4}E_{CM}^2 (1 + \cos\theta)^2. \qquad (6.40)$$

Integrating over angles gives

$$\sigma_{tot}(\bar{\nu}_e e^-) = \frac{1}{3}\sigma_{tot}(\nu_e e^-). \qquad (6.41)$$

Neutrino–electron scattering is not, of course, a very practical reaction to study experimentally, but these ideas may be taken over to deep inelastic neutrino scattering from nucleons, where the latter are assumed to be composed of constituent quarks whose masses may be neglected at high energies. This will enable us to extend the discussion of Section 5.7 for charged leptons. In this case the neutrino is assumed to interact with a single quark within the nucleon (this is again the spectator model) and we must take account of all relevant quarks and antiquarks. In practice we can neglect interactions with s and $\bar{s}$ quarks as these will be suppressed by the Cabibbo factor. So, taking into account only the u and d quarks and their antiparticles, we can generalize Equations (6.39) and (6.41) to give

$$\sigma_{tot}(\nu_e N) = \frac{M_N c^2 G_F^2 E_\nu}{\pi(\hbar c)^4}\left(H + \frac{1}{3}\overline{H}\right) \qquad (6.42a)$$

and

$$\sigma_{tot}(\bar{\nu}_e N) = \frac{M_N c^2 G_F^2 E_\nu}{\pi(\hbar c)^4}\left(\frac{1}{3}H + \overline{H}\right), \qquad (6.42b)$$

for scattering from an isoscalar nucleus, i.e. one with an equal number of neutrons and protons, where M_N is the mass of the nucleon. The quantities H and $\overline{H}$ are given by

$$H \equiv \int_0^1 x[u(x) + d(x)]dx \quad \text{and} \quad \overline{H} \equiv \int_0^1 x[\bar{u}(x) + \bar{d}(x)]dx, \qquad (6.43)$$

---

[12]This behaviour has arisen because of the approximation Equation (6.35). It cannot of course continue indefinitely. At very high values of Q^2 the full form of the propagator would have to be taken into account and this would introduce an energy dependence in the denominator of Equation (6.39).

where $u(x)$ etc. are the quark densities defined in Section 5.7 and the integral is over the scaling variable x.

Setting $y = \overline{H}/H$, we have from Equations (6.42)

$$R \equiv \frac{\sigma(\bar{\nu}_e N)}{\sigma(\nu_e N)} = \frac{1 + 3y}{3 + y}. \tag{6.44}$$

Some data for R are shown in Figure 6.15 from an experiment using muon–neutrinos. These show that R is approximately constant, as predicted by Equation (6.44), and has a value of about 0.51, which implies $y \approx 0.2$, i.e. antiquarks exist in the nucleon at the level of about 20 per cent. Other experiments yield similar results in the range 15–20 per cent.

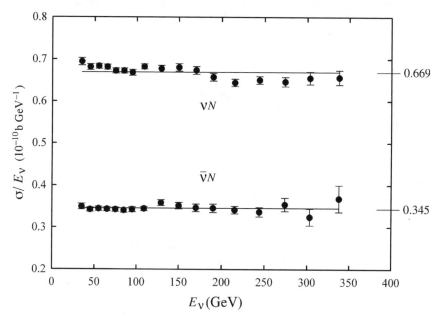

Figure 6.15 Neutrino and antineutrino total cross-sections (data from Se97)

6.6 Neutral Meson Decays

Neutral mesons are of particular interest not only because they enable very sensitive tests of *CP*-conservation to be made, but also because the application of basic quantum mechanics leads to surprising effects that, for example, allow the symmetry between particles and antiparticles to be tested with extraordinary precision. In both cases the crucial ingredient is the phenomenon of particle mixing that we have met before in connection with the mixing of neutrino flavour states. Because most work has been done on the neutral kaons, we will mainly discuss this system as an example. The equivalent formalisms for B- and D-decays are similar. We start with a discussion of CP violation.

6.6.1 CP violation

We have seen that there are two neutral kaon states

$$K^0(498) = d\bar{s} \quad \text{and} \quad \overline{K}^0(498) = s\bar{d}, \qquad (6.45)$$

which have strangeness $S = +1$ and $S = -1$ respectively. However, because strangeness is not conserved in weak interactions, these states can be converted into each other by higher-order weak processes like those shown in Figure 6.16.

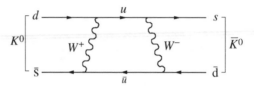

Figure 6.16   Example of a process that can convert a K^0 state to a $\overline{K}^0$ state

This is in marked contrast to most other particle–antiparticle systems, for which such transitions are forbidden, because the particle and its antiparticle differ by quantum numbers that are conserved in all known interactions. For example, the π^+ and π^- have opposite electric charges, and the neutron and antineutron have opposite baryon numbers. For neutral kaons, however, there is no conserved quantum number to distinguish the K^0 and $\overline{K}^0$ states when weak interactions are taken into account and the observed physical particles correspond not to the K^0 and $\overline{K}^0$ states themselves, but to linear combinations of them. Similar mixing can occur between $B^0 - \overline{B}^0$ and $D^0 - \overline{D}^0$ states. We have met the idea that observed states can be linear combinations of other states in the CKM mixing scheme for quarks above and earlier when we discussed neutrino oscillations in the absence of lepton number conservation in Chapter 3. In the present case it leads to the phenomena of $K^0 - \overline{K}^0$ mixing, and *strangeness oscillations*.

We start by assuming that CP-invariance is exact and that neutral kaons are eigenstates of the combined CP operator. In this case, using the standard phase convention, we can define

$$C|K^0, \mathbf{p}\rangle = -|\overline{K}^0, \mathbf{p}\rangle, \quad C|\overline{K}^0, \mathbf{p}\rangle = -|K^0, \mathbf{p}\rangle, \qquad (6.46)$$

where $|K^0, \mathbf{p}\rangle$ denotes a K^0 state with momentum $\mathbf{p}$, etc.. Since kaons have negative intrinsic parity, we also have for $\mathbf{p} = \mathbf{0}$

$$P|K^0, \mathbf{0}\rangle = -|K^0, \mathbf{0}\rangle, \quad P|\overline{K}^0, \mathbf{0}\rangle = -|\overline{K}^0, \mathbf{0}\rangle, \qquad (6.47)$$

so that

$$CP|K^0, \mathbf{0}\rangle = |\overline{K}^0, \mathbf{0}\rangle, \quad CP|\overline{K}^0, \mathbf{0}\rangle = |K^0, \mathbf{0}\rangle. \tag{6.48}$$

Thus CP eigenstates $K^0_{1,2}$ are

$$|K^0_{1,2}, \mathbf{0}\rangle = \frac{1}{\sqrt{2}} \left\{ |K^0, \mathbf{0}\rangle \pm |\overline{K}^0, |\mathbf{0}\rangle \right\} \quad (CP = \pm 1). \tag{6.49}$$

If CP is conserved, then K^0_1 should decay entirely to states with $CP = 1$ and K^0_2 should decay entirely into states with $CP = -1$. We examine the consequences of this for decays leading to pions in the final state.

Consider the state $\pi^0 \pi^0$. Since the kaon has spin-0, by angular momentum conservation the pion pair must have zero orbital angular momentum in the rest frame of the decaying particle. Its parity is therefore given by [cf. Equation (1.14)]

$$P = P_\pi^2 (-1)^L = 1, \tag{6.50}$$

where $P_\pi = -1$ is the intrinsic parity of the pion. The C-parity is given by

$$C = (C_{\pi^0})^2 = 1, \tag{6.51}$$

where $C_{\pi^0} = 1$ is the C-parity of the neutral pion. Combining these results gives $CP = 1$. The same result holds for the $\pi^+ \pi^-$ final state.

The argument for three-pion final states $\pi^+ \pi^- \pi^0$ and $\pi^0 \pi^0 \pi^0$ is more complicated, because there are two orbital angular momenta to consider, If we denote by $\mathbf{L}_{12}$ the orbital angular momentum of one pair (either $\pi^+ \pi^-$ or $\pi^0 \pi^0$) in their mutual centre-of-mass frame, and $\mathbf{L}_3$ is the orbital angular momentum of the third pion about the centre-of-mass of the pair in the overall centre-of-mass frame, then the total orbital angular momentum $\mathbf{L} \equiv \mathbf{L}_{12} + \mathbf{L}_3 = \mathbf{0}$, since the decaying particle has spin-0. This can only be satisfied if $L_{12} = L_3$. This implies that the parity of the final state is

$$P = P_\pi^3 (-1)^{L_{12}} (-1)^{L_3} = -1. \tag{6.52}$$

For the $\pi^0 \pi^0 \pi^0$ final state, the C-parity is

$$C = (C_{\pi^0})^3 = 1 \tag{6.53}$$

and combining these results gives $CP = -1$ overall. The same result can be shown to hold for the $\pi^+ \pi^- \pi^0$ final state.

The experimental position is that two neutral kaons are observed, called K^0-*short* and K^0-*long*, denoted K^0_S and K^0_L, respectively. They have almost equal masses of about 499 MeV/c^2, but very different lifetimes and decay modes. The K^0_S has a lifetime of 0.89×10^{-10} s and decays overwhelmingly to two pions; the longer-lived K_L^0 has a lifetime of 0.52×10^{-7}s with a significant branching ratio to three pions, but not to two. In view of the CP analysis above, this immediately suggests the identification

$$K_S^0 = K_1^0, \quad \text{and} \quad K_L^0 = K_2^0. \tag{6.54}$$

However, in 1964 it was discovered that the K_L^0 also decayed to two pions[13]

$$K_L^0 \rightarrow \pi^+ + \pi^-, \tag{6.55}$$

but with a very small branching ratio of the order of 10^{-3}. This result is clear evidence of CP violation. This was confirmed in later experiments on the decay $K^0 \rightarrow \pi^0\pi^0$.

Because CP is not conserved, the physical states K_S^0 and K_L^0 need not correspond to the CP-eigenstates K_1^0 and K_2^0, but can contain small components of states with the opposite CP, i.e. we may write

$$|K_S^0, \mathbf{0}\rangle = \frac{1}{(1 + |\varepsilon|^2)^{1/2}} \left[|K_1^0, \mathbf{0}\rangle - \varepsilon|K_2^0, \mathbf{0}\rangle\right] \tag{6.56a}$$

and

$$|K_L^0, \mathbf{0}\rangle = \frac{1}{(1 + |\varepsilon|^2)^{1/2}} \left[\varepsilon|K_1^0, \mathbf{0}\rangle + |K_2^0, \mathbf{0}\rangle\right], \tag{6.56b}$$

where ε is a small complex parameter. (The factor in front of the brackets is to normalize the states.) The CP-violating decays can then occur in two different ways: either (a) the CP-forbidden K_1^0 component in the K_L^0 decays via a CP-allowed processe, giving a contribution proportional to the probability $|\varepsilon|^2[1 + |\varepsilon|^2]^{-1} \approx |\varepsilon|^2$ of finding a K_1^0 component in the K_L^0; or (b) the CP-allowed K_2^0 component in the K_L^0 decays via a CP-violating reaction. A detailed analysis of the data for the $\pi\pi$ decay modes[14] shows that it is the former mechanism that dominates, with $|\varepsilon| \approx 2.2 \times 10^{-3}$.

This is confirmed in the semileptonic decays

$$K^0 \rightarrow \pi^- + e^+ + \nu_e \tag{6.57a}$$

---

[13]The experiment was led by James Cronin and Val Fitch. They received the 1980 Nobel Prize in Physics for their discovery.
[14]See, for example, Ei04.

and

$$\bar{K}^0 \rightarrow \pi^+ + e^- + \bar{\nu}_e. \tag{6.57b}$$

For example, if we start with a beam of K^0 particles, with initially equal amounts of K_S^0 and K_L^0, then after a time that is large compared with the K_S^0 lifetime, the K_S^0 component will have decayed leaving just the K_L^0 component, which itself will be an equal admixture of K^0 and $\bar{K}^0$ components. We would therefore expect to observe identical numbers of electrons (N^-) and positrons (N^+) from the decays of Equations (6.57). However, if K_L^0 is not an eigenstate of CP, then there will be an asymmetry in these numbers, which will depend on the relative strengths of the K^0 and $\bar{K}^0$ components in K_L^0. The asymmetry is given by $2\mathrm{Re}\,\varepsilon$, where ε is the CP-violating parameter defined in Equation (6.56).

Figure 6.17 shows data on the asymmetry $(N^+ - N^-)/(N^+ + N^-)$ as a function of proper time. After the initial oscillations there is seen to be an asymmetry whose value is $2\mathrm{Re}\,\varepsilon \approx 3.3 \times 10^{-3}$, which is consistent with the value of ε obtained from the $\pi\pi$ modes. Thus CP-violation in K-decay occurs mainly, though not entirely, by the mixing of the CP-eigenstates in the physical states rather than by direct CP-violating decays, both of which are allowed in the CKM mixing scheme.

What do these results mean for the CKM mixing scheme? The CKM matrix is a 3×3 matrix and in general contains nine complex elements. However, the unitary nature of the matrix implies that there are relations between the elements, such as

$$V_{ud}V_{ub}^* + V_{cd}V_{cb}^* + V_{td}V_{tb}^* = 0. \tag{6.58}$$

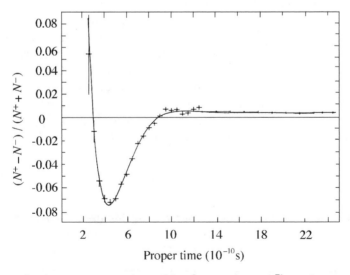

Figure 6.17 The charge asymmetry observed for $K^0 \rightarrow \pi^- e^+ \nu_e$ and $\bar{K}^0 \rightarrow \pi^+ e^- \bar{\nu}_e$ as a function of proper time, for a beam that is initially predominantly K^0 (adapted from Gj74, copyright Elsevier, with permission)

Using these and exploiting the freedom to define the phases of the basic quark states, the matrix may be parameterized by just four quantities. A number of different parameterizations are used, but an approximate form that is commonly used to discuss *CP* violation is

$$V = \begin{pmatrix} 1 - \frac{1}{2}\lambda^2 & \lambda & A\lambda^3(\rho - i\eta) \\ -\lambda & 1 - \frac{1}{2}\lambda^2 & A\lambda^2 \\ A\lambda^3(1 - \rho - i\eta) & -A\lambda^2 & 1 \end{pmatrix} + O(\lambda^4), \tag{6.59}$$

with parameters A, λ, ρ and η. The quantity $\lambda = |V_{us}| \approx 0.22$ plays the role of an expansion parameter in this approximation and a non-zero value of η would be indicative of *CP* violation.

The parameter ε in Equations (6.56) is just one *CP*-violating parameter that may be measured in various *K*-decay modes. We will not pursue this further, but note that by combining the values of the parameters with information on other elements of the CKM matrix, a value of the *CP*-violating parameter η may be deduced and used to predict the size of *CP*-violating effects in other decays. There are very few other places where such mixing effects can occur, but in principle they should be possible in the $D^0\bar{D}^0$ and $B^0\bar{B}^0$ systems, which are analogues of the *K*-mesons, but with a strange quark replaced by a charmed and bottom quark, respectively. Mixing in the $B^0\bar{B}^0$ states due to $B^0\bar{B}^0$ oscillations has in fact been observed and also very recently direct *CP* violation. The latter was established by comparing the decay $B^0 \to K^+\pi^-$ with the decay $\bar{B}^0 \to K^-\pi^+$. Moreover, the size of the effect is much stronger than in neutral kaon decays and this is in agreement with the predictions of the CKM mixing scheme.

There is still much to be done in studying *CP* violation. For example, the cleanest measurement of the *CP*-violating parameter would be from the decays $B^0/\bar{B} \to (J/\psi)K_S^0$, where J/ψ is the $^3S_1$ ground state of charmonium, but the present limits on these decays are orders of magnitude from those required to test the predictions. On the theoretical side, although the CKM mixing model accounts for all *CP*-violating data to date, it fails by several orders of magnitude to account for the observed matter–antimatter asymmetry observed in the universe (which will be discussed in Chapter 9) and so there is probably a *CP*-violating mechanism beyond the standard model awaiting to be discovered.

6.6.2   Flavour oscillations

One interesting consequence of flavour mixing for the $K^0 - \bar{K}^0$ system is the phenomenon of *strangeness oscillation*, which occurs whenever a neutral kaon is produced in a strong interaction process. For example, the neutral kaon produced in the strong interaction

$$\pi^- + p \to K^0 + \Lambda^0$$
$$S = 0 \qquad 0 \qquad 1 \qquad -1 \tag{6.60}$$

must necessarily be a K^0 state with $S = 1$, in order to conserve strangeness. However, if the produced particle is allowed to travel through free space and its strangeness is measured, one finds that it no longer has a definite strangeness $S = 1$, but has components with both $S = 1$ and $S = -1$ whose intensities oscillate with time. These are called strangeness oscillations. The phenomenon is very similar mathematically to that describing the flavour oscillations of neutrinos we met in Chapter 3 and enables the mass difference between K_S^0 and K_L^0 particles to be measured with extraordinary precision, as we will now show.

In what follows, we shall measure time in the rest frame of the produced particle, and define $t = 0$ as the moment when it is produced. If we ignore the very small CP violations, the initial state produced in the $\pi^- p$ reaction above is

$$|K^0, \mathbf{p}\rangle = \frac{1}{\sqrt{2}} \{|K_S^0, \mathbf{p}\rangle + |K_L^0, \mathbf{p}\rangle\}. \tag{6.61}$$

At later times, however, this will become

$$\frac{1}{\sqrt{2}} \{a_S(t)|K_S^0, \mathbf{p}\rangle + a_L(t)|K_L^0, \mathbf{p}\rangle\}, \tag{6.62}$$

where

$$a_\alpha(t) = e^{-im_\alpha t} e^{-\Gamma_\alpha t/2} \quad (\alpha = S, L) \tag{6.63}$$

and m_α and Γ_α are the mass and decay rate of the particle concerned. Here the first exponential factor is the usual oscillating time factor e^{-iEt} associated with any quantum mechanical stationary state, evaluated in the rest frame of the particle. The second exponential factor reflects the fact that the particles decay, and it ensures that the probability

$$\left| \frac{1}{\sqrt{2}} a_\alpha(t) \right|^2 = \frac{1}{2} e^{-\Gamma_\alpha t} \quad (\alpha = S, L) \tag{6.64}$$

of finding a K_S^0 or K_L^0 decreases exponentially with a mean lifetime $\tau_\alpha = \Gamma_\alpha^{-1}$ ($\alpha = S$, L). Because $\tau_S \ll \tau_L$, for times t such that $\tau_S \ll t \lesssim \tau_L$ only the K_L^0 component survives, implying equal intensities for the K^0 and $\bar{K}^0$ components. Here we are interested in the intensities of the K^0 and $\bar{K}^0$ components at shorter times, and to deduce these we rewrite the expression

$$\frac{1}{\sqrt{2}} \{a_S(t)|K_S^0, \mathbf{p}\rangle + a_L(t)|K_L^0, \mathbf{p}\rangle\} \tag{6.65}$$

in the form

$$\{A_0(t)|K^0, \mathbf{p}\rangle + \bar{A}_0(t)|\bar{K}^0, \mathbf{p}\rangle\}, \tag{6.66}$$

where

$$A_0(t) = \frac{1}{2}[a_S(t) + a_L(t)] \quad \text{and} \quad \bar{A}_0(t) = \frac{1}{2}[a_S(t) - a_L(t)]. \tag{6.67}$$

The intensities of the two components are then given by

$$I(K^0) \equiv |A_0(t)|^2 = \frac{1}{4}\left[e^{-\Gamma_S t} + e^{-\Gamma_L t} + 2e^{-(\Gamma_S + \Gamma_L)t/2}\cos(\Delta m t)\right] \tag{6.68a}$$

and

$$I(\bar{K}^0) \equiv |\bar{A}_0(t)|^2 = \frac{1}{4}\left[e^{-\Gamma_S t} + e^{-\Gamma_L t} - 2e^{-(\Gamma_S + \Gamma_L)t/2}\cos(\Delta m t)\right] \tag{6.68b}$$

where $\Delta m \equiv |m_S - m_L|$ and we have used Equation (6.63) explicitly to evaluate the amplitudes.

The variation of the $\bar{K}^0$ intensity $I(\bar{K}^0)$ with time can be determined experimentally by measuring the rate of production of *hyperons* (baryons with non-zero strangeness) in strangeness-conserving strong interactions such as

$$\begin{aligned} \bar{K}^0 + p &\to \pi^+ + \Lambda^0 \\ &\to \pi^0 + \Sigma^+ \end{aligned} \tag{6.69}$$

as a function of the distance from the K^0 source. The data are then fitted by Equations (6.68) with Δm as a free parameter and the predictions are in good agreement with the experiments for a mass difference

$$\Delta m = (3.522 \pm 0.016) \times 10^{-12} \text{ MeV/c}^2. \tag{6.70}$$

The states K_S^0 and K_L^0 are not antiparticles but the K^0 and $\bar{K}^0$ are, of course, and the mass difference Δm can be shown to arise solely from the possibility of transitions $K^0 \leftrightarrow \bar{K}^0$, whose magnitude can be calculated from diagrams like that shown in Figure 6.16. We shall not discuss this further, but merely note that the resulting agreement between the predicted and measured values confirms the equality $m_{K^0} = m_{\bar{K}^0}$ to better than one part in 10^{17}. (This should be compared with the next most precisely tested particle–antiparticle mass relation $m_{e^+} = m_{e^-}$ which only verified to within an experimental error of the order of one part in 10^7.) This equality is a prediction of the so-called *CPT theorem*, which states that under very general conditions any relativistic quantum theory will be invariant under the combined operations of C, P and T.

6.7   Neutral Currents and the Unified Theory

Neutral current reactions are those that involve the emission, absorption or exchange of Z^0 bosons. The unified electroweak theory predicted the existence

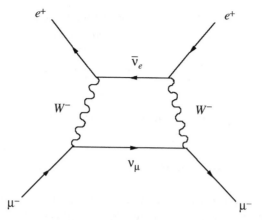

Figure 6.18 Higher order contribution to the reaction $e^+\mu^- \rightarrow e^+\mu^-$ from the exchange of two W-bosons

of such reactions before their discovery in 1973. This theory[15] was proposed mainly in order to solve problems associated with Feynman diagrams in which more than one W boson was exchanged, like that shown in Figure 6.18, which contributes to the reaction $e^+\mu^- \rightarrow e^+\mu^-$.

Such contributions are expected to be small because they are higher order in the weak interaction and this appears to be confirmed by experimental data, which are in good agreement with theoretical predictions that neglect them entirely. (For example, in the experimentally accessible reaction $e^+ + e^- \rightarrow \mu^+ + \mu^-$.) However, when these higher-order contributions are explicitly calculated, they are found to be proportional to divergent integrals, i.e. they are infinite. In the unified theory, this problem is solved when diagrams involving the exchange of Z^0 bosons and photons are taken into account. These also give infinite contributions, but when all the diagrams of a given order are added together the divergences cancel (!), giving a well-defined and finite contribution overall.[16] This cancellation is not accidental, but is a consequence of a fundamental symmetry relating the weak and electromagnetic interactions. Here we will simply comment on some phenomenological consequences of the theory.

---

[15]The formulation of the theory is in terms of four massless vector bosons arranged as multiplets of 'weak isospin' and 'weak hypercharge'. Specifically, three states are a weak isospin triplet and the fourth is a weak isospin singlet. The fact that they all have zero masses ensures that gauge invariance is satisfied. These fields then interact with additional scalar fields associated with new postulated particles called Higgs bosons, which we have mentioned elsewhere. This process, known as 'spontaneous symmetry breaking' generates the observed masses of the W, Z and γ bosons, while still preserving gauge invariance. (For further details see, for example, Section 8.4 of Pe00.) The originators of this theory, Sheldon Glasow, Abdus Salam and Steven Weinberg, shared the 1979 Nobel Prize in Physics for their contributions to formulating the electroweak theory and the prediction of weak neutral currents.

[16]The first people to demonstrate that this occurred were Gerardus 't Hooft and Martinus Veltman. They shared the 1999 Nobel Prize in Physics for this discovery.

The first is that to ensure the cancellation, the theory requires a relation between the weak and electromagnetic couplings, called the *unification condition*. This is

$$\frac{e}{2\sqrt{2}\varepsilon_0^{1/2}} = g_w \sin\theta_W = g_z \cos\theta_W, \tag{6.71}$$

where the *weak mixing angle* θ_W (also called the *Weinberg angle* after one of the authors of the theory) is given by

$$\cos\theta_W \equiv M_W/M_Z \quad (0 < \theta < \pi/2) \tag{6.72}$$

and g_z is a coupling constant which characterizes the strength of the neutral current vertices. The unification condition relates the strengths of the various interactions to the W and Z masses, and historically was used to predict the latter from the former before the $W^\pm$ and Z^0 bosons were discovered.

Secondly, just as all the charged current interactions of leptons can be understood in terms of the basic $W^\pm$-lepton vertices, in the same way all known neutral current interactions can be accounted for in terms of basic Z^0-lepton vertices shown in Figures 6.19(a) and 16.19(b). The corresponding quark vertices can be obtained from the lepton vertices by using lepton–quark symmetry and quark mixing, in the same way that $W^\pm$-quark vertices are obtained from the $W^\pm$-lepton vertices. Thus, making the replacements

$$\nu_e \to u, \quad \nu_\mu \to c, \quad e^- \to d', \quad \mu^- \to s' \tag{6.73}$$

in the lepton vertices

$$\nu_e\nu_e Z^0, \quad \nu_\mu\nu_\mu Z^0, \quad e^-e^- Z^0, \quad \mu^-\mu^- Z^0, \tag{6.74}$$

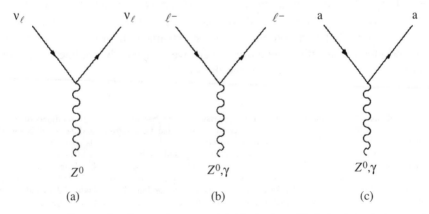

Figure 6.19 Z^0 and γ couplings to leptons and quarks in the unified electroweak theory, where $\ell = e$, μ and τ and a denotes a quark

leads to the quark vertices

$$uuZ^0, \quad ccZ^0, \quad d'd'Z^0, \quad s's'Z^0. \tag{6.75}$$

Finally, we interpret the latter two of these using Equations (6.30). Thus, for example,

$$d'd'Z^0 = (d \cos \theta_C + s \sin \theta_C)(d \cos \theta_C + s \sin \theta_C)Z^0$$
$$= ddZ^0 \cos^2 \theta_C + ssZ^0 \sin^2 \theta_C + (dsZ^0 + sdZ^0) \sin \theta_C \cos \theta_C \tag{6.76}$$

When all the terms in Expression (6.75) are evaluated, ones obtains a set of vertices equivalent to

$$uuZ^0, \quad ccZ^0, \quad ddZ^0, \quad ssZ^0, \tag{6.77}$$

which are shown in Figure 6.19(c).

One important difference from charged current reactions that follows from Figure 6.19 is that neutral current interactions *conserve* individual quark numbers. Thus, for example, strangeness-changing weak neutral current reactions are forbidden. An example is the decay $K^0 \rightarrow \mu^+\mu^-$ and indeed this is not seen experimentally, although nothing else forbids it.

It follows from the above that in any process in which a photon is exchanged, a Z^0 boson can be exchanged as well. At energies that are small compared with the Z^0 mass, the Z^0-exchange contributions can be neglected compared to the corresponding photon exchange contributions, and these reactions can be regarded as purely electromagnetic to a high degree of accuracy. However, at very high energy and momentum transfers, Z^0-exchange contributions become comparable with those of photon exchange and we are therefore dealing with genuinely electroweak processes which involve both weak and electromagnetic interactions to a comparable degree.

These points are clearly illustrated by the cross-section for the muon pair production reaction $e^+ + e^- \rightarrow \mu^+ + \mu^-$. If we assume that the energy is large enough for the lepton masses to be neglected, then the centre-of-mass energy E is the only quantity in the system that has dimensions. Because a cross-section has the dimensions of area, on dimensional grounds the electromagnetic cross-section for one-photon exchange is of the form $\sigma_\gamma \approx \alpha^2 (\hbar c)^2 / E^2$. For Z^0-exchange with $E \ll M_Z c^2$, a similar argument gives for the weak interaction cross-section $\sigma_Z \approx \alpha_Z^2 E^2 (\hbar c)^2 / (M_Z c^2)^4$. (The factor in the denominator comes from the propagator of the Z^0-boson.) Thus the one-photon exchange diagram dominates at low energies, and the cross-section falls as E^{-2}. This is in agreement with the observed behaviour shown in Figure 6.20 and justifies our neglect of the Z^0-exchange contribution at low energies. However, the relative importance of the Z^0-exchange contribution increases rapidly with energy and at beam energies of about 35 GeV it

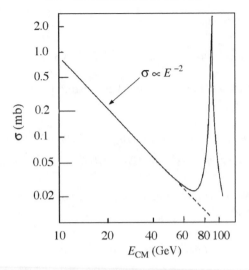

Figure 6.20   Total cross-section for the reaction $e^+e^- \rightarrow \mu^+\mu^-$

begins to make a significant contribution to the total cross-section. At still higher energies, the cross-section is dominated by a very large peak at an energy corresponding to the Z^0 mass, as illustrated in Figure 6.20. At this energy the low-energy approximation is irrelevant and Figure 6.20 corresponds to the formation of physical Z^0 bosons in the process $e^+ + e^- \rightarrow Z^0$ followed by the subsequent decay $Z^0 \rightarrow \mu^+ + \mu^-$ to give the final-state muons. Finally, beyond the peak we once again regain the electroweak regime where both contributions are comparable.

If the exchange of a Z^0 boson always accompanies the exchange of a photon, then there will also in principle be parity-violating effects in reactions that at first sight we would expect to be purely electromagnetic. Their observation would be further unambiguous evidence for electroweak unification. This was first tested in 1978 by scattering polarized electrons from a deuterium target and measuring the parity-violating asymmetry

$$A_{PV} \equiv \frac{\sigma_R - \sigma_L}{\sigma_R + \sigma_L}, \tag{6.78}$$

where $\sigma_R (\sigma_L)$ is the cross-section for incident right (left)-handed electrons. To produce polarized electrons is a complicated multistage process that starts with linearly polarized photons from a laser source that are then converted to states with circular polarization. Finally these are used to pump a GaAs crystal (photocathode) to produce the require electrons. Polarizations of about 80 per cent are obtained by this means. The asymmetry is very small and in this experiment A_{PV} is predicted to be only a few parts per million. Nevertheless, a non-zero value was definitely established. Moreover, A_{PV} was also measured as a function of the fractional energy loss of the initial electron. This is a function of the weak mixing angle and a value was found in agreement with other determinations, e.g. from deep inelastic neutrino scattering. A later experiment confirmed the effect in polarized electron–proton scattering.

A very recent experiment (2004) has measured A_{PV} for e^-e^- scattering. This was done using electrons of about 50 GeV primary energy from the SLAC linear accelerator in Stanford, USA, and scattering them from a liquid hydrogen target. The experiment was able to distinguish final-state electrons scattered from the atomic electrons from those scattered from protons. Taking account of all sources of error, the measured value was $A_{PV} = (-175 \pm 40) \times 10^{-9}$ (note the exponent – parts per billion) and the experiment also yielded a value of $\sin^2 \theta_W$ consistent with other determinations. These remarkable experiments provide unambiguous evidence for parity violation in 'electromagnetic' processes at the level predicted by theory and hence for the electroweak unification as specified in the standard model.[17]

It should also in principle be possible to detect parity violating effects in atomic physics, where the electromagnetic interactions of the electrons dominate. For example, measurements have been made of the slight change in the polarization angle of light passing through a vapour of metallic atoms. In this case the rotation angle is extremely small ($\sim 10^{-7}$ rad), but very sensitive experiments can measure the effect to an accuracy of $\sim$1 per cent. However, to predict the size of the effects requires a detailed knowledge of the atomic theory of the atom and in all cases to date the uncertainties on the predictions are such that a null effect cannot be ruled out. Thus at present, atomic physics does not compete with particle physics experiments in detecting parity-violating effects and measuring $\sin^2 \theta_W$, although this could change in the future.

Problems

6.1 Define charged and neutral current reactions in weak interactions and give an example of each in symbol form. How do they differ in respect of conservation of the strangeness quantum number? Why does observation of the process $\bar{\nu}_\mu + e^- \rightarrow \bar{\nu}_\mu + e^-$ constitute unambiguous evidence for weak neutral currents, whereas the observation of $\bar{\nu}_e + e^- \rightarrow \bar{\nu}_e + e^-$ does not?

6.2 The reaction $e^+e^- \rightarrow \mu^+\mu^-$ is studied using colliding beams each of energy 7 GeV and at these energies the reaction is predominantly electromagnetic. Draw its lowest order Feynman diagram. The differential cross-section is given by

$$\left(\frac{d\sigma}{d\Omega}\right) = \frac{\alpha^2 \hbar^2 c^2}{4E_{CM}^2} (1 + \cos^2 \theta),$$

---

[17]Incidentally, all these experiments are of the fixed-target type, showing that this type of experiment still has a lot to offer.

where E_{CM} is the total centre-of-mass energy and θ is the scattering angle with respect to the beam direction. Calculate the total cross-section in nanobarns at this energy.

The weak interaction also contributes to this process. Draw the corresponding lowest-order Feynman diagram. The weak interaction adds an *additional* term to the differential cross-section of the form

$$\left(\frac{d\sigma}{d\Omega}\right) = \frac{\alpha^2\hbar^2c^2}{4E_{CM}^2}C_{wk}\cos\theta.$$

The constant C_{wk} may be determined experimentally by measuring the 'forward–backward' asymmetry, defined by

$$A_{FB} = \frac{\sigma_F - \sigma_B}{\sigma_F + \sigma_B},$$

where $\sigma_F(\sigma_B)$ is the total cross-section for scattering in the forward (backward) hemisphere, i.e. $0 \leq \cos\theta \leq 1$ ($-1 \leq \cos\theta \leq 0$). Derive a relation between C_{wk} and A_{FB}.

6.3 Draw a Feynman diagram at the quark level for the decay $\Lambda \to p + \pi^-$. If nature were to double the weak coupling constant and decrease the mass of the W boson by a factor of four, what would be the effect on the decay rate $\Gamma(\Lambda \to p + \pi^-)$?

6.4 Neglecting the electron mass, the energy spectrum for the electrons emitted in muon decay is give by

$$\frac{d\omega}{dE_e} = \frac{2G_F^2(m_\mu c^2)^2 E_e^2}{(2\pi)^3(\hbar c)^6}\left(1 - \frac{4E_e}{3m_\mu c^2}\right).$$

What is the most probable energy for the electron? Draw a diagram showing the orientation of the momenta of the three outgoing particles and their helicities for the case when $E_e \approx m_\mu c^2/2$. Show also the helicity of the muon. Integrate the energy spectrum to obtain an expression for the total decay width of the muon. Hence calculate the muon lifetime in seconds $(G_F/(\hbar c)^3 = 1.166 \times 10^{-5}\,\text{GeV}^{-2})$.

6.5 Use lepton universality and lepton–quark symmetry to estimate the branching ratios for (a) the decays $b \to c + e^- + \bar{\nu}_e$ (where the b and c quarks are bound in hadrons) and (b) $\tau^- \to e^- + \bar{\nu}_e + \nu_\tau$. Ignore final states that are Cabibbo-suppressed relative to the lepton modes.

6.6 The couplings of the Z^0 to right-handed (R) and left-handed (L) fermions are given by

$$g_R(f) = -q_f\sin^2\theta_W, \quad g_L(f) = \pm 1/2 - q_f\sin^2\theta_W,$$

where q_f is the electric charge of the fermion f in units of e and θ_W is the weak mixing angle. The positive sign in g_L is used for neutrinos and the $q = u, c, t$

quarks; the negative sign is used for charged leptons and the $q = d,\ s,\ b$ quarks. If the partial width for $Z^0 \to f\bar{f}$ is given by

$$\Gamma_f = \frac{G_F M_Z^3 c^6}{3\pi\sqrt{2}(\hbar c)^3}\left[g_R^2(f) + g_L^2(f)\right],$$

calculate the partial widths to neutrinos Γ_ν and to $q\bar{q}$ pairs Γ_q and explain the relation of Γ_q to the partial width to hadrons Γ_{hadron}.

The widths to hadrons and to charged leptons are measured to be $\Gamma_{\text{had}} = (1738 \pm 12)$ MeV and $\Gamma_{\text{lep}} = (250 \pm 2)$ MeV, and the total width to all final states is measured to be $\Gamma_{\text{tot}} = (2490 \pm 7)$ MeV. Use these experimental results and your calculated value for the decay width to neutrinos to show that there are only three generations of neutrinos with masses $M_\nu < M_Z/2$.

6.7 Explain, with the aid of Feynman diagrams, why the decay $D^0 \to K^- + \pi^+$ can occur as a charged-current weak interaction at lowest order, but the decay $D^+ \to K^0 + \pi^+$ cannot.

6.8 Why is the decay rate of the charged pion much smaller than that of the neutral pion? Draw Feynman diagrams to illustrate your answer.

6.9 Draw the lowest-order Feynman diagrams for the decays $\pi^- \to \mu^- + \bar{\nu}$ and $K^- \to \mu^- + \bar{\nu}_\mu$. Use lepton–quark symmetry and the Cabibbo hypothesis with the Cabibbo angle $\theta_C = 12°$ to estimate the ratio

$$R \equiv \frac{\text{Rate}(K^- \to \mu^- + \bar{\nu}_\mu)}{\text{Rate}(\pi^- \to \mu^- + \bar{\nu}_\mu)},$$

ignoring all kinematic and spin effects. Comment on your result.

6.10 Estimate the ratio of decay rates

$$R \equiv \frac{\Gamma(\Sigma^- \to n + e^- + \bar{\nu}_e)}{\Gamma(\Sigma^- \to \Lambda + e^- + \bar{\nu}_e)}$$

and explain why the decay $\Gamma(\Sigma^+ \to n + e^+ + \nu_e)$ has never been seen.

6.11 One way of looking for the Higgs boson H is in the reaction $e^+e^- \to Z^0 H$. If this reaction is studied at a centre-of-mass energy of 500 GeV in a collider operating for 10^7 s per year and the cross-section at this energy is 60 fb, what instantaneous luminosity (in units of $\text{cm}^{-2}\,\text{s}^{-1}$) would be needed to collect 2000 events in one year if the detection efficiency is 10 per cent. For a Higgs boson with mass $M_H < 120$ GeV, the branching ratio for $H \to b\bar{b}$ is predicted to be 85 per cent. Why will looking for b quarks help distinguish $e^+e^- \to Z^0 H$ from the background reaction $e^+e^- \to Z^0 Z^0$?

6.12 Hadronic strangeness-changing weak decays approximately obey the so-called '$\Delta I = \frac{1}{2}$ rule', i.e. the total isospin changes by $\frac{1}{2}$ in the decay. By assuming a fictitious strangeness zero $I = \frac{1}{2}$ particle S^0 in the initial state, find the prediction of this rule for the ratio

$$R \equiv \frac{\Gamma(\Xi^- \to \Lambda + \pi^-)}{\Gamma(\Xi^0 \to \Lambda + \pi^0)}.$$

Assume that the state $|\Xi^0, S^0\rangle$ is an equal mixture of states with $I = 0$ and $I = 1$.

6.13   The charged-current differential cross-sections for ν and $\bar{\nu}$ scattering from a spin-$\frac{1}{2}$ target are given by generalizations of Equations (6.37) and (6.40) and may be written

$$\frac{d\sigma^{CC}(\nu)}{dy} = \frac{1}{\pi} \frac{G^2 Hs}{(\hbar c)^4}, \quad \frac{d\sigma^{CC}(\bar{\nu})}{dy} = \frac{d\sigma^{CC}(\nu)}{dy}(1-y)^2,$$

where $s = E_{CM}^2$, $y = \frac{1}{2}(1 - \cos\theta)$ and H is the integral of the quark density for the target (cf. Equation (6.43)). The corresponding cross-sections for neutral current scattering are

$$\frac{d\sigma^{NC}(\nu)}{dy} = \frac{d\sigma^{CC}(\nu)}{dy}[g_L^2 + g_R^2(1-y)^2],$$

$$\frac{d\sigma^{NC}(\bar{\nu})}{dy} = \frac{d\sigma^{CC}(\nu)}{dy}[g_L^2(1-y)^2 + g_R^2],$$

where the right- and left-hand couplings to u and d quarks are given by

$$g_L(u) = \frac{1}{2} - \frac{2}{3}\sin^2\theta_W, \quad g_R(u) = -\frac{2}{3}\sin^2\theta_W,$$

$$g_L(d) = -\frac{1}{2} + \frac{1}{3}\sin^2\theta_W, \quad g_R(d) = \frac{1}{3}\sin^2\theta_W.$$

Derive expressions for the ratios $\sigma^{NC}(\nu)/\sigma^{CC}(\nu)$ and $\sigma^{NC}(\bar{\nu})/\sigma^{CC}(\bar{\nu})$ in the case of an isoscalar target consisting of valence u and d quarks only.

7
Models and Theories of Nuclear Physics

Nuclei are held together by the strong nuclear force between nucleons, so we start this chapter by looking at the form of this, which is more complicated than that generated by simple one-particle exchange. Much of the phenomenological evidence comes from low-energy nucleon–nucleon scattering experiments which we will simply quote, but we will interpret the results in terms of the fundamental strong interaction between quarks. The rest of the chapter is devoted to various models and theories that are constructed to explain nuclear data in particular domains.

7.1 The Nucleon – Nucleon Potential

The existence of stable nuclei implies that overall the net nucleon–nucleon force must be *attractive* and much stronger than the Coulomb force, although it cannot be attractive for all separations, or otherwise nuclei would collapse in on themselves. So at very short ranges there must be a repulsive core. However, the repulsive core can be ignored in low-energy nuclear structure problems because low-energy particles cannot probe the short-distance behaviour of the potential. In lowest order, the potential may be represented dominantly by a central term (i.e. one that is a function only of the radial separation of the particles), although there is also a smaller non-central part. We know from proton–proton scattering experiments[1] that the nucleon–nucleon force is *short-range*, of the same order as the size of the nucleus, and thus does not correspond to the exchange of gluons, as in the fundamental strong interaction. A schematic diagram of the resulting potential is shown in Figure 7.1. In practice of course this strong

---

[1]For reviews see, for example, Chapter 7 of Je90 and Chapter 14 of Ho97.

Nuclear and Particle Physics B. R. Martin
© 2006 John Wiley & Sons, Ltd

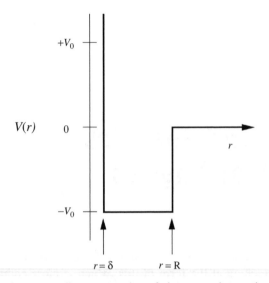

Figure 7.1   Idealized square well representation of the strong interaction nucleon--nucleon potential. The distance R is the range of the nuclear force and $\delta \ll R$ is the distance at which the short-range repulsion becomes important. The depth V_0 is approximately 40 MeV

interaction potential must be combined with the Coulomb potential in the case in the case of protons.

A comparison of *nn* and *pp* scattering data (after allowing for the Coulomb interaction) shows that the nuclear force is *charge-symmetric* ($pp = nn$) and almost *charge-independent* ($pp = nn = pn$).[2] We have commented in Chapter 3 that there is also evidence for this from nuclear physics. Charge-symmetry is seen in comparisons of the energy levels of mirror nuclei (see, for example, Figure 3.9) and evidence for charge-independence comes from the energy levels of triplets of related nuclei with the same A values. Nucleon–nucleon forces are, however, *spin-dependent*. The force between a proton and neutron in an overall spin-1 state (i.e. with spins parallel) is strong enough to support a weakly bound state (the *deuteron*), whereas the potential corresponding to the spin-0 state (i.e. spins antiparallel) has no bound states. Finally, nuclear forces *saturate*. This describes that fact that a nucleon in a typical nucleus experiences attractive interactions only with a limited number of the many other nucleons and is a consequence of the short-range nature of the force. The evidence for this is the form of the nuclear binding energy curve and was discussed in Chapter 2.

Ideally one would like to be able to interpret the nucleon–nucleon potential in terms of the fundamental strong quark–quark interactions. It is not yet possible to give a complete explanation along these lines, but it is possible to go some way in this direction. If we draw an analogy with atomic and molecular structure, with

---

[2]For a discussion of these data see, for example, the references in Footnote 1.

quarks playing the role of electrons, then possibilities are: an ionic-type bond, a van der Waals type of force, or a covalent bond.[3] The first can be ruled out because the confining forces are too strong to permit a quark to be 'lent' from one nucleon to another and the second can also be ruled out because the resulting two-gluon exchange is too weak. This leaves a covalent bond due to the sharing of single quarks between the nucleons analogous to the covalent bond that binds the hydrogen molecule. However, nucleons have to remain 'colourless' during this process and so the shared quark from one nucleon has to have the same colour as the shared quark from the other nucleon. The effect of this is to reduce the effective force (because there are three possible colour states) and by itself it is unable to explain the depth of the observed potential. In addition to the three (valence) quarks within the nucleon there are also present quark–antiquark pairs due to vacuum fluctuations.[4] Such pairs can be colourless and so can also be shared between the nucleons. These quarks actually play a greater role in generating the nuclear strong interaction than single quarks. The lightest such diquarks will be pions and this exchange gives the largest contribution to the attractive part of the nucleon–nucleon force (see, for example, the Feynman diagram Figure 1.4).

In principle, the short-range repulsion could be due to the exchange of heavier diquarks (i.e. mesons), possibly also in different overall spin states. Experiment provides many suitable meson candidates, in agreement with the predictions of the quark model, and each exchange would give rise to a specific contribution to the overall nucleon–nucleon potential, by analogy with the Yukawa potential resulting from the exchange of a spin-0 meson, as discussed in Chapter 1. It is indeed possible to obtain excellent fits to nucleon–nucleon scattering data in a model with several such exchanges.[5] Thus this approach can yield a satisfactory potential model, but is semi-phenomenological only, as it requires the couplings of each of the exchanged particles to be found by fitting nucleon–nucleon scattering data. (The couplings that result broadly agree with values found from other sources.) Boson-exchange models therefore cannot give a fundamental explanation of the repulsion. The reason for the repulsion at small separations in the quark model lies in the spin dependence of the quark–quark strong interaction, which like the phenomenological nucleon–nucleon interaction, is strongly spin-dependent. We have discussed this in the context of calculating hadron masses in Section 3.3.3. When the two nucleons are very close, the wavefunction is effectively that for a 6-quark system with zero angular momentum between the quarks, i.e. a symmetric spatial wave function. Since the colour wave function is antisymmetric, (recall the discussion of Chapter 5), it follows that the spin wavefunction is symmetric. However, the

---

[3]Recall from chemistry that in ionic bonding, electrons are permanently transferred between constituents to form positive and negative ions that then bind by electrostatic attraction; in covalent bonding the constituents share electrons; and the van der Waals force is generated by the attraction between temporary charges induced on the constituents by virtue of slight movements of the electrons.

[4]These are the 'sea' quarks mentioned in connection with the quark model in Chapter 3 and which are probed in deep inelastic lepton scattering that was discussed in Chapter 6.

[5]This approach is discussed in, for example, Chapter 3 of Co01 and also in the references given in Footnote1.

potential energy increases if all the quarks remain in the $L = 0$ state with spins aligned.[6] The two-nucleon system will try to minimize its 'chromomagnetic' energy, but this will compete with the need to have a symmetric spin wavefunction. The optimum configuration at small separations is when one pair of quarks is in an $L = 1$ state, although the excitation energy is comparable to the decrease in chromomagnetic energy, so there will still be a net increase in energy at small separations.

Some tantalizing clues exist about the role of the quark–gluon interaction in nuclear interactions, such as the small nuclear effects in deep inelastic lepton scattering mentioned in Chapter 5. There is also a considerable experimental programme in existence (for example at CEBAF, the superconducting accelerator facility at the Jefferson Laboratory, Virginia, USA, mentioned in Chapter 4) to learn more about the nature of the strong nucleon–nucleon force in terms of the fundamental quark–gluon strong interaction and further progress in this area may well result in the next few years. Meanwhile, in the absence of a fundamental theory to describe the nuclear force, specific models and theories are used to interpret the phenomena in different areas of nuclear physics. In the remainder of this chapter we will discuss a number of such approaches.

7.2 Fermi Gas Model

In this model, the protons and neutrons that make up the nucleus are assumed to comprise two independent systems of nucleons, each freely moving inside the nuclear volume subject to the constraints of the Pauli principle. The potential felt by every nucleon is the superposition of the potentials due to all the other nucleons. In the case of neutrons this is assumed to be a finite-depth square well; for protons, the Coulomb potential modifies this. A sketch of the potential wells in both cases is shown in Figure 7.2.

For a given ground state nucleus, the energy levels will fill up from the bottom of the well. The energy of the highest level that is completely filled is called the *Fermi level* of energy E_F and has a momentum $p_F = (2ME_F)^{1/2}$, where M is the mass of the nucleon. Within the volume V, the number of states with a momentum between p and $p + dp$ is given by the *density of states factor*

$$n(p)dp = dn = \frac{4\pi V}{(2\pi\hbar)^3}p^2dp, \qquad (7.1)$$

---

[6]Compare the mass of the $\Delta(1232)$ resonance, where all three quarks spins are aligned, to that of the lighter nucleon, where one pair of quarks spins is anti-aligned to give a total spin of zero. This is discussed in detail in Section 3.3.3.

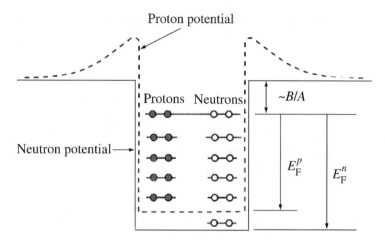

Figure 7.2 Proton and neutron potentials and states in the Fermi gas model

which is derived in Appendix A. Since every state can contain two fermions of the same species, we can have (using $n = 2 \int_0^{p_F} dn$)

$$N = \frac{V(p_F^n)^3}{3\pi^2\hbar^3} \quad \text{and} \quad Z = \frac{V(p_F^p)^3}{3\pi^2\hbar^3} \tag{7.2}$$

neutrons and protons, respectively, with a nuclear volume

$$V = \frac{4}{3}\pi R^3 = \frac{4}{3}\pi R_0^3 A, \tag{7.3}$$

where experimentally $R_0 = 1.21$ fm, as we have seen from electron and hadron scattering experiments discussed in Chapter 2. Assuming for the moment that the depths of the neutron and proton wells are the same, we find for a nucleus with $Z = N = A/2$, the Fermi momentum

$$p_F = p_F^n = p_F^p = \frac{\hbar}{R_0}\left(\frac{9\pi}{8}\right)^{1/3} \approx 250\,\text{MeV}/c. \tag{7.4}$$

Thus the nucleons move freely within the nucleus with quite large momenta.
  The Fermi energy is

$$E_F = \frac{p_F^2}{2M} \approx 33\,\text{MeV}. \tag{7.5}$$

The difference between the top of the well and the Fermi level is constant for most heavy nuclei and is just the average binding energy per nucleon

$\tilde{B} \equiv B/A = 7\text{–}8$ MeV. The depth of the potential and the Fermi energy are to a good approximation independent of the mass number A:

$$V_0 = E_F + \tilde{B} \approx 40 \, \text{MeV}. \tag{7.6}$$

Heavy nuclei generally have a surplus of neutrons. Since the Fermi levels of the protons and neutrons in a stable nucleus have to be equal (otherwise the nucleus can become more stable by β-decay) this implies that the depth of the potential well for the neutron gas has to be deeper than for the proton gas, as shown in Figure 7.2. Protons are therefore on average less tightly bound in nuclei than are neutrons.

We can use the Fermi gas model to give a theoretical expression for some of the dependence of the binding energy on the surplus of neutrons, as follows. First, we define the average kinetic energy per nucleon as

$$\langle E_{\text{kin}} \rangle \equiv \left[\int_0^{p_F} E_{\text{kin}} p^2 \mathrm{d}p \right] \left[\int_0^{p_F} p^2 \mathrm{d}p \right]^{-1}. \tag{7.7}$$

Evaluating the integrals gives

$$\langle E_{\text{kin}} \rangle = \frac{3}{5} \frac{p_F^2}{2M} \approx 20 \, \text{MeV}. \tag{7.8}$$

The total kinetic energy of the nucleus is then

$$E_{\text{kin}}(N, Z) = N \langle E_n \rangle + Z \langle E_p \rangle = \frac{3}{10 M} [N(p_F^n)^2 + Z(p_F^p)^2], \tag{7.9}$$

which may be re-expressed as

$$E_{\text{kin}}(N, Z) = \frac{3}{10 M} \frac{\hbar^2}{R_0^2} \left(\frac{9\pi}{4} \right)^{2/3} \left[\frac{N^{5/3} + Z^{5/3}}{A^{2/3}} \right], \tag{7.10}$$

where again we have taken the radii of the proton and neutron wells to be equal. This expression is for fixed A but varying N and has a minimum at $N = Z$. Hence the binding energy gets smaller for $N \neq Z$. If we set $N = (A + \Delta)/2$, $Z = (A - \Delta)/2$, where $\Delta \equiv N - Z$, and expand Equation (7.10) as a power series in Δ/A, we obtain

$$E_{\text{kin}}(N, Z) = \frac{3}{10 M} \frac{\hbar^2}{R_0^2} \left(\frac{9\pi}{8} \right)^{2/3} \left[A + \frac{5}{9} \frac{(N - Z)^2}{A} + \ldots \right], \tag{7.11}$$

which gives the dependence on the neutron excess. The first term contributes to the volume term in the semi-empirical mass formula (SEMF), while the second describes the correction that results from having $N \neq Z$. This is a contribution to the asymmetry term we have met before in the SEMF and grows as the square of the neutron excess. Evaluating this term from Equation (7.11) shows that its contribution to the asymmetry coefficient defined in Equation (2.51) is about 44 MeV/c^2, compared with the empirical value of about 93 MeV/c^2 given in Equation (2.54). In practice, to reproduce the actual term in the SEMF accurately we would have to take into account the change in the potential energy for $N \neq Z$.

7.3 Shell Model

The nuclear shell model is based on the analogous model for the orbital structure of atomic electrons in atoms. In some areas it gives more detailed predictions than the Fermi gas model and it can also address questions that the latter model cannot. Firstly, we recap the main features of the atomic case.

7.3.1 Shell structure of atoms

The binding energy of electrons in atoms is due primarily to the central Coulomb potential. This is a complicated problem to solve in general because in a multi-electron atom we have to take account of not only the Coulomb field of the nucleus, but also the fields of all the other electrons. Analytic solutions are not usually possible. However, many of the general features of the simplest case of hydrogen carry over to more complicated cases, so it is worth recalling the former.

Atomic energy levels are characterized by a quantum number $n = 1, 2, 3, 4, \ldots$. called the *principal quantum number*. This is defined so that it determines the energy of the system.[7] For any n there are energy-degenerate levels with *orbital angular momentum quantum numbers* given by

$$\ell = 0, 1, 2, 3, \ldots, (n - 1) \tag{7.12}$$

(this restriction follows from the form of the Coulomb potential) and for any value of ℓ there are $(2\ell + 1)$ sub-states with different values of the projection of orbital angular momentum along any chosen axis (the *magnetic quantum number*):

$$m_\ell = -\ell, -\ell + 1, \ldots, 0, 1, \ldots, \ell - 1, \ell. \tag{7.13}$$

Due to the rotational symmetry of the Coulomb potential, all such sub-states are degenerate in energy. Furthermore, since electrons have spin-$\frac{1}{2}$, each of the above

---

[7]In nuclear physics we are not dealing with the same simple Coulomb potential, so it would be better to call n the *radial node quantum number*, as it still determines the form of the radial part of the wavefunction.

states can be occupied by an electron with spin 'up' or 'down', corresponding to the *spin-projection quantum number*

$$m_s = \pm 1/2. \tag{7.14}$$

Again, both these states will have the same energy. So finally, any energy eigenstate in the hydrogen atom is labelled by the quantum numbers (n, ℓ, m_ℓ, m_s) and for any n, there will be n_d degenerate energy states, where

$$n_d = 2 \sum_{\ell=0}^{n-1} (2\ell + 1) = 2n^2. \tag{7.15}$$

The high degree of degeneracy can be broken if there is a preferred direction in space, such as that supplied by a magnetic field, in which case the energy levels could depend on m_ℓ and m_s. One such interaction is the spin–orbit coupling, which is the interaction between the magnetic moment of the electron (due to its spin) and the magnetic field due to the motion of the nucleus (in the electron rest frame). This leads to corrections to the energy levels called *fine structure*, the size of which are determined by the electromagnetic fine structure constant α.

In atomic physics the fine-structure corrections are small and so, if we ignore them for the moment, in hydrogen we would have a system with electron orbits corresponding to shells of a given n, with each shell having degenerate sub-shells specified by the orbital angular momentum ℓ. Going beyond hydrogen, we can introduce the electron–electron Coulomb interaction. This leads to a splitting in any energy level n according to the ℓ value. The degeneracies in m_l and m_s are unchanged. It is straightforward to see that if a shell or sub-shell is filled, then we have

$$\sum m_s = 0 \quad \text{and} \quad \sum m_\ell = 0, \tag{7.16}$$

i.e. there is a strong pairing effect for closed shells. In these cases it can be shown that the Pauli principle implies

$$\mathbf{L} = \mathbf{S} = \mathbf{0} \quad \text{and} \quad \mathbf{J} = \mathbf{L} + \mathbf{S} = \mathbf{0}. \tag{7.17}$$

For any atom with a closed shell or a closed sub-shell structure, the electrons are paired off and thus no valence electrons are available. Such atoms are therefore chemically inert. It is straightforward to work out the atomic numbers at which this occurs. These are

$$Z = 2, 10, 18, 36, 54. \tag{7.18}$$

For example, the inert gas argon $Ar(Z = 18)$ has closed shells corresponding to $n = 1, 2$ and closed sub-shells corresponding to $n = 3, \ell = 0, 1$. These values of Z are called the *atomic magic numbers*.

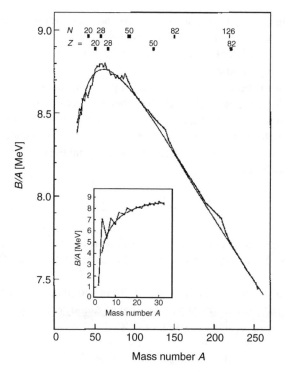

Figure 7.3 Binding energy per nucleon for even values of A: the solid curve is the SEMF (from Bo69)

7.3.2 Nuclear magic numbers

In nuclear physics, there is also evidence for magic numbers, i.e. values of Z and N at which the nuclear binding is particularly strong. This can been seen from the B/A curves of Figure 2.10 where at certain values of N and Z the data lie above the SEMF curve. This is also shown in Figure 7.3, where the inset shows the low-A region magnified. (The figure only shows results for even values of the mass number A.)

The *nuclear magic numbers* are found from experiment to be

$$N = 2, 8, 20, 28, 50, 82, 126$$
$$Z = 2, 8, 20, 28, 50, 82$$

(7.19)

and correspond to one or more closed shells, plus eight nucleons filling the s and p sub-shells of a nucleus with a particular value of n. Nuclei with both N and Z having one of these values are called *doubly magic*, and have even greater stability. An example is the helium nucleus, the α-particle.

Shell structure is also suggested by a number of other phenomena. For example: 'magic' nuclei have many more stable isotopes than other nuclei; they have very small electric dipole moments, which means they are almost spherical, the most tightly bound shape; and neutron capture cross-sections show sharp drops compared with neighbouring nuclei. However, to proceed further we need to know something about the effective potential.

A simple Coulomb potential is clearly not appropriate and we need some form that describes the effective potential of all the other nucleons. Since the strong nuclear force is short-ranged we would expect the potential to follow the form of the density distribution of nucleons in the nucleus. For medium and heavy nuclei, we have seen in Chapter 2 that the Fermi distribution fits the data and the corresponding potential is called the *Woods–Saxon* form

$$V_{\text{central}}(r) = \frac{-V_0}{1 + e^{(r-R)/a}}. \tag{7.20}$$

However, although these potentials can be shown to offer an explanation for the lowest magic numbers, they do not work for the higher ones. This is true of all purely central potentials.

The crucial step in understanding the origin of the magic numbers was taken in 1949 by Mayer and Jensen who suggested that by analogy with atomic physics there should also be a spin–orbit part, so that the total potential is

$$V_{\text{total}} = V_{\text{central}}(r) + V_{\ell s}(r)\mathbf{L} \cdot \mathbf{S}, \tag{7.21}$$

where $\mathbf{L}$ and $\mathbf{S}$ are the orbital and spin angular momentum operators for a single nucleon and $V_{\ell s}(r)$ is an arbitrary function of the radial coordinate.[8] This form for the total potential is the same as that used in atomic physics except for the presence of the function $V_{\ell s}(r)$. Once we have coupling between $\mathbf{L}$ and $\mathbf{S}$ then m_ℓ and m_s are no longer 'good' quantum numbers and we have to work with eigenstates of the total angular momentum vector $\mathbf{J}$, defined by $\mathbf{J} = \mathbf{L} + \mathbf{S}$. Squaring this, we have

$$\mathbf{J}^2 = \mathbf{L}^2 + \mathbf{S}^2 + 2\mathbf{L} \cdot \mathbf{S}, \tag{7.22}$$

i.e.

$$\mathbf{L} \cdot \mathbf{S} = \frac{1}{2}(\mathbf{J}^2 - \mathbf{L}^2 - \mathbf{S}^2) \tag{7.23}$$

and hence the expectation value of $\mathbf{L} \cdot \mathbf{S}$, which we write as $\langle \ell s \rangle$, is

$$\langle \ell s \rangle = \frac{\hbar^2}{2}[j(j+1) - \ell(\ell+1) - s(s+1)] = \begin{cases} \ell/2 & \text{for} \quad j = \ell + \frac{1}{2} \\ -(\ell+1)/2 & \text{for} \quad j = \ell - \frac{1}{2} \end{cases}. \tag{7.24}$$

---

[8] For their work on the shell structure of nuclei. Maria Goeppert-Mayer and J. Hans Jensen were awarded a half share of the 1963 Nobel Prize in Physics. (They shared the prize with Wigner, mentioned in Chapter 1 for his development of the concept of parity.)

(We are always dealing with a single nucleon, so that $s = \frac{1}{2}$.) The splitting between the two levels is thus

$$\Delta E_{ls} = \frac{2\ell + 1}{2}\hbar^2 \langle V_{\ell s}\rangle. \tag{7.25}$$

Experimentally, it is found that $V_{\ell s}(r)$ is negative, which means that the state with $j = \ell + \frac{1}{2}$ has a lower energy than the state with $j = \ell - \frac{1}{2}$. This is the opposite of the situation in atoms. Also, the splittings are substantial and increase linearly with ℓ. Hence for higher ℓ, crossings between levels can occur. Namely, for large ℓ, the splitting of any two neighbouring degenerate levels can shift the $j = \ell - \frac{1}{2}$ state of the initial lower level to lie above the $j = \ell + \frac{1}{2}$ level of the previously higher level.

An example of the resulting splittings up to the $1G$ state is shown in Figure 7.4, where the usual atomic spectroscopic notation has been used, i.e. levels are written $n\ell_j$ with $S, P, D, F, G, \ldots$ used for $\ell = 0, 1, 2, 3, 4, \ldots$. Magic numbers occur when there are particularly large gaps between groups of levels. Note that there is no restriction on the values of ℓ for a given n because, unlike in the atomic case, the strong nuclear potential is not Coulomb-like.

The *configuration* of a real nuclide (which of course has both neutrons and protons) describes the filling of its energy levels (sub-shells), for protons and for neutrons, in order, with the notation $(n\ell_j)^k$ for each sub-shell, where k is the occupancy of the given sub-shell. Sometimes, for brevity, the completely filled

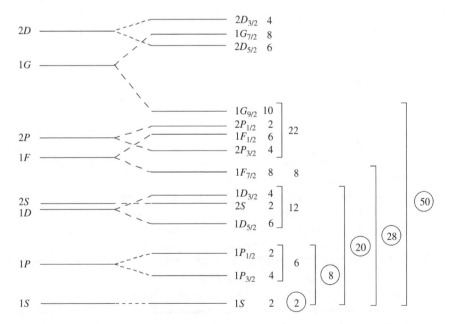

Figure 7.4 Low-lying energy levels in a single-particle shell model using a Woods--Saxon potential plus spin--orbit term; circled integers correspond to nuclear magic numbers

sub-shells are not listed, and if the highest sub-shell is nearly filled, k can be given as a negative number, indicating how far from being filled that sub-shell is. Using the ordering diagram above, and remembering that the maximum occupancy of each sub-shell is $2j + 1$, we predict, for example, the configuration for $^{17}_{8}O$ to be:

$$(1s_{\frac{1}{2}})^2(1p_{\frac{3}{2}})^4(1p_{\frac{1}{2}})^2 \quad \text{for the protons} \tag{7.26a}$$

and

$$(1s_{\frac{1}{2}})^2(1p_{\frac{3}{2}})^4(1p_{\frac{1}{2}})^2(1d_{\frac{5}{2}})^1 \quad \text{for the neutrons.} \tag{7.26b}$$

Notice that all the proton sub-shells are filled, and that all the neutrons are in filled sub-shells except for the last one, which is in a sub-shell on its own. Most of the ground state properties of $^{17}_{8}O$ can therefore be found from just stating the neutron configuration as $(1d_{\frac{5}{2}})^1$.

7.3.3   Spins, parities and magnetic dipole moments

The nuclear shell model can be used to make predictions about the *spins* of ground states. A filled sub-shell must have zero total angular momentum, because j is always an integer-plus-a-half, so the occupancy of the sub-shell, $2j + 1$, is always an even number. This means that in a filled sub-shell, for each nucleon of a given $m_j(= j_z)$ there is another having the opposite m_j. Thus the pair have a combined m_j of zero and so the complete sub-shell will also have zero m_j. Since this is true whatever axis we choose for z, the total angular momentum must also be zero. Since magic number nuclides have closed sub-shells, such nuclides are predicted to have zero contribution to the nuclear spin from the neutrons or protons or both, whichever are magic numbers. Hence magic-Z/magic-N nuclei are predicted to have zero nuclear spin. This is indeed found to be the case experimentally.

   In fact it is found that *all* even-Z/even-N nuclei have zero nuclear spin. We can therefore make the hypothesis that for ground state nuclei, pairs of neutrons and pairs of protons in a given sub-shell *always* couple to give a combined angular momentum of zero, even when the sub-shell is not filled. This is called the *pairing hypothesis*. We can now see why it is the last proton and/or last neutron that determines the net nuclear spin, because these are the only ones that may not be paired up. In odd-A nuclides there is only one unpaired nucleon, so we can predict precisely what the nuclear spin will be by referring to the filling diagram. For even-A/odd-Z/odd-N nuclides, however, we will have both an unpaired proton and an unpaired neutron. We cannot then make a precise prediction about the net spin because of the vectorial way that angular momenta combine; all we can say is that the nuclear spin will lie in the range $|j_p - j_n|$ to $(j_p + j_n)$.

   Predictions can also be made about nuclear *parities*. First, recall the following properties of parity: (1) parity is the transformation $\mathbf{r} \rightarrow -\mathbf{r}$; (2) the wavefunction of a single-particle quantum state will contain an angular part proportional to the spherical harmonic $Y_m^l(\theta, \phi)$, and under the parity transformation

$$PY_m^l(\theta, \phi) = (-)^\ell Y_m^l(\theta, \phi);\qquad(7.27)$$

(3) a single-particle state will also have an *intrinsic parity*, which for nucleons is defined to be positive. Thus the parity of a single-particle nucleon state depends exclusively on the orbital angular momentum quantum number with $P = (-1)^\ell$. The total parity of a multiparticle state is the *product* of the parities of the individual particles. A pair of nucleons with the same ℓ will therefore always have a combined parity of $+1$. The pairing hypothesis then tells us that the total parity of a nucleus is found from the product of the parities of the last proton and the last neutron. So we can predict the parity of *any* nuclide, including the odd/odd ones, and these predictions are in agreement with experiment.

Unless the nuclear spin is zero, we expect nuclei to have *magnetic (dipole) moments*, since both the proton and the neutron have intrinsic magnetic moments, and the proton is electrically charged, so it can produce a magnetic moment when it has orbital motion. The shell model can make predictions about these moments. Using a notation similar to that used in atomic physics, we can write the nuclear magnetic moment as

$$\mu = g_j\, j\mu_N,\qquad(7.28)$$

where μ_N is the *nuclear magneton* that was used in the discussion of hadron magnetic moments in Section 3.3.3, g_j is the *Landé g-factor* and j is the nuclear spin quantum number. For brevity we can write simply $\mu = g_j\, j$ nuclear magnetons.

We will find that the shell model does not give very accurate predictions for magnetic moments, even for the even–odd nuclei where there is only a single unpaired nucleon in the ground state. We will therefore not consider at all the much more problematic case of the odd–odd nuclei having an unpaired proton and an unpaired neutron.

For the even–odd nuclei, we would expect all the paired nucleons to contribute zero net magnetic moment, for the same reason that they do not contribute to the nuclear spin. Predicting the nuclear magnetic moment is then a matter of finding the correct way to combine the orbital and intrinsic components of magnetic moment of the single unpaired nucleon. We need to combine the spin component of the moment, $g_s s$, with the orbital component, $g_\ell \ell$ (where g_s and g_ℓ are the g-factors for spin and orbital angular momentum.) to give the total moment $g_j\, j$. The general formula for doing this is[9]

$$g_j = \frac{j(j+1) + \ell(\ell+1) - s(s+1)}{2j(j+1)}g_\ell + \frac{j(j+1) - \ell(\ell+1) + s(s+1)}{2j(j+1)}g_s,\qquad(7.29)$$

---

[9] See, for example, Section 6.6 of En66.

which simplifies considerably because we always have $j = \ell \pm \frac{1}{2}$. Thus

$$jg_j = g_\ell \ell + g_s/2 \quad \text{for} \quad j = \ell + \frac{1}{2} \tag{7.30a}$$

and

$$jg_j = g_\ell j \left(1 + \frac{1}{2\ell + 1}\right) - g_s j \left(\frac{1}{2\ell + 1}\right) \quad \text{for} \quad j = \ell - \frac{1}{2}. \tag{7.30b}$$

Since $g_\ell = 1$ for a proton and 0 for a neutron, and g_s is approximately $+5.6$ for the proton and -3.8 for the neutron, Equations (7.30) yield the results (where $g_{\text{proton(neutron)}}$ is the g-factor for nuclei with an odd proton(neutron))

$$jg_{\text{proton}} = \ell + \frac{1}{2} \times 5.6 = j + 2.3 \quad \text{for} \quad j = \ell + \frac{1}{2}$$

$$jg_{\text{proton}} = j \left(1 + \frac{1}{2\ell + 1}\right) - 5.6 \times j \left(\frac{1}{2\ell + 1}\right) = j - \frac{2.3j}{j + 1} \quad \text{for} \quad j = \ell - \frac{1}{2}$$

$$\tag{7.31}$$

$$jg_{\text{neutron}} = -\frac{1}{2} \times 3.8 = -1.9 \quad \text{for} \quad j = j = \ell + \frac{1}{2}$$

$$jg_{\text{neutron}} = 3.8 \times j \left(\frac{1}{2\ell + 1}\right) = \frac{1.9j}{j + 1} \quad \text{for} \quad j = \ell - \frac{1}{2}.$$

Accurate values of magnetic dipole moments are available for a wide range of nuclei and plots of a sample of measured values for a range of odd-Z and odd-N nuclei across the whole periodic table are shown in Figure 7.5. It is seen that for a given j, the measured moments usually lie somewhere between the $j = \ell - \frac{1}{2}$ and the $j = \ell + \frac{1}{2}$ values (the so-called *Schmidt lines*), but beyond that the model does not predict the moments accurately. The only exceptions are a few low-A nuclei where the numbers of nucleons are close to magic values.

Why should the shell model work so well when predicting nuclear spins and parities, but be poor for magnetic moments? There are several likely problem areas, including the possibility that protons and neutrons inside nuclei may have effective intrinsic magnetic moments that are different to their free-particle values, because of their very close proximity to one another.

7.3.4 Excited states

In principle, the shell model's energy level structure can be used to predict nuclear excited states. This works quite well for the first one or two excited states when there is only one possible configuration of the nucleus. However, for higher states the

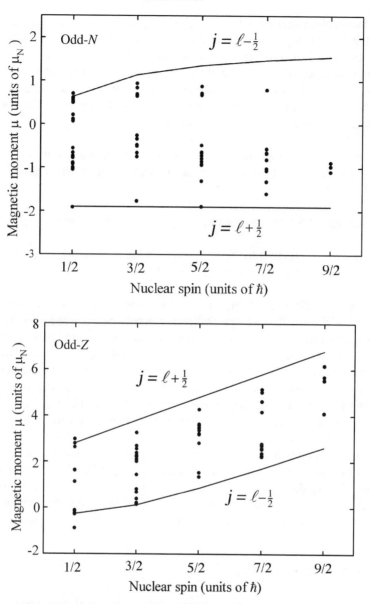

Figure 7.5 Magnetic moments for odd-*N*, even-*Z* nuclei (upper diagram) and odd-*Z*, even-*N* (lower diagram) as functions of nuclear spin compared with the predictions of the single-particle shell model (the Schmidt lines)

spectrum becomes very complicated because several nucleons can be excited simultaneously into a superposition of many different configurations to produce a given nuclear spin and parity. When trying to predict the first one or two excited states using a filling diagram like Figure 7.4, we are looking for the configuration that is nearest to the ground state configuration. This will normally involve *either* moving an unpaired nucleon to the next highest level, *or* moving a nucleon from the sub-shell below the unpaired nucleon up one level to pair with it. Thus it is necessary to consider levels just above and below the last nucleons (protons and neutrons).

As an example, consider the case of $^{17}_{8}O$. Its ground-state configuration is given in Equations (7.26). All the proton sub-shells are filled, and all the neutrons are in filled sub-shells except for the last one, which is in a sub-shell on its own. There are three possibilities to consider for the first excited state:

1. promote one of the $1p_{\frac{1}{2}}$ protons to $1d_{\frac{5}{2}}$, giving a configuration of $(1p_{\frac{1}{2}})^{-1}(1d_{\frac{5}{2}})^{1}$, where the superscript -1 means that the shell is one particle short of being filled;

2. promote one of the $1p_{\frac{1}{2}}$ neutrons to $1d_{\frac{5}{2}}$, giving a configuration of $(1p_{\frac{1}{2}})^{-1}(1d_{\frac{5}{2}})^{2}$;

3. promote the $1d_{\frac{5}{2}}$ neutron to the next level, which is probably $2s_{\frac{1}{2}}$ (or the nearby $1d_{\frac{3}{2}}$), giving a configuration of $(1s_{\frac{1}{2}})^{1}$ or $(1d_{\frac{3}{2}})^{1}$.

Following the diagram of Figure 7.4, the third of these possibilities would correspond to the smallest energy shift, so it should be favoured over the others. The next excited state might involve moving the last neutron up a further level to $1d_{\frac{3}{2}}$, or putting it back where it was and adopting configurations (1) or (2). Option (2) is favoured over (1) because it keeps the excited neutron paired with another, which should have a slightly lower energy than creating two unpaired protons. When comparing these predictions with the observed excited levels it is found that the expected excited states do exist, but not necessarily in precisely the order predicted.

The shell model has many limitations, most of which can be traced to its fundamental assumption that the nucleons move independently of one another in a spherically symmetric potential. The latter, for example, is only true for nuclei that are close to having doubly-filled magnetic shells and predicts zero electric quadruple moments, whereas in practice many nuclei are deformed and quadruple moments are often substantial. We discuss this important observation in the next section.

7.4   Non-Spherical Nuclei

So far we have discussed only spherical nuclei, but with non-sphericity new phenomena are allowed, including additional modes of excitation and the possibility of an electric quadrupole moment.

7.4.1   Electric quadrupole moments

The charge distribution in a nucleus is described in terms of electric multipole moments and follows from the ideas of classical electrostatics. If we have a localized classical charge distribution with charge density $\rho(\mathbf{x})$ within a volume τ, then the first moment that can be non-zero is the electric quadrupole Q, defined by

$$eQ \equiv \int \rho(\mathbf{x})(3z^2 - r^2)\mathrm{d}^3\mathbf{x}, \qquad (7.32)$$

where we have taken the axis of symmetry to be the z-axis. The analogous definition in quantum theory is

$$Q = \frac{1}{e}\sum_i \int \psi^* q_i (3z_i^2 - r^2)\psi\, \mathrm{d}^3\mathbf{x}, \qquad (7.33)$$

where ψ is the nuclear wavefunction and the sum is over all relevant nucleons, each with charge q_i.[10] The quadrupole moment is zero if $|\psi|^2$ is spherically symmetric and so a non-zero value of Q would be indicative of a non-spherical nuclear charge distribution.

If we consider a spheroidal distribution with semi-axes defined as in Figure 2.14, then evaluation of Equation (7.32) leads to the result

$$Q_{\text{intrinsic}} = \frac{2}{5}Ze(a^2 - b^2), \qquad (7.34)$$

where $Q_{\text{intrinsic}}$ is the value of the quadrupole moment for a spheroid at rest and Ze is its total charge. For small deformations,

$$Q_{\text{intrinsic}} \approx \frac{6}{5}ZeR^2\varepsilon, \qquad (7.35)$$

where ε is defined in Equation (2.70) and R is the nuclear radius. Thus, for a prolate distribution $(a > b)$, $Q > 0$ and for an oblate distribution $(a < b)$, $Q < 0$, as illustrated in Figure 7.6. The same results hold in the quantum case.

If the nucleus has a spin J and magnetic quantum number M, then Q will depend on M because it depends on the shape and hence the orientation of the charge distribution. The quadrupole moment is then defined as the value of Q for which M has its maximum value projected along the z-axis. This may be evaluated from Equation (7.33) in the single-particle shell model and without proof we state the resulting prediction that for odd-A, odd-Z nuclei with a single proton having a total angular moment j outside closed sub-shells, the value of Q is given by

$$Q \approx -R^2 \frac{(2j-1)}{2(j+1)}. \qquad (7.36)$$

---

[10]The electric dipole moment $d_z = \frac{1}{e}\sum_i \int \psi^* q_i z_i \psi\, \mathrm{d}\tau$ vanishes because it will contain a sum of terms of the form $\langle \psi_i | z_i | \psi_i \rangle$, all of which are zero by parity conservation.

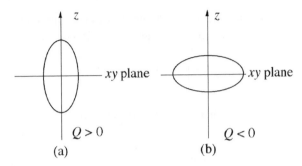

Figure 7.6 Shapes of nuclei leading to (a) $Q > 0$ (prolate), and (b) $Q < 0$ (oblate)

Thus, $Q = 0$ for $j = \frac{1}{2}$. For odd-A, odd-N nuclei with a single neutron outside closed sub-shells Q is predicted to be zero because the neutron has zero electric charge, as will all even-Z, odd-N nuclei because of the pairing effect.

Unlike magnetic dipole moments, electric quadrupole moments are not always well measured and the quoted experimental errors are often far larger than the differences between the values obtained in different experiments. Significant (and difficult to apply) corrections also need to be made to the data to extract the quadrupole moment and this is not always done. The compilation of electric dipole moment data shown in Figure 7.7 is therefore representative. The solid lines are simply to guide the eye and

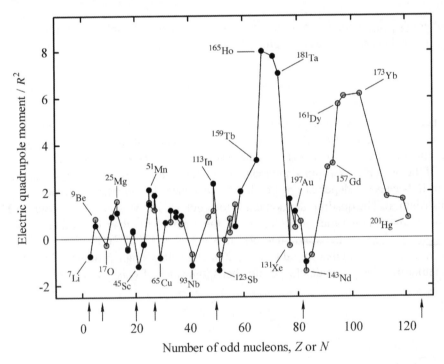

Figure 7.7 Some measured electric quadrupole moments for odd-A nuclei, normalized by dividing by R^2, the squared nuclear radius: grey circles denote odd-N nuclei and black circles odd-Z nuclei; the solid lines have no theoretical significance and the arrows denote the position of closed shells

have no theoretical significance. The arrows indicate the positions of major closed shells. A change of sign of Q at these points is expected because a nucleus with one proton less than a closed shell behaves like a closed-shell nucleus with a negatively charged proton (a proton hole) and there is some evidence for this in the data.

Two features emerge from this diagram. Firstly, while odd-A, odd-Z nuclei with only a few nucleons outside a closed shell do have moments of order $-R^2$, in general the measured moments are larger by factors of two to three and for some nuclei the discrepancy can be as large as a factor of 10. Secondly, odd-A, odd-N nuclei also have non-zero moments, contrary to expectations and, moreover, there is little difference between these and the moments for odd-A, odd-Z nuclei, except that the former tend to be somewhat smaller. These results strongly suggest that for some nuclei it is not a good approximation to assume spherical symmetry and that these nuclei must be considered to have non-spherical mass distributions.

The first attempt to explain the measured electric quadrupole moments in terms of non-spherical nuclei was due to Rainwater. His approach can be understood using the model we discussed in Chapter 2 when considering fission and used above to derive the results of Equations (7.34) and (7.35). There the sphere was deformed into an ellipsoid (see Figure 2.14) with axes parameterized in terms of a small parameter ε via Equation (2.70). The resulting change in the binding energy ΔE_B was found to be

$$\Delta E_B = -\alpha\varepsilon^2, \tag{7.37}$$

where

$$\alpha = \frac{1}{5}(2a_s A^{\frac{2}{3}} - a_c Z^2 A^{-\frac{1}{3}}) \tag{7.38}$$

and the coefficients a_s and a_c are those of the SEMF with numerical values given in Equation (2.54). Rainwater assumed that this expression only held for closed-shell nuclei, but not for nuclei with nucleons in unfilled shells. In the latter cases he showed that distortion gives rise to an additional term in ΔE_B that is linear in ε, so that the total change in binding energy is

$$\Delta E_B = -\alpha\varepsilon^2 - \beta\varepsilon, \tag{7.39}$$

where β is a parameter that could be calculated from the Fermi energy of the nucleus. This form has a minimum value $\beta^2/4\alpha$ where $\varepsilon = -\beta/2\alpha$. The ground state would therefore be deformed and not spherical.

Finally, once the spin of the nucleus is taken into account in quantum theory, the measured electric quadrupole moment for ground states is predicted to be

$$Q = \frac{j(2j-1)}{(j+1)(2j+1)} Q_{\text{intrinsic}}. \tag{7.40}$$

This model gives values for Q that are of the correct sign, but overestimates them by typically a factor of two. Refined variants of the model are capable of bringing the predictions into agreement with the data by making better estimates of the parameter β.

7.4.2 Collective model

The Rainwater model is equivalent to assuming an *aspherical* liquid drop and Aage Bohr (the son of Neils Bohr) and Mottelson showed that many properties of heavy nuclei could be ascribed to the surface motion of such a drop. However, the single-particle shell model cannot be abandoned because it explains many general features of nuclear structure. The problem was therefore to reconcile the shell model with the liquid-drop model. The outcome is the *collective model*.[11]

This model views the nucleus as having a hard core of nucleons in filled shells, as in the shell model, with outer valence nucleons that behave like the surface molecules of a liquid drop. The motions of the latter introduce non-sphericity in the core that in turn causes the quantum states of the valence nucleons to change from the unperturbed states of the shell model. Such a nucleus can both rotate and vibrate and these new degrees of freedom give rise to rotational and vibrational energy levels. For example, the rotational levels are given by $E_J = J(J + 1)\hbar^2/2I$, where I is the moment of inertia and J is the spin of the nucleus. The predictions of this simple model are quite good for small J, but overestimate the energies for larger J. Vibrational modes are due predominantly to *shape oscillations*, where the nucleus oscillates between prolate and oblate ellipsoids. Radial oscillations are much rarer because nuclear matter is relatively incompressible. The energy levels are well approximated by a simple harmonic oscillator potential with spacing $\Delta E = \hbar\omega$, where ω is the oscillator frequency.

In practice, the energy levels of deformed nuclei are very complicated, because there is often coupling between the various modes of excitation, but nevertheless many predictions of the collective model are confirmed experimentally.[12]

7.5   Summary of Nuclear Structure Models

The shell model is based upon the idea that the constituent parts of a nucleus move independently. The liquid-drop model implies just the opposite, since in a drop of incompressible liquid, the motion of any constituent part is correlated with the motion of all the neighbouring pairs. This emphasizes that *models* in physics have a limited domain of applicability and may be unsuitable if applied to a different set of phenomena. As knowledge evolves, it is natural to try and incorporate more

---

[11]For their development of the collective model, Aage Bohr, Ben Mottelson and Leo Rainwater shared the 1975 Nobel Prize in Physics.
[12]The details are discussed, for example, in Section 2.3 of Je90 and Chapter 17 or Ho97.

phenomena by modifying the model to become more general, until (hopefully) we have a model with firm theoretical underpinning which describes a very wide range of phenomena, i.e. a *theory*. The collective model, which uses the ideas of both the shell and liquid drop models, is a step in this direction. We will conclude this section with a brief summary of the assumptions of each of the nuclear models we have discussed and what each can tell us about nuclear structure.

Liquid-drop model

This model assumes that all nuclei have similar mass densities, with binding energies approximately proportional to their masses, just as in a classical charged liquid drop. The model leads to the SEMF, which gives a good description of the average masses and binding energies. It is largely classical, with some quantum mechanical terms (the asymmetry and pairing terms) inserted in an *ad hoc* way. Input from experiment is needed to determine the coefficients of the SEMF.

Fermi gas model

The assumption here is that nucleons move independently in a net nuclear potential. The model uses quantum statistics of a Fermi gas to predict the depth of the potential and the asymmetry term of the SEMF.

Shell model

This is a fully quantum mechanical model that solves the Schrödinger equation with a specific spherical nuclear potential. It makes the same assumptions as the Fermi gas model about the potential, but with the addition of a strong spin–orbit term. It is able to successfully predict nuclear magic numbers, spins and parities of ground-state nuclei and the pairing term of the SEMF. It is less successful in predicting magnetic moments.

Collective model

This is also a fully quantum mechanical model, but in this case the potential is allowed to undergo deformations from the strictly spherical form used in the shell model. The result is that the model can predict magnetic dipole and electric quadrupole magnetic moments with some success. Additional modes of excitation, both vibrational and rotational, are possible and are generally confirmed by experiment.

It is clear from the above that there is at present no universal nuclear model. What we currently have is a number of models and theories that have limited domains of applicability and even within which they are not always able to explain all the observations. For example, the shell model, while able to give a convincing account of the spins and parities of the ground states of nuclei, is unable to predict the spins of excited states with any real confidence. And of course the shell model has absolutely nothing to say about whole areas of nuclear physics phenomena. Some attempt has been made to combine features of different models, such as is done in the collective model, with some success. A more fundamental theory will require the full apparatus of many-body theory applied to interacting nucleons and some progress has been made in this direction for light nuclei, as we will mention in Chapter 9. A theory based on interacting quarks is a more distant goal.

7.6 α-Decay

To discuss α-decays, we could return to the semiempirical mass formula of Chapter 2 and by taking partial derivatives with respect to A and Z find the limits of α-stability, but the result is not very illuminating. To get a very rough idea of the stability criteria, we can write the SEMF in terms of the binding energy B. Then α-decay is energetically allowed if

$$B(2,4) > B(Z,A) - B(Z-2,A-4). \tag{7.41}$$

If we now make the *approximation* that the line of stability is $Z = N$ (the actual line of stability deviates from this, see Figure 2.7), then there is only one independent variable. If we take this to be A, then

$$B(2,4) > B(Z,A) - B(Z-2,A-4) \approx 4\frac{dB}{dA}, \tag{7.42}$$

and we can write

$$4\frac{dB}{dA} = 4\left[A\frac{d(B/A)}{dA} + \frac{B}{A}\right]. \tag{7.43}$$

From the plot of B/A (Figure 2.2), we have $d(B/A)/dA \approx -7.7 \times 10^{-3}$ MeV for $A \geq 120$ and we also know that $B(2,4) = 28.3$ MeV, so we have

$$28.3 \approx 4[B/A - 7.7 \times 10^{-3}A], \tag{7.44}$$

which is a straight line on the B/A versus A plot which cuts the plot at $A \approx 151$. Above this value of A, Equation (7.41) is satisfied by most nuclei and α-decay becomes energetically possible.

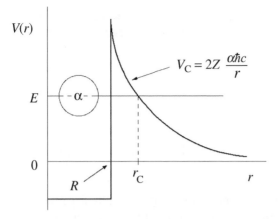

Figure 7.8 Schematic diagram of the potential energy of an α-particle as a function of its distance r from the centre of the nucleus

Lifetimes of α-emitters span an enormous range, and examples are known from 10 ns to 10^{17} years. The origin of this large spread lies in the quantum mechanical phenomenon of *tunelling*. Individual protons and neutrons have binding energies in nuclei of about 8 MeV, even in heavy nuclei (see Figure 2.2), and so cannot in general escape. However, a bound group of nucleons can sometimes escape because its binding energy increases the total energy available for the process. In practice, the most significant decay process of this type is the emission of an α-particle, because unlike systems of two and three nucleons it is very strongly bound by 7 MeV/ nucleon. Figure 7.8 shows the potential energy of an α-particle as a function of r, its distance from the centre of the nucleus.

Beyond the range of the nuclear force, $r > R$, the α-particle feels only the Coulomb potential

$$V_{\mathrm{C}}(r) = \frac{2Z\alpha\hbar c}{r}, \tag{7.45}$$

where we now use Z to be the atomic number of the *daughter* nucleus. Within the range of the nuclear force, $r < R$, the strong nuclear potential prevails, with its strength characterized by the depth of the well. Since the α-particle can escape from the nuclear potential, $E_\alpha > 0$. It is this energy that is released in the decay. Unless E_α is larger than the Coulomb barrier (in which case the decay would be so fast as to be unobservable) the only way the α-particle can escape is by quantum mechanical tunelling through the barrier.

The probability T for transmission through a barrier of height V and thickness Δr by a particle of mass m with energy E_α is given approximately by

$$T \approx \mathrm{e}^{-2\kappa\Delta r}, \tag{7.46}$$

where $\hbar\kappa = [2m|V_C - E_\alpha|]^{1/2}$. Using this result, we can model the Coulomb barrier as a succession of thin barriers of varying height. The overall transmission probability is then

$$T = e^{-G}, \tag{7.47}$$

where the *Gamow factor* G is

$$G = \frac{2}{\hbar} \int_R^{r_C} [2m|V_C(r) - E_\alpha|]^{1/2} dr, \tag{7.48}$$

with $\beta = v/c$ and v is the velocity of the emitted particle.[13] This assumes that the orbital angular momentum of the α-particle is zero, i.e. we ignore possible centrifugal barrier corrections.[14] Since r_C is the value of r where $E_\alpha = V_C(r_C)$,

$$r_C = 2Ze^2/4\pi\varepsilon_0 E_\alpha \tag{7.49}$$

and hence

$$V_C(r) = 2Ze^2/4\pi\varepsilon_0 r = r_C E_\alpha/r. \tag{7.50}$$

So, substituting into Equation (7.48) gives

$$G = \frac{2(2mE_\alpha)^{1/2}}{\hbar} \int_R^{r_C} \left[\frac{r_C}{r} - 1\right]^{1/2} dr, \tag{7.51}$$

where m is the reduced mass of the α-particle and the daughter nucleus, i.e. $m = m_\alpha m_D/(m_\alpha + m_D) \approx m_\alpha$. Evaluating the integral in Equation (7.51) gives

$$G = 4Z\alpha \left(\frac{2mc^2}{E_\alpha}\right)^{1/2} \left[\cos^{-1}\sqrt{\frac{R}{r_C}} - \sqrt{\frac{R}{r_C}\left(1 - \frac{R}{r_C}\right)}\right]. \tag{7.52}$$

Finally, since E_α is typically 5 MeV and the height of the barrier is typically 40 MeV, $r_C \gg R$ and from (7.52), $G \approx 4\pi\alpha Z/\beta$, where $\beta = v_\alpha/c$ and v_α is the velocity of the alpha particle within the nucleus.

The probability per unit time λ of the α-particle escaping from the nucleus is proportional to the product of: (a) the probability $w(\alpha)$ of finding the α-particle in the nucleus; (b) the frequency of collisions of the α-particle with the barrier (this

---

[13]These formulae are derived in Appendix A.

[14]The existence of an angular momentum barrier will suppress the decay rate (i.e. increase the lifetime) compared with a similar nucleus without such a barrier. Numerical estimates of the suppression factors, which increase rapidly with angular momentum, have been calculated by Blatt and Weisskopf and are given in their book B152.

is $v_\alpha/2R$); and (c) the transition probability. Thus, combining these factors, λ is given by

$$\lambda = w(\alpha)\frac{v_\alpha}{2R}e^{-G} \tag{7.53}$$

and since

$$G \propto \frac{Z}{\beta} \propto \frac{Z}{\sqrt{E_\alpha}}, \tag{7.54}$$

small differences in E_α have strong effects on the lifetime.
To examine this further we can take logarithms of Equation (7.53) to give

$$\log_{10} t_{\frac{1}{2}} = a + bZE_\alpha^{-\frac{1}{2}}, \tag{7.55}$$

where $t_{\frac{1}{2}}$ is the half-life. The quantity a depends on the probability $w(\alpha)$ and so is a function of the nucleus, whereas b is a constant that may be estimated from the above equations to be about 1.7. Equation (7.55) is a form of a relation that was found empirically by Geiger and Nuttall in 1911 long before its theoretical derivation in 1928. It is therefore called the *Geiger-Nuttall relation*. It predicts that for fixed Z, the log of the half-life of α-emitters varies linearly with $E_\alpha^{-\frac{1}{2}}$.

Figure 7.9 shows lifetime data for the isotopes of four nuclei. The very strong variation with α-particle energy is evident; changing E_α by a factor of about 2.5 changes the lifetime by 20 orders of magnitude. In all cases the agreement with the Geiger–Nuttall relation is very reasonable and the slopes are compatible with the

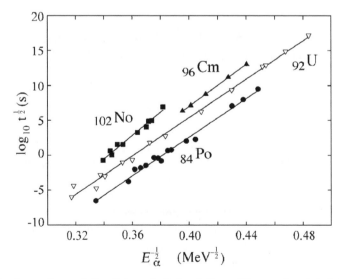

Figure 7.9 Comparison of the Geiger--Nuttall relation with experimental data for some α-emitters

estimate for b above. Thus the simple barrier penetration model is capable of explaining the very wide range of lifetimes of nuclei decaying by α-emission.

7.7 β-Decay

In Chapter 2 we discussed in some detail the phenomenology of β-decay using the SEMF. In this section we return to these decays and examine their theoretical interpretation.

7.7.1 Fermi theory

The first successful theory of nuclear β-decay was proposed in the 1930s by Fermi, long before the W and Z bosons were known and the quark model formulated. He therefore had to construct a theory based on very general principles, working by analogy with the quantum theory of electromagnetic processes (QED), the only successful theory known at the time for quantum particles.

The general equation for electron β-decay is

$$_Z^A X \rightarrow {}_{Z+1}^A Y + e^- + \bar{\nu}_e. \tag{7.56}$$

In Chapter 2, we interpreted this reaction as the decay of a bound neutron, i.e. $n \rightarrow p + e^- + \bar{\nu}_e$, and in Chapter 3 we gave the quark interpretation of this decay. In general, it is possible for the internal state of the nucleus to change in other ways during the transition, but we will simplify matters by considering just the basic neutron decay process.

We have also met the Second Golden Rule, which enables transition rates to be calculated provided the interaction is relatively weak. We will write the Golden Rule as

$$\omega = \frac{2\pi}{\hbar} |M_{fi}|^2 n(E), \tag{7.57}$$

where ω is the *transition rate* (probability per unit time), M_{fi} is the *transition amplitude* (also called the *matrix element* because it is one element of a matrix whose elements are all the possible transitions from the initial state i to different final states f of the system) and $n(E)$ is the *density of states*, i.e. the number of quantum states available to the final system per unit interval of total energy. The density-of-states factor can be calculated from purely kinematical factors, such as energies, momenta, masses and spins where appropriate.[15] The *dynamics* of the process is contained in the matrix element.

---

[15]This is done explicitly in Appendix A.

The matrix element can in general be written in terms of five basic Lorentz invariant interaction operators, $\hat{O}$:

$$M_{fi} = \int \Psi_f^* (g\hat{O}) \Psi_i \, d^3\mathbf{x}, \qquad (7.58)$$

where Ψ_f and Ψ_i are total wavefunctions for the final and initial states, respectively, g is a dimensionless coupling constant, and the integral is over three-dimensional space The five categories are called *scalar* (S), *pseudo-scalar* (P), *vector* (**V**), *axial-vector* (**A**), and *tensor* (T); the names having their origin in the mathematical transformation properties of the operators. (We have met the **V** and **A** forms previously in Chapter 6 on the electroweak interaction.) The main difference between them is the effect on the spin states of the parent and daughter nuclei. When there are no spins involved, and at low energies, $(g\hat{O})$ is simply the interaction potential, i.e. that part of the potential that is responsible for the change of state of the system.

Fermi guessed that $\hat{O}$ would be of the vector type, since electromagnetism is a vector interaction, i.e. it is transmitted by a spin-1 particle – the photon. (Decays of the vector type are called *Fermi transitions*.) We have seen from the work of Chapter 6 that we now know that the weak interaction violates parity conservation and is correctly written as a mixture of both vector and axial-vector interactions (the latter are called *Gamow–Teller transitions* in nuclear physics), but as long as we are not concerned with the spins of the nuclei, this does not make much difference, and we can think of the matrix element in terms of a classical weak interaction potential, like the Yukawa potential. Applying a bit of modern insight, we can assume the potential is of extremely short range (because of the large mass of the W boson), in which case we have seen that we can approximate the interaction by a point-like form and the matrix element then becomes simply a constant, which we write as

$$M_{fi} = \frac{G_F}{V}, \qquad (7.59)$$

where G_F is the Fermi coupling constant we met in Chapter 6. It has dimensions [energy][length]^3 and is related to the charged current weak interaction coupling α_W by

$$G_F = \frac{4\pi(\hbar c)^3 \alpha_W}{(M_W c^2)^2}. \qquad (7.60)$$

In Equation (7.59) it is convenient to factor out an arbitrary volume V, which is used to normalize the wavefunctions. (It will eventually cancel out with a factor in the density-of-states term.)

In nuclear theory, the Fermi coupling constant G_F is taken to be a universal constant and with appropriate corrections for changes of the nuclear state this assumption is also used in β-decay. Experimental results are consistent with the theory under this assumption. However, the theory goes beyond nuclear β-decay, and can be applied to any process mediated by the W boson, provided the energy is not too great. In Chapter 6, for example, we used the same ideas to discuss neutrino scattering. The best process to determine the value of G_F is one not complicated by hadronic (nuclear) effects and muon decay is usually used. The lifetime of the muon τ_μ is given to a very good approximation by (ignoring the Cabbibo correction)

$$\frac{1}{\tau_\mu} = \frac{(m_\mu c^2)^5}{192\pi^3 \hbar (\hbar c)^6} G_F^2, \tag{7.61}$$

from which we can deduce that the value of G_F is about 90 eV fm^3. It is usually quoted in the form $G_F/(\hbar c)^3 = 1.166 \times 10^{-5}$ GeV^{-2}.

7.7.2   Electron momentum distribution

We see from Equation (7.58) that the transition rate (i.e. β-decay lifetime) depends essentially on kinematical factors arising through the density-of-states factor, $n(E)$. To simplify the evaluation of this factor, we consider the neutron and proton to be 'heavy', so that they have negligible kinetic energy, and all the energy released in the decay process goes into creating the electron and neutrino and in giving them kinetic energy. Thus we write

$$E = E_e + E_\nu, \tag{7.62}$$

where E_e is the total (relativistic) energy of the electron, E_ν is the total energy of the neutrino, and E is the total energy released. (This equals $(\Delta m)c^2$, if Δm is the neutron–proton mass difference, or the change in mass of the decaying nucleus.)

The transition rate w can be measured as a function of the electron momentum, so we need to obtain an expression for the spectrum of β-decay electrons. Thus we will fix E_e and find the differential transition rate for decays where the electron has energy in the range E_e to $E_e + dE_e$. From the Golden Rule, this is

$$d\omega = \frac{2\pi}{\hbar} |M|^2 n_\nu (E - E_e) n_e (E_e) dE_e, \tag{7.63}$$

where n_e and n_ν are the density of states factors for the electron and neutrino, respectively. These may be obtained from our previous result:

$$n(p_e)dp_e = \frac{V}{(2\pi\hbar)^3} 4\pi p_e^2 dp_e, \tag{7.64}$$

with a similar expression for n_ν, by changing variables using

$$\frac{dp}{dE} = \frac{E}{pc^2},$$ (7.65)

so that

$$n(E_e)dE_e = \frac{4\pi V}{(2\pi\hbar)^3 c^2} p_e E_e dE_e,$$ (7.66)

with a similar expression for $n(E_\nu)$. Using these in Equation (7.57) and setting $M = G_F/V$, gives

$$\frac{d\omega}{dE_e} = \frac{G_F^2}{2\pi^3\hbar^7 c^4} p_e E_e p_\nu E_\nu$$ (7.67)

where in general

$$p_\nu c = \sqrt{E_\nu^2 - m_\nu^2 c^4} = \sqrt{(E - E_e)^2 - m_\nu^2 c^4}.$$ (7.68)

Finally, it is useful to change the variable to p_e by writing

$$\frac{d\omega}{dp_e} = \frac{dE_e}{dp_e}\frac{d\omega}{dE_e} = \frac{G_F^2}{2\pi^3\hbar^7 c^2} p_e^2 p_\nu E_\nu.$$ (7.69)

If we take the antineutrino to be *precisely* massless, then $p_\nu = E_\nu/c$ and Equation (7.69) reduces to

$$\frac{d\omega}{dp_e} = \frac{G_F^2 p_e^2 p_\nu^2}{2\pi^3\hbar^7 c} = \frac{G_F^2 p_e^2 E_\nu^2}{2\pi^3\hbar^7 c^3} = \frac{G_F^2 p_e^2 (E - E_e)^2}{2\pi^3\hbar^7 c^3}.$$ (7.70)

This expression gives rise to a bell-shaped electron momentum distribution, which rises from zero at zero momentum, reaches a peak and falls to zero again at an electron energy equal to E, as illustrated in the curve labelled $Z = 0$ in Figure 7.10. Studying the precise shape of the distribution near its upper end-point is one way in principle of finding a value for the antineutrino mass. If the neutrino has zero mass, then the gradient of the curve approaches zero at the end-point, whereas any non-zero value results in an end-point that falls to zero with an asymptotically infinite gradient. We will return to this later.

There are several factors that we have ignored or over-simplified in deriving this momentum distribution. The principal ones are to do with the possible changes of nuclear spin of the decaying nucleus, and the electric force acting between the β-particle (electron or positron) and the nucleus. In the first case, when the electron–antineutrino carry away a combined angular momentum of 0 or 1, the

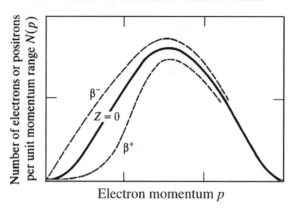

Figure 7.10   Predicted electron spectra: $Z = 0$, without Fermi screening factor; $\beta^{\pm}$, with Fermi screening factor

above treatment is essentially correct: these are the so-called *'allowed transitions'*. However, the nucleus may change its spin by more than 1 unit, and then the simplified short-range potential approach to the matrix element is inadequate. The decay rate in these cases is generally suppressed, although not completely forbidden, despite these being traditionally known as the *'forbidden transitions'*.[16] In the second case, the electric potential between the positive nucleus and a positive β-particle will cause a shift of the low end of its momentum spectrum to the right, since it is propelled away by electrostatic repulsion. Conversely, the low end of the negative β-spectrum is shifted to the left (see Figure 7.10). The precise form of these effects is difficult to calculate, and requires quantum mechanics, but the results are published in tables of a factor $F(Z, E_e)$, called the *Fermi screening factor*, to be applied to the basic β-spectrum.

7.7.3   Kurie plots and the neutrino mass

The usual way of experimentally testing the form of the electron momentum spectrum given by the Fermi theory is by means of a *Kurie plot*. From Equation (7.70), with the Fermi screening factor included, we have

$$\frac{d\omega}{dp_e} = \frac{F(Z, E_e)G_{\mathrm{F}}^2 p_e^2 (E - E_e)^2}{2\pi^3 \hbar^7 c^3}, \tag{7.71}$$

which can be written as

$$H(E_e) \equiv \left[\left(\frac{d\omega}{dp_e} \right) \frac{1}{p_e^2 K(Z, p_e)} \right]^{\frac{1}{2}} = E - E_e, \tag{7.72}$$

---

[16]For a discussion of forbidden transitions see, for example, Co01.

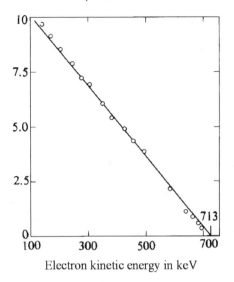

Figure 7.11 Kurie plot for the β-decay of ^{36}Cl (the y-axis is proportional to the function $H(E_e)$ above)

where $K(Z, p_e)$ includes $F(Z, E_e)$ and all the fixed constants in Equation (7.71). A plot of the left-hand side of this equation – using the measured $d\omega/dp_e$ and p_e, together with the calculated value of $K(Z, p_e)$ – against the electron energy E_e should then give a straight line with an intercept of E. An example is shown in Figure 7.11.

If the neutrino mass is not exactly zero then it is straightforward to repeat the above derivation and to show that the left-hand side of the Kurie plot is proportional to

$$\{(E - E_e)[(E - E_e)^2 - m_\nu^2 c^4]^{\frac{1}{2}}\}^{\frac{1}{2}}. \tag{7.73}$$

This will produce a *very* small deviation from linearity extremely close to the end-point of the spectrum and the straight line will curve near the end point and cut the axis vertically at $E_0' = E_0 - m_\nu c^2$. In order to have the best conditions for measuring the neutrino mass, it is necessary to use a nucleus where a non-zero mass would have a maximum effect, i.e. the maximum energy release $E = E_0$ should only be a few keV. Also at such low energies atomic effects have to be taken into account, so the initial and final atomic states must be very well understood. The most suitable case is the decay of tritium,

$$^3\text{H} \rightarrow \, ^3\text{He} + e^- + \bar{\nu}_e, \tag{7.74}$$

where $E_0 = 18.6 \, \text{keV}$. The predicted Kurie plot very close to the end-point is shown in Figure 7.12.

Since the counting rate near E_0 is vanishingly small, the experiment is extremely difficult. In practice, the above formula is fitted to data close to the end-point of the spectrum and extrapolated to E_0. The best experiments are consistent with a zero

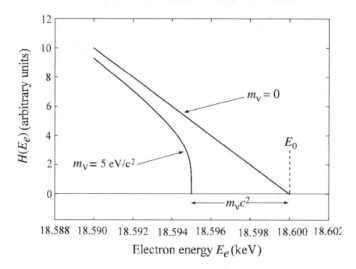

Figure 7.12   Expected Kurie plot for tritium decay very close to the end-point of the electron energy spectrum for two cases: $m_\nu = 0$ and $m_\nu = 5\,\text{eV}/c^2$

neutrino mass, but when experimental and theoretical uncertainties are taken into account, an upper limit of about 2–3 eV/c^2 results.

7.8   γ-Emission and Internal Conversion

In Chapter 2 we mentioned that excited states of nuclei frequently decay to lower states (often the ground state) by the emission of photons in the energy range appropriate to γ-rays and that in addition it is possible for the nucleus to de-excite by ejecting an electron from a low-lying atomic orbit. We shall discuss this only briefly because a proper treatment requires using a quantized electromagnetic radiation field and is beyond the scope of this book. Instead, we will outline the results, without proof.

7.8.1   Selection rules

Gamma emission is a form of electromagnetic radiation and like all such radiation is caused by a changing electric field inducing a magnetic field. There are two possibilities, called electric (E) radiation and magnetic (M) radiation. These names derive from the semiclassical theory of radiation, in which the radiation field arises because of the time variation of charge and current distributions. The classification of the resulting radiation is based on the fact that total angular momentum and parity are conserved in the overall reaction, the latter because it is an electromagnetic process.

The photon carries away a total angular momentum, given by a quantum number L[17], which must include the fact that the photon is a spin-1 vector meson. The minimum value is $L = 1$. This is because a real photon has two possible polarization states corresponding, for example, to $L_z = \pm 1$. Thus in the transition, there must be a change of L_z for the emitting nucleus of ± 1 and this cannot happen if $L = 0$. Hence, if the spins of the initial and final nuclei states are denoted by $\mathbf{J}_i$ and $\mathbf{J}_f$ respectively, the transition $\mathbf{J}_i = \mathbf{0} \rightarrow \mathbf{J}_f = \mathbf{0}$ is strictly forbidden. In general, the photons are said to have a multipolarity L and we talk about multipole radiation; transitions are called dipole $(L = 1)$, quadrupole $(L = 2)$, octupole $(L = 3)$ etc.. Thus, for example, M2 stands for magnetic quadrupole radiation. The allowed values of L are restricted by the conservation equation relating the photon total angular momentum $\mathbf{L}$ and the spins of the initial and final nuclei states, i.e.

$$\mathbf{J}_i = \mathbf{J}_f + \mathbf{L}. \tag{7.75}$$

Thus, L may lie in the range

$$J_i + J_f \geq L \geq |J_i - J_f|. \tag{7.76}$$

It is also necessary to take account of parity. In classical physics, an electric dipole $q\mathbf{r}$ is formed by having two equal and opposite charges q separated by a distance $\mathbf{r}$. It therefore has negative parity under $\mathbf{r} \rightarrow -\mathbf{r}$. Similarly, a magnetic dipole is equivalent to a charge circulating with velocity $\mathbf{v}$ to form a current loop of radius $\mathbf{r}$. The magnetic dipole is then of the form $q\mathbf{r} \times \mathbf{v}$, which does not change sign under a parity inversion and thus has positive parity. The general result, which we state without proof, is that electric multipole radiation has parity $(-1)^L$, whereas magnetic multipole radiation has parity $(-1)^{L+1}$. We thus are led to the selection rules for γ emission shown in Table 7.1. Using this table we can determine which radiation types are allowed for any specific transition. Some examples are shown in Table 7.2.

Table 7.1 Selection rules for γ emission

Multipolarity	Dipole		Quadrupole		Octupole	
Type of radiation	E1	M1	E2	M2	E3	M3
L	1	1	2	2	3	3
ΔP	Yes	No	No	Yes	Yes	No

Although transitions $\mathbf{J}_i = \mathbf{0} \rightarrow \mathbf{J}_f = \mathbf{0}$ are forbidden because the photon is a real particle, such transitions could occur if a *virtual* photon is involved, provided parity does not change. The reason for this is that a virtual photon does not have the

---

[17]As this is the total angular momentum, logically it would be better to employ the symbol $\mathbf{J}$. However, as $\mathbf{L}$ is invariably used in the literature, it will be used in what follows.

Table 7.2 Examples of nuclear electromagnetic transitions

$J_i^{P_i}$	$J_f^{P_f}$	ΔP	L	Allowed transitions
0^+	0^+	No	–	None
$\frac{1}{2}^+$	$\frac{1}{2}^-$	Yes	1	E1
1^+	0^+	No	1	M1
2^+	0^+	No	2	E2
$\frac{3}{2}^-$	$\frac{1}{2}^+$	Yes	1, 2	E1, M2
2^+	1^+	No	1, 2, 3	M1, E2, M3
$\frac{3}{2}^-$	$\frac{5}{2}^+$	Yes	1, 2, 3, 4	E1, M2, E3, M4

restriction on its states of polarization that a real photon does. In practice, the energy of the virtual photon can be transferred to an orbital atomic electron that can thereby be ejected. This is the process of *internal conversion*. There is another possibility whereby the virtual photon can create an internal e^+e^- pair. This is referred to as *internal pair production*.

7.8.2 Transition rates

In semi-classical radiation theory, the transition probability per unit time, i.e. the emission rate, is given by[18]

$$T_{fi}^{E,M}(L) = \frac{1}{4\pi\varepsilon_0} \frac{8\pi(L+1)}{L[(2L+1)!!]^2} \frac{1}{\hbar}\left(\frac{E_\gamma}{\hbar c}\right)^{2L+1} B_{fi}^{E,M}(L), \tag{7.77}$$

where E_γ is the photon energy, E and M refer to electric and magnetic radiation, and for odd-n, $n!! \equiv n(n-2)(n-4)\ldots 3.1$. The function $B_{fi}^{E,M}(L)$ is the so-called *reduced transition probability* and contains all the nuclear information. It is essentially the square of the matrix element of the appropriate operator causing the transition producing photons with multipolarity L, taken between the initial and final nuclear state wave functions. For electric transitions, B is measured in units of $e^2\,\mathrm{fm}^{2L}$ and for magnetic transitions in units of $(\mu_N/c)^2\mathrm{fm}^{2L-2}$ where μ_N is the nuclear magneton.

To go further requires knowledge of the nuclear wave functions. An approximation due to Weisskopf is based on the single-particle shell model. This approach assumes that the radiation results from the transition of a single proton from an initial orbital state of the shell model to a final state of zero angular momentum. In this model the general formulas reduce to

$$B^E(L) = \frac{e^2}{4\pi}\left(\frac{3R^L}{L+3}\right)^2 \tag{7.78a}$$

---

[18]See, for example, Chapter 16 of Ja75.

for electric radiation and

$$B^M(L) = 10\left(\frac{\hbar}{m_p cR}\right)^2 B^E(L) \tag{7.78b}$$

for magnetic radiation, where R is the nuclear radius and m_p is the mass of the proton. Finally, from the work in Chapter 2 on nuclear sizes, we can substitute $R = R_0 A^{1/3}$, with $R_0 = 1.21$ fm, to give the final results:

$$B^E(L) = \frac{e^2}{4\pi}\left(\frac{3}{L+3}\right)^2 (R_0)^{2L} A^{2L/3} \tag{7.79a}$$

and

$$B^M(L) = \frac{10}{\pi}\left(\frac{e\hbar}{2m_p c}\right)^2 \left(\frac{3}{L+3}\right)^2 (R_0)^{2L-2} A^{(2L-2)/3}. \tag{7.79b}$$

Figure 7.13 shows an example of the transition rates $T^{E,M}$ calculated from Equation (7.77) using the approximations of Equations (7.79). Although these are only approximate predictions, they do confirm what is observed experimentally: for a given transition there is a very substantial decrease in decay rates with increasing L, and electric transitions have decay rates about two orders of magnitude higher than the corresponding magnetic transitions.

Finally, it is often useful to have simple formulas for *radiative widths* Γ_γ. These follow from Equations (7.77), (7.78) and (7.79) and for the lowest multipole transitions may be written

$$\Gamma_\gamma(E1) = 0.068E_\gamma^3 A^{2/3}; \quad \Gamma_\gamma(M1) = 0.021E_\gamma^3; \quad \Gamma_\gamma(E2) = (4.9 \times 10^{-8})E_\gamma^5 A^{4/3},$$
$$\tag{7.80}$$

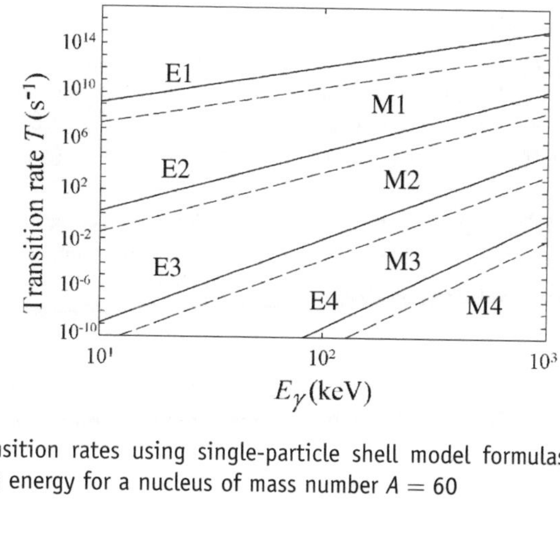

Figure 7.13 Transition rates using single-particle shell model formulas of Weisskopf as a function of photon energy for a nucleus of mass number $A = 60$

where Γ_γ is measured in eV, the transition energy E_γ is measured in MeV and A is the mass number of the nucleus. These formulae are based on the single-particle approximation and in practice collective effects often give values that are much greater than those predicted by Equations (7.80).

Problems

7.1 Assume that in the shell model the nucleon energy levels are ordered as shown in Figure 7.4. Write down the shell-model configuration of the nucleus $^7_3$Li and hence find its spin, parity and magnetic moment (in nuclear magnetons). Give the two most likely configurations for the first excited state, assuming that only protons are excited.

7.2 A certain odd-parity shell-model state can hold up to a maximum of 16 nucleons; what are its values of j and ℓ?

7.3 The ground state of the radioisotope $^{17}_9$F has spin-parity $j^P = \frac{5}{2}^+$ and the first excited state has $j^P = \frac{1}{2}^-$. By reference to Figure 7.4, suggest two possible configurations for the latter state.

7.4 What are the configurations of the ground states of the nuclei $^{93}_{41}$Nb and $^{33}_{16}$S and what values are predicted in the single-particle shell model for their spins, parities and magnetic dipole moments?

7.5 Show explicitly that a uniformly charged ellipsoid at rest with total charge Ze and semi-axes defined in Figure 2.14, has a quadrupole moment $Q = \frac{2}{3}Ze(a^2 - b^2)$.

7.6 The ground state of the nucleus $^{165}_{67}$Ho has an electric quadrupole moment $Q \approx 3.5\,$b. If this is due the fact that the nucleus is a deformed ellipsoid, use the result of Question 7.5 to estimate the sizes of its semi-major and semi-minor axes.

7.7 The decay $^{244}_{98}$Cf$(0^+) \rightarrow {}^{240}_{96}Cm(0^+) + \alpha$ has a Q-value of 7.329 MeV and a half-life of 19.4 mins. If the frequency and probability of forming α-particles (see Equation (7.53)) for this decay are the same as those for the decay $^{228}_{90}$Th$(0^+) \rightarrow {}^{224}_{88}R(0^+) + \alpha$, estimate the half-life for the α-decay of $^{228}_{90}$Th, given that its Q-value is 5.520 MeV.

7.8 The hadrons Σ^0 and Δ^0 can both decay via photon emission:
$\Sigma^0(1193) \rightarrow \Lambda(1116) + \gamma$ (branching ratio ~ 100 per cent); $\Delta^0(1232) \rightarrow n + \gamma$ (branching ratio 0.56 per cent). If the lifetime of the Δ^0 is 0.6×10^{-23} s, estimate the lifetime of the Σ^0.

7.9 The reaction ^{34}S$(p, n)^{34}$Cl has a threshold proton laboratory energy of 6.45 MeV. Calculate non-relativistically the upper limit of the positron energy in the β-decay of ^{34}Cl, given that the mass difference between the neutron and the hydrogen atom is 0.78 MeV.

7.10 To determine the mass of the electron neutrino from the β-decay of tritium requires measurements of the electron energy spectrum very close to the end-point where there is a paucity of events (see Figure 7.12.). To see the nature of the problem, estimate the fraction of electrons with energies within 10 eV of the end-point.

7.11 The electron energy spectra of β-decays with very low-energy end-points E_0 may be approximated by $d\omega/dE = E^{1/2}(E_0 - E)^2$. Show that in this case the mean energy is $\frac{1}{3}E_0$.

7.12 The ground state of $^{35}_{73}$Br has $J^P = \frac{1}{2}^-$ and the first two excited states have $J^P = \frac{5}{2}^-$ (26.92 keV) and $J^P = \frac{3}{2}^-$ (178.1 keV). List the possible γ-transitions between these levels and estimate the lifetime of the $\frac{3}{2}^-$ state.

7.13 Use the Weisskopf formulas of Equations (7.79) to calculate the radiative width $\Gamma_\gamma(E3)$ expressed in a form analogous to Equations (7.80).

8

Applications of Nuclear Physics

Nuclear physics impinges on our everyday lives[1] in a way that particle physics does not, at least not yet. A minor example of this is radioactive dating of historical artefacts which we discussed in Chapter 2. It is appropriate, therefore, to discuss some of these applications. For reasons of space, we will consider just three important areas: fission, fusion and biomedical applications, concentrating in the latter on medical imaging and the therapeutic use of radiation.

8.1   Fission

Fission was discussed in Chapter 2 in the context of the semi-empirical mass formula and among other things we showed that spontaneous fission only occurs for very heavy nuclei. In this section we discuss fission in more detail in the context of its use in the production of energy.

8.1.1   Induced fission – fissile materials

In Chapter 2 we saw that for a nucleus with $A \approx 240$, the Coulomb barrier, which inhibits spontaneous fission, is between 5 and 6 MeV. If a neutron with zero kinetic energy enters a nucleus to form a compound nucleus, the latter will have an excitation energy above its ground state equal to the neutron's binding energy in that state. For example, a zero-energy neutron entering a nucleus of ^{235}U forms a state of ^{236}U with excitation energy of 6.5 MeV. This energy is well above the fission barrier and the compound nucleus quickly undergoes fission, with decay products similar to those found in the spontaneous decay of ^{236}U. To induce fission

---

[1]This is literally true, because we shall see that the energy of the Sun has its origins in nuclear reactions.

Nuclear and Particle Physics   B. R. Martin
© 2006 John Wiley & Sons, Ltd

in ^{238}U on the other hand, requires a neutron with kinetic energy of at least 1.2 MeV. The binding energy of the last neutron in ^{239}U is only 4.8 MeV and an excitation energy of this size is below the fission threshold of ^{239}U.

The differences in the binding energies of the last neutron in even-A and odd-A nuclei are given by the pairing term in the semi-empirical mass formula. Examination of the value of this term (see Equation (2.52)) leads to the explanation of why the odd-A nuclei

$$^{233}_{92}\text{U}, \quad ^{235}_{92}\text{U}, \quad ^{239}_{94}\text{Pu}, \quad ^{241}_{94}\text{Pu} \tag{8.1}$$

are 'fissile', i.e. fission may be induced by even zero-energy neutrons, whereas the even-A (even-Z/even-N) nuclei

$$^{232}_{90}\text{Th}, \quad ^{238}_{92}\text{U}, \quad ^{240}_{94}\text{Pu}, \quad ^{242}_{94}\text{Pu} \tag{8.2}$$

require an energetic neutron to induce fission.

A *nuclear reactor* is a device designed to produce useful energy. The most commonly used fuel in reactors is uranium, so we will focus on this element. Natural uranium consists of 99.3 per cent ^{238}U and only 0.7 per cent ^{235}U. The total and fission cross-sections, σ_{tot} and σ_{fission}, respectively, for neutrons incident on ^{235}U and ^{238}U are shown in Figure 8.1.

The most important features of these figures are (cf. the discussion of nuclear reactions in Section 2.9) as follows.

1. At energies below 0.1 eV, σ_{tot} for ^{235}U is much larger than that for ^{238}U and the fission fraction is large ($\sim$84 per cent). (The other 16 per cent is mainly radiative capture with the formation of an excited state of ^{236}U, plus one or more photons.)

2. In the region between 0.1 eV and 1 keV, the cross-sections for both isotopes show prominent peaks corresponding to neutron capture into resonances. The widths of these states are $\sim$0.1 eV and thus their lifetimes are of the order of $\tau_f \approx \hbar/\Gamma_f \approx 10^{-14}$ s. In the case of ^{235}U these compound nuclei lead to fission, whereas in the case of ^{238}U, neutron capture leads predominantly to radiative decay of the excited state.

3. Above 1 keV, the ratio $\sigma_{\text{fission}}/\sigma_{\text{tot}}$ for ^{235}U is still significant, although smaller than at very low energies. In both isotopes, σ_{tot} is mainly due to contributions from elastic scattering and inelastic excitation of the nucleus.

The fission fragments (which are not unique – several final states are possible) carry away about 90 per cent of the energy of the primary fission reaction. The accompanying neutrons in the primary fission process (referred to as *prompt*

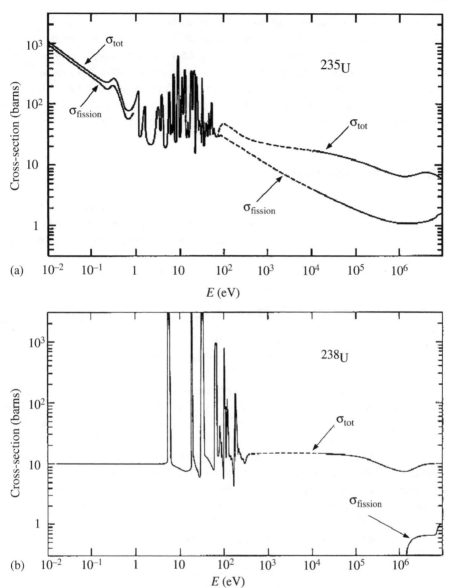

Figure 8.1 Total cross-section σ_{tot} and fission cross-section $\sigma_{fission}$ as functions of energy for neutrons incident on (a) ^{235}U and (b) ^{238}U (adapted from Ga76, Courtesy of Brookhaven National Laboratory)

neutrons) carry away only about 2 per cent of the energy. For ^{235}U, the average number of prompt neutrons per fisson is $n \approx 2.5$, with the value depending a little on the incident neutron energy and they have an average energy of about 2 MeV.

In addition to the neutrons produced in the primary fission, the decay products will themselves decay by chains of β-decays and some of the resulting nuclei will themselves give off further neutrons. This *delayed* component constitutes about 13 per cent of the energy release in the fission of ^{235}U. Although the mean delay is about 13 s, some components have very long lifetimes and may not decay until many years later. One consequence of this is that a reactor will still produce heat even after it has ceased to be used for power production and another is that the delayed component may be emitted after the fuel has been used and removed from the reactor, leading to the biological hazard of radioactive waste.[2]

8.1.2   Fission chain reactions

We have seen in Chapter 2 that in each fission reaction a large amount of energy is produced, which of course is what is needed for power production. However, just as important is the fact that the fission decay products contain other neutrons. For example, we have said that in the case of fission of ^{235}U, on average $n = 2.5$ neutrons are produced. Since neutrons can induce fission, the potential exists for a sustained chain reaction, although a number of conditions have to be fulfilled for this to happen in practice. If we define

$$k \equiv \frac{\text{number of neutrons produced in the } (n+1)\text{ th stage of fission}}{\text{number of neutrons produced in the } n\text{th stage of fission}}, \quad (8.3)$$

then for $k = 1$ the process is said to be *critical* and a sustained reaction can occur. This is the ideal situation for the operation of a power plant based on nuclear fission. If $k < 1$, the process is said to be *subcritical* and the reaction will die out; if $k > 1$, the process is *supercritical* and the energy will grow very rapidly, leading to an uncontrollable explosion, i.e. a nuclear fission bomb.

Again we will focus on uranium as the fissile material and consider the length and timescales for a chain reaction to occur. If we assume that the uranium is a mixture of the two isotopes ^{235}U and ^{238}U with an average neutron total cross-section $\bar{\sigma}_{\text{tot}}$, then the mean free path, i.e. the mean distance the neutron travels between interactions (see Chapter 4), is given by

$$\ell = 1/(\rho_{\text{nucl}}\bar{\sigma}_{\text{tot}}), \quad (8.4)$$

where $\rho_{\text{nucl}} = 4.8 \times 10^{28}$ nuclei/m^3 is the nuclei density of uranium metal. For example, the average energy of a prompt neutron from fission is 2 MeV and at this energy we can see from Figure 8.1 that $\bar{\sigma}_{\text{tot}} \approx 7$ barns, so that $\ell \approx 3$ cm. A 2 MeV neutron will travel this distance in about 1.5×10^{-9} s.

Consider first the case of the *explosive release* of energy in a nuclear bomb, using the highly enriched isotope ^{235}U (for simplicity we will assume a sample of

---

[2]We will return to the effect of radiation on living tissue later in this chapter, in Section 8.3.1.

100 per cent ^{235}U). From Figure 8.1, we see that a neutron with energy of 2 MeV has a probability of about 18 per cent to induce fission in an interaction with a ^{235}U nucleus. Otherwise it will scatter and lose energy, so that the probability for a further interaction will be somewhat increased (because the cross-section increases with decreasing energy). If the probability of inducing fission in a collision is p, the probability that a neutron has induced fission after n collisions is $p(1 - p)^{n-1}$ and the mean number of collisions to induce fission will be

$$\bar{n} = \sum_{n=1}^{\infty} np(1 - p)^{n-1}, \tag{8.5}$$

provided the neutron does not escape outside the target. The value of $\bar{n}$ can be estimated using the measured cross-sections and is about six. Thus the neutron will move a linear (net) distance of $3\sqrt{6}\,\text{cm} \approx 7\,\text{cm}$ in a time $t_p \approx 10^{-8}\,\text{s}$ before inducing a further fission and being replaced on average by 2.5 new neutrons with average energy of 2 MeV.[3]

The above argument suggests that the critical mass of uranium ^{235}U that would be necessary to produce a nuclear explosion is a sphere of radius about 7 cm. However, not all neutrons will be available to induce fission. Some will escape from the surface and some will undergo radiative capture. If the probability that a newly created neutron induces fission is q, then each neutron will on average lead to the creation of $(nq - 1)$ additional neutrons in the time t_p. If there are $N(t)$ neutrons present at time t, then at time $t + \delta t$ there will be

$$N(t + \delta t) = N(t) \left[1 + (nq - 1)\left(\delta t / t_p\right)\right], \tag{8.6}$$

neutrons and hence

$$\frac{N(t + \delta t)}{\delta t} = \frac{N(t)\,(nq - 1)}{t_p}. \tag{8.7}$$

In the limit as $\delta t \to 0$, this gives

$$\frac{\text{d}N}{\text{d}t} = \frac{(nq - 1)}{t_p} N(t), \tag{8.8}$$

---

[3]The square root appears because we are assuming that at each collision the direction changes randomly, i.e. the neutron executes a *random walk*. Thus if the distance travelled in the ith collision is l_i, the displacement vector $\mathbf{d}$ after n collisions will be $\mathbf{d} = \sum_{i=1}^{n} \mathbf{l}_i$ and the net distance travelled d will be given by

$$d^2 = \sum_{i=1}^{n} \sum_{j=1}^{n} (\mathbf{l}_i \cdot \mathbf{l}_j) = l_1^2 + l_2^2 + l_3^2 + \cdots + l_n^2 + 2(\mathbf{l}_1 \cdot \mathbf{l}_2 + \mathbf{l}_1 \cdot \mathbf{l}_3 + \cdots).$$

When the average is taken over several collisions, the scalar products will cancel because the direction of each step is random. Finally, setting $l_i = l$, the mean distance travelled per collision, gives $d = l\sqrt{n}$.

and hence by integrating Equation (8.8)

$$N(t) = N(0) \exp\left[(nq - 1)t/t_p\right].$$ (8.9)

Thus the number increases or decreases exponentially, depending on whether $nq > 1$ or $nq < 1$. For ^{235}U, the number increases if $q > 1/n \approx 0.4$ (recall that $n \approx 2.5$). Clearly if the dimensions of the metal are substantially less than 7 cm, q will be small and the chain reaction will die out exponentially. However, a sufficiently large mass brought together at $t = 0$ will have $q > 0.4$. There will always be some neutrons present at $t = 0$ arising from spontaneous fission and since $t_p \approx 10^{-8}$ s, an explosion will occur very rapidly. For a simple sphere of ^{235}U the critical radius at which $nq = 1$ is actually close to 9 cm and the critical mass is about 50 kg.

Despite the above simple analysis, it is not easy to make a nuclear bomb! This is because the thermal energy released as the assembly becomes critical will produce an outward pressure that is sufficient to blow apart the fissile material unless special steps are taken to prevent this. In early 'atom bombs', a sub-critical mass was assembled and a small plug fired into a prepared hollow in the material so that the whole mass became supercritical. In later devices, the fissile material was a sub-critical sphere of ^{239}Pu surrounded by chemical explosives. These were specially designed ('shaped') so that when they exploded, the resulting shock wave *imploded* the plutonium, which as a result became supercritical.

8.1.3 Nuclear power reactors

The production of power in a controlled way for peaceful use is carried out in a *nuclear reactor* and is just as complex as producing a bomb. There are several distinct types of reactor available. We will discuss just one of these, the *thermal reactor*, which uses uranium as the fuel and low-energy neutrons to establish a chain reaction. The discussion will concentrate on the principles operating such a reactor and not on practical details.

A schematic diagram of the main elements of a generic example of a thermal reactor is shown in Figure 8.2. The most important part is the core, shown schematically in Figure 8.3. This consists of fissile material (fuel elements), control rods and the moderator. The roles of the control rods and the moderator will be explained later. The most commonly used fuel is uranium and many thermal reactors use natural uranium, even though it has only 0.7 per cent of ^{235}U. Because of this, a neutron is much more likely to interact with a nucleus of ^{238}U. However, a 2 MeV neutron from the primary fission has very little chance of inducing fission in a nucleus of ^{238}U. Instead it is much more likely to scatter inelastically, leaving the nucleus in an excited state and after a couple of such collisions the energy of the neutron will be below the threshold of 1.2 MeV for inducing fission in ^{238}U.

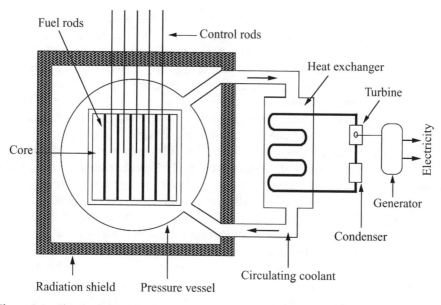

Fuel rods

Control rods

Heat exchanger

Turbine

Core

Electricity

Generator

Condenser

Circulating coolant

Radiation shield   Pressure vessel

Figure 8.2   Sketch of the main elements of a thermal reactor -- the components are not to scale (after Li01, Copyright, John Wiley & Sons)

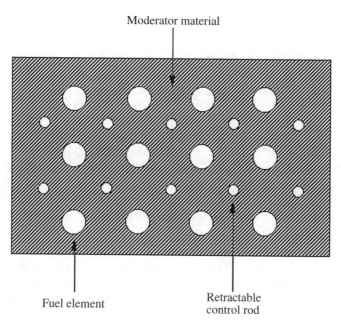

Moderator material

Fuel element

Retractable
control rod

Figure 8.3   Sketch of the elements of the core of a reactor

A neutron with its energy so reduced will have to find a nucleus of ^{235}U if it is to induce fission, but its chances of doing this are very small unless its energy has been reduced to very low energies below 0.1 eV, where the cross-section is large (see Figure 8.1). Before that happens it is likely to have been captured into one of the ^{238}U resonances with the emission of photons. Thus, to sustain a chain reaction, either the fuel must be enriched with a greater fraction of ^{235}U (2–3 per cent is common in some types of commercial reactor), or if natural uranium is to be used, some method must be devised to overcome this problem.

This is where the moderator comes in. This surrounds the fuel elements and its volume is much greater than that of the fuel elements themselves. Its main purpose is to slow down fast neutrons produced in the fission process. Fast neutrons will escape from the fuel rods into the moderator and are reduced to very low energies by elastic collisions. In this way the absorption into resonances of ^{238}U is avoided. The moderator must therefore be a material with a negligible cross-section for absorption and ideally should also be inexpensive. In practice, heavy water (a form of water where the hydrogen atoms are replaced by atoms of deuterium), or carbon (in the form of graphite), are the moderators of choice in many thermal reactors using natural uranium. For enriched reactors, ordinary water may be used.

Consider now the stability of the chain reaction. This is where the control rods play their part. They are usually made of cadmium, which has a very high absorption cross-section for neutrons. By mechanically manipulating the control rods, i.e. by retracting or inserting them, the number of neutrons available to induce fission can be regulated. This mechanism is the key to maintaining a constant k value of unity and therefore a constant power output. However, safe working of the reactor is not possible with prompt neutrons alone. To see this, we return to Equation (8.9) and set $nq - 1 = k - 1$, so that

$$N(t) = N(0) \exp\left[(k - 1)t/t_p\right]. \tag{8.10}$$

The value of t_p is determined by the mean free path for neutron absorption and unlike the case of pure ^{235}U we considered in Section 8.1.2, is given approximately by $t_p \sim 10^{-3}$ s. Thus, for example, if we take $k = 1.001$, i.e. an increase of only 0.1 per cent, the reactor flux would increase by $e^{60} \approx 10^{26}$ in only one minute. Clearly a much smaller rate of increase has to be achieved for safe manipulation of the control rods if a disaster is to be averted. This is where the delayed neutrons play a crucial role.

In a nuclear weapon, the delayed neutrons are of no consequence, because the explosion will have taken place long before they would have been emitted, but in a power reactor they are vital for reactor safety. Taking account of delayed neutrons, each fission leads to $[(n + \delta n)q - 1]$ additional neutrons, where we have defined δn as the number of delayed neutrons per fission. In practice $\delta n \sim 0.02$. In the steady-state operation, with constant energy output, the neutron density must remain constant (i.e. $k = 1$ in Equation (8.3)). Thus q must satisfy the critical condition

$$(n + \delta n)q - 1 = 0. \tag{8.11}$$

Equation (8.10) is now modified to have an additional term that depends on the mean time t_d of the delayed neutrons, which is about 13 s. Provided $n(k - 1) \ll \delta n$, it is the latter term that dominates and (without proof) the modified form of Equation (8.10) is given approximately by

$$N(t) \approx N(0) \exp\left\{\frac{n(k - 1)t}{[\delta n - n(k - 1)]t_d}\right\}. \tag{8.12}$$

Thus the timescale to manipulate the control rods is determined by that of the delayed neutrons. For example, using $n = 2.5, \delta n = 0.02, k = 1.001$ and $t_d = 13$ s in Equation (8.12) gives an increase in reactor flux of less than a factor two in one minute. Clearly, the precise increase is sensitive to the parameters chosen, but factors of this size are manageable. The reactor design therefore ensures that $nq - 1 < 0$ always, so that the reactor can only become critical in the presence of delayed neutrons.

This simple discussion has ignored many practical details that will modify the real formulas used in reactor dynamics, such as the fact that the fuel and moderator are not uniformly distributed throughout the core and that some of the fission products themselves have appreciable cross-sections for neutron absorption and will therefore reduce the flux of neutrons available to sustain the chain reaction.[4]

Returning to Figure 8.2, the core is surrounded by a coolant (often water), which removes the heat generated in the core from the energy deposited by the fission fragments. A thick concrete shield to prevent radiation leaks surrounds the entire set-up. At start-up, the value of k is set slightly higher than unity and is kept at that value until the desired power output is achieved and the operating temperature is reached, after which the value of k is lowered by adjusting the control rods. It is very important for safety reasons that $dq/dT < 0$, so that an increase in temperature T leads to a fall in reaction rate. The rest of the plant is conventional engineering. Thus, the heated coolant gives up its heat in a heat exchanger and is used to boil water and drive a steam turbine, which in turn produces electricity.

It is worth calculating the efficiency with which one can expect to produce energy in a nuclear reactor. We can use the SEMF to calculate the energy released during fission, by finding the binding energies of the two fission products and comparing their sum to the binding energy of the decaying nucleus. For the fission of a single ^{235}U nucleus this is $\sim$200 MeV or 3.2×10^{-11} J. (As we have mentioned above, about 90 per cent of this is in the form of 'prompt' energy.) We also know

---

[4]More details of reaction dynamics are discussed in, for example, Section 10.3 of Li01. In Section 10.6 of this reference there is also a discussion of several other types of commercial reactor.

that 1 g of any element contains N_A/A atoms, where N_A is Avogadro's number. Thus 1g of ^{235}U has about $6 \times 10^{23}/235 \approx 3 \times 10^{21}$ atoms and if fission were complete would yield a total energy of about 10^{11} J, or 1 MW-day. This is about 3×10^6 times greater than the yield obtained by burning (chemical combustion) 1 g of coal. In practice only about 1 per cent of the energy content of natural uranium can be extracted (the overall efficiency is greatly reduced by the conventional engineering required to produce electricity via steam turbines), but this can be increased significantly in another type of reactor, called a *fast breeder* discussed briefly below.

We can also calculate the power output from an ideal thermal reactor for a given mass of uranium. From Equation (1.44) of Chapter 1, the reaction rate for fission W_f is given by

$$W_f = J N \sigma_{\text{fission}}, \tag{8.13}$$

where J is the flux, N is the number of nuclei undergoing fission and σ_{fission} is the fission cross-section. Consider, for example, a reactor containing 100 tonnes of natural uranium, generating a neutron flux of 10^{13} cm^{-2} s^{-1} and with a fission cross-section for ^{235}U of 580 b at the appropriate energy (see Figure 8.1). Since the fraction of ^{235}U in natural uranium is 0.072 per cent, the number of ^{235}U nuclei undergoing fission is given by

$$N = \frac{100 \times 10^3 \times 0.0072 \times N_A}{A} = 1.82 \times 10^{27}, \tag{8.14}$$

where $A = 238.03$ is the mass number of natural uranium. The power generated is thus

$$P = W_f E, \tag{8.15}$$

where $E = 200$ MeV is the total energy released per fission (see above). Evaluating Equation (8.15) gives $P \approx 340$ MW. In addition to causing fission, neutrons will be absorbed by ^{235}U without causing fission. If the total absorption cross-section σ_a is 680 b, then the number of ^{235}U nuclei that will be consumed per second will be $N J \sigma_a$, i.e. 1.24×10^{19} s^{-1}. Since we started with 1.82×10^{27} nuclei, the fuel will be used at the rate of about 1.8 per cent per month.

We turn now to the fast breeder reactor mentioned above. In this reactor there is no large volume of moderator and no large density of thermal neutrons is established. In such a reactor, the proportion of fissile material is increased to about 20 per cent and fast neutrons are used to induce fission. The fuel used is ^{239}Pu rather than ^{235}U, the plutonium being obtained by chemical separation from the spent fuel rods of a thermal reactor. This is possible because some ^{238}U nuclei in the latter will have captured neutrons to produce ^{239}U, which subsequently decays via a chain of β-decays to plutonium. The whole sequence is as follows:

$$n + {}^{238}U \rightarrow {}^{239}U \ (23\,\text{mins}) \rightarrow {}^{239}Np \ (2.4\,\text{days}) \rightarrow {}^{239}Pu \ (2.4 \times 10^4\,\text{years}). \quad (8.16)$$

The mean number of neutrons produced in the fission of ^{239}Pu is 2.96, so this nucleus is very suitable for use in a fast reactor. In practice, the core is a mixture of 20 per cent ^{239}Pu and 80 per cent ^{238}U surrounded by a blanket of more ^{238}U, where more plutonium is made. The ^{238}U obtained from spent fuel rods in thermal reactors is called *depleted* uranium. Such a reactor can produce more fissile ^{239}Pu than it consumes, hence the name 'breeder'. In principle such a reactor can consume all the energy content of natural uranium, rather than the 1 per cent used in thermal reactors, although in practice there are limits to its efficiency.

Whatever type of reactor is used, a major problem is the generation of radioactive waste, including transuranic elements and long-lived fission fragments, which in some cases may have to be stored safely for hundreds of years.[5] Much effort has been expended on this problem, but a totally satisfactory solution is still not available. Short-lived waste with low activity (for example, consumables such as protective clothing) is simply buried in the ground. One idea for long-lived waste with high activity is to 'glassify' it into stable forms that can be stored underground without risk of spillage, leakage into the water table, etc..

A particularly ingenious idea is to 'defuse' long-lived fission fragments by using the resonance capture of neutrons to convert them to short-lived, or even stable, nuclei. For example, ^{99}Tc (technetium), which concentrates in several organs of the body and also in the blood, has a very long half-life. However, it has a large resonant cross-section for neutron capture to a completely stable isotope ^{100}Ru (ruthenium) and in principle this reaction could be used to 'neutralize' ^{99}Tc. Needless to say, the problems to be overcome are far from trivial. First, the amount of radioactive waste is very large, so one problem is to find a source of neutrons capable of handling it. (Reactors themselves are one possible source!) Secondly, the neutron energy has to be matched to the particular waste material, which therefore would have to be separated and prepared before being bombarded by the neutrons. All this would take energy and would increase the overall cost of energy production by nuclear power, which is already more expensive than conventional burning of fossil fuels. Nevertheless, there is considerable interest in the principle of this method and proposals have been made to exploit it without the attendant drawbacks above. We will return to this in Chapter 9, where we discuss some of the outstanding problems in nuclear physics and their possible future solutions. However, until such time as this, or some other, method is realized in practice, the safe long-term disposal of radioactive waste remains a serious unsolved problem.

---

[5]In principle, there would be no such problem with fast breeder reactors, but in practice the ideal is not realized.

8.2   Fusion

We have seen that the plot of binding energy per nucleon (Figure 2.2) has a maximum at $A \approx 56$ and slowly decreases for heavier nuclei. For lighter nuclei, the decrease is much quicker, so that with the exception of magic nuclei, lighter nuclei are less tightly bound than medium size nuclei. Thus, in principle, energy could be produced by two light nuclei fusing to produce a heavier and more tightly bound nucleus – the inverse process to fission. Just as for fission, the energy released comes from the difference in the binding energies of the initial and final states. This process is called *nuclear fusion*. Since light nuclei contain fewer nucleons than heavier nuclei, the energy released per fusion is smaller than in fission. However, as a potential source of power, this is more than balanced by the far greater abundance of stable light nuclei in nature than very heavy nuclei. Thus fusion offers enormous potential for power generation, if the huge practical problems could be overcome.

8.2.1   Coulomb barrier

The practical problem to obtaining fusion, whether in power production or more generally, has its origin in the Coulomb repulsion, which inhibits two nuclei getting close enough together to fuse. This is given by the Coulomb potential

$$V_{\text{C}} = \frac{1}{4\pi\varepsilon_0} \frac{ZZ'e^2}{R + R'}, \tag{8.17}$$

where Z and Z' are the atomic numbers of the two nuclei and R and R' are their effective radii. The quantity $(R + R')$ is therefore classically the distance of closest approach. Recalling, from the work on nuclear structure in Chapter 2, that for medium and heavy nuclei $R = 1.2A^{\frac{1}{3}}$ fm, we have

$$V_{\text{C}} = \left(\frac{e^2}{4\pi\varepsilon_0 \hbar c}\right) \frac{\hbar c\, Z Z'}{1.2 \left[A^{1/3} + (A')^{1/3}\right] \text{fm}} = 1.198 \frac{ZZ'}{A^{1/3} + (A')^{1/3}} \text{MeV}. \tag{8.18}$$

If, for illustration, we set $A \approx A' \approx 2Z \approx 2Z'$, then

$$V_{\text{C}} \approx 0.15 A^{\frac{5}{3}} \text{ MeV}. \tag{8.19}$$

Thus, for example, with $A \approx 8$, $V_{\text{C}} \approx 4.8\,\text{MeV}$ and this energy has to be supplied to overcome the Coulomb barrier.

This is a relatively small amount of energy to supply and it might be thought that it could be achieved by simply colliding two accelerated beams of light nuclei, but in practice nearly all the particles would be elastically scattered. The only practical way is to heat a confined mixture of the nuclei to supply enough thermal energy to overcome the Coulomb barrier. The temperature necessary may be estimated from the relation $E = kT$, where k_B is Boltzmann's constant, given by $k_B = 8.6 \times 10^{-5} \, \text{eV K}^{-1}$. For an energy of 4.8 Mev, this implies a temperature of 5.6×10^{10} K. This is well above the typical temperature of 10^8 K found in stellar interiors.[6] It is also the major hurdle to be overcome in achieving a controlled fusion reaction in a reactor, as we shall see later.

Fusion actually occurs at a lower temperature than this estimate due to a combination of two reasons. The first and most important is the phenomenon of quantum tunnelling, which means that the full height of the Coulomb barrier does not have to be overcome. In Chapter 7 we discussed a similar problem in the context of α-decay, and we can draw on that analysis here. The probability of barrier penetration depends on a number of factors, but the most important is the Gamow factor, which is a function of the relative velocities and the charges of the reaction products. In particular, the probability is proportional to $\exp[-G(E)]$, where $G(E)$ is a generalization of the Gamow factor of Chapter 7. This may be written as $G = \sqrt{E_G/E}$, where again, generalizing the equations in Chapter 7,

$$E_G = 2mc^2(\pi\alpha Z_1 Z_2)^2. \tag{8.20}$$

Here, m is the reduced mass of the two fusing nuclei and they have electric charges $Z_1 e$ and $Z_2 e$. Thus the probability of barrier penetration increases as E increases. Nevertheless, the probability of fusion is still extremely small. For example, if we consider the fusion of two protons (which we will see below is an important ingredient of the reactions that power the Sun), at a typical stellar temperature of 10^7 K, we find $E_G \approx 490 \, \text{keV}$ and $E \approx 1 \, \text{keV}$. Hence the probability of fusion is proportional to $\exp[-(E_G/E)^{1/2}] \approx \exp(-22) \approx 10^{-9.6}$ which is a very large suppression factor and so the actual fusion rate is still extremely slow.

The other reason that fusion occurs at a lower temperature than expected is that a collection of nuclei at a given mean temperature, whether in stars or elsewhere, will have a Maxwellian distribution of energies about the mean and so there will be some with energies substantially higher than the mean energy. Nevertheless, even a stellar temperature of 10^8 K corresponds to an energy of only about 10 keV, so the fraction of nuclei with energies of order 1 MeV in such a star would only be of the order of $\exp(-E/kT) \sim \exp(-100) \sim 10^{-43}$, a minute amount. We will return to these questions in more detail in Section. 8.2.3.

8.2.2 Stellar fusion

The energy of the Sun comes from nuclear fusion reactions, foremost of which is the so-called *proton–proton cycle*. This has more than one branch, but one of these,

---

[6]Because of this, many scientists refused to accept that fusion occurred in stars when the suggestion was first made. When challenged on this, Eddington's reposte was simple: '.... go and find a hotter place'.

the PPI cycle, is dominant. This starts with the fusion of hydrogen nuclei to produce deuterium via the weak interaction:

$$^1H + {}^1H \rightarrow {}^2H + e^+ + \nu_e + 0.42\,\text{MeV}. \tag{8.21}$$

The deuterium then fuses with more hydrogen to produce ^{3}He via the electromagnetic interaction:

$$^1H + {}^2H \rightarrow {}^3He + \gamma + 5.49\,\text{MeV} \tag{8.22}$$

and finally, two ^{3}He nuclei fuse to form ^{4}He via the nuclear strong interaction:

$$^3He + {}^3He \rightarrow {}^4He + 2({}^1H) + 12.86\,\text{MeV}. \tag{8.23}$$

The relatively large energy release in the last reaction is because ^{4}He is a doubly magic nucleus and so is very tightly bound. The first of these reactions, being a weak interaction, proceeds at an extremely slow rate and sets the scale for the long lifetime of the Sun. Combining these equations, we have overall

$$4({}^1H) \rightarrow {}^4He + 2e^+ + 2\nu_e + 2\gamma + 24.68\,\text{MeV}. \tag{8.24}$$

Because the temperature of the Sun is $\sim 10^7$ K, all its material is fully ionized. Matter in this state is referred to as a *plasma*. The positrons produced above will annihilate with electrons in the plasma to release a further 1.02 MeV of energy per positron and so the total energy released is 26.72 MeV. However, of this each neutrino will carry off 0.26 MeV on average, which is lost into space.[7] Thus on average, 6.55 MeV of electromagnetic energy is radiated from the Sun for every proton consumed in the PPI chain.

The PPI chain is not the only fusion cycle contributing to the energy output of the Sun, but it is the most important. Another interesting cycle is the carbon, or CNO chain. Although this contributes only about 3 per cent of the energy output of the Sun, it plays an important role in the evolution of other stellar objects. In the presence of any of the nuclei $^{12}_{6}$C, $^{13}_{6}$C, $^{14}_{7}$N or $^{15}_{7}$N, hydrogen will catalyse burning via the reactions

$$\begin{aligned}^{12}C + {}^1H &\rightarrow {}^{13}N + \gamma + 1.95\text{ MeV}\\ {}^{13}N &\rightarrow {}^{13}C + e^+ + \nu_e + 1.20\text{ MeV}\end{aligned} \tag{8.25}$$

$$^{13}C + {}^1H \rightarrow {}^{14}N + \gamma + 7.55\text{ MeV} \tag{8.26}$$

$$\begin{aligned}^{14}N + {}^1H &\rightarrow {}^{15}O + \gamma + 7.34\text{ MeV}\\ {}^{15}O &\rightarrow {}^{15}N + e^+ + \nu_e + 1.68\text{ MeV}\end{aligned} \tag{8.27}$$

---

[7]These are the main contributors to the neutrino flux observed at the surface of the Earth that was discussed in Chapter 3.

and

$$^{15}\text{N} + {}^{1}\text{H} \rightarrow {}^{12}\text{C} + {}^{4}\text{He} + 4.96 \text{ MeV} \tag{8.28}$$

Thus, overall in the CNO cycle we have

$$4\,({}^{1}\text{H}) \rightarrow {}^{4}\text{He} + 2e^{+} + 2\nu_{e} + 3\gamma + 24.68 \text{ MeV}. \tag{8.29}$$

These and other fusion chains all produce electron neutrinos as final-state products and using detailed models of the Sun, the flux of such neutrinos at the surface of the Earth can be predicted.[8] The actual count rate is far lower than the theoretical expectation. This is the *solar neutrino problem* that we met in Section 3.1.4. The solution to this problem is almost certainly neutrino oscillations, where some ν_{e} are converted to neutrinos of other flavours in their passage from the Sun to the Earth. We saw in Chapter 3 that this is only possible if neutrinos have mass, so a definitive measurement of neutrino masses would be an important piece of evidence to finally resolve the solar neutrino problem. Such measurements should be available in a few years.

The process whereby heavier elements (including the ^{12}C required in the CNO cycle) are produced by fusion of lighter ones can continue beyond the reactions above. For example, when the hydrogen content is depleted, at high temperatures helium nuclei can fuse to form an equilibrium mixture with ^{8}Be via the reaction

$$^{4}\text{He} + {}^{4}\text{He} \rightleftharpoons {}^{8}\text{Be} \tag{8.30}$$

and the presence of ^{8}Be allows the rare reaction

$$^{4}\text{He} + {}^{8}\text{Be} \rightarrow {}^{12}\text{C}^{*} \tag{8.31}$$

to occur, where C^{*} is an excited state of carbon. A very small fraction of the latter will decay to the ground state, so that overall we have[9]

$$3({}^{4}\text{He}) \rightarrow {}^{12}\text{C} + 7.27 \text{ MeV}. \tag{8.32}$$

---

[8] The expectations are based on a detailed model of the Sun known as the standard solar model that we met in Chapter 3.

[9] The occurrence of this crucial reaction depends critically on the existence of a particular excited state of ^{12}C. For a discussion of this and the details of the other reactions mentioned below see, for example, Section 4.3 of Ph94. Very recent experiments (2005) have found evidence for other nearby excited states that change the accepted energy dependence (or equivalently the temperature dependence) of this reaction which will have implications for theories of stellar evolution.

The presence of ^{12}C enables another series of fusion reactions to occur, in addition to the CNO cycle. Thus ^{16}O can be produced via the reaction

$$^4\text{He} + {}^{12}\text{C} \rightarrow {}^{16}\text{O} + \gamma \tag{8.33}$$

and the production of neon, sodium and magnesium is possible via the reactions

$$^{12}\text{C} + {}^{12}\text{C} \rightarrow {}^{20}\text{Ne} + {}^4\text{He}, \quad {}^{23}\text{Na} + p, \quad {}^{23}\text{Mg} + n. \tag{8.34}$$

Fusion processes continue to synthesize heavier elements until the core of the stellar object is composed mainly of nuclei with $A \approx 56$, i.e. the peak of the binding energy per nucleon curve. Heavier nuclei are produced in supernova explosions, but this is properly the subject of astrophysics and we will not pursue it further here, although we will return to it briefly in Chapter 9.

8.2.3   Fusion reaction rates

We have mentioned in Section 8.2.1 that quantum tunnelling and the Maxwellian distribution of energies combine to enable fusion to occur at a lower temperature than might at first be expected. The product of the increasing barrier penetration factor with energy and the Maxwellian decreasing exponential actually means that in practice fusion takes place over a rather narrow range of energies. To see this we will consider the fusion between two types of nuclei, a and b, having number densities n_a and n_b (i.e. the number of particles per unit volume) and at a temperature T. We assume that the temperature is high enough so that the nuclei form a plasma, with uniform values of number densities and temperature. We also assume that the velocities of the two nuclei are given by the Maxwell–Boltzmann distribution, so that the probability of having two nuclei with a relative speed v in the range v to $v + dv$ is

$$P(v)\, dv = \left(\frac{2}{\pi}\right)^{1/2} \left(\frac{m}{kT}\right)^{3/2} \exp\left[\frac{-mv^2}{2kT}\right] v^2\, dv, \tag{8.35}$$

where m is the reduced mass of the pair. The fusion reaction rate per unit volume is then

$$R_{ab} = n_a n_b \langle \sigma_{ab} v \rangle, \tag{8.36}$$

where σ_{ab} is the fusion cross-section[10] and the brackets denote an average, i.e.

$$\langle \sigma_{ab} v \rangle \equiv \int_0^\infty \sigma_{ab} v P(v)\, dv. \tag{8.37}$$

---

[10]The product $n_A n_B$ is the number of pairs of nuclei that can fuse. If the two nuclei are of the same type, with $n_A = n_B = n$, then the product must be replaced by $\frac{1}{2}n(n-1) \approx \frac{1}{2}n^2$, because in quantum theory such nuclei are indistinguishable.

The fusion cross-section may be written

$$\sigma_{ab}(E) = \frac{S(E)}{E} \exp\left[-\left(\frac{E_G}{E}\right)^{1/2}\right], \tag{8.38}$$

where the exponential follows from the previous discussion of quantum tunnelling and $S(E)$ contains the details of the nuclear physics. The term $1/E$ is conveniently factored out because many nuclear cross-sections have this behaviour at low energies. Using (8.35) and (8.38) in (8.37) gives, from (8.36):

$$R_{ab} = n_a n_b \left(\frac{8}{\pi m}\right)^{1/2} \left(\frac{1}{kT}\right)^{3/2} \int_0^\infty S(E) \exp\left[-\frac{E}{kT} - \left(\frac{E_G}{E}\right)^{1/2}\right] dE. \tag{8.39}$$

Because the factor $1/E$ has been taken out of the expression for $\sigma(E)$, the quantity $S(E)$ is slowly varying and the behaviour of the integrand is dominated by the behaviour of the exponential term. The falling exponential of the Maxwellian energy distribution combines with the rising exponential of the quantum tunnelling effect to produce a maximum in the integrand at $E = E_0$ where

$$E_0 = \left[\frac{1}{4} E_G (kT)^2\right]^{1/3} \tag{8.40}$$

and fusion takes place over a relatively narrow range of energies $E_0 \pm \Delta E_0$ where

$$\Delta E_0 = \frac{4}{3^{1/2} 2^{1/3}} E_G^{1/6} (kT)^{5/6}. \tag{8.41}$$

The importance of the temperature and the Gamow energy $E_G = 2mc^2 (\pi \alpha Z_a Z_b)^2$ is clear. A schematic illustration of the interplay between these two effects is shown in Figure 8.4.

As a real example, consider the pp reaction (Equation (8.21)), at a temperature of 2×10^7 K. We have $E_G = 493\,\text{keV}$ and $kT = 1.7\,\text{keV}$, so that fusion is most likely at $E_0 = 7.2\,\text{keV}$ and the half-width of the distribution is $\Delta E_0/2 = 4.1\,\text{keV}$. The resulting function $\exp[-E/kT - (E_G/E)^{1/2}]$ is shown in Figure 8.5.

In the approximation where $S(E)$ is taken as a constant $S(E_0)$, the integral in Equation (8.39) may be done and gives

$$\langle \sigma_{ab} v \rangle \approx \frac{8}{9} S(E_0) \left(\frac{2}{3mE_G}\right)^{1/2} \tau^2 \exp[-\tau], \tag{8.42}$$

where $\tau = 3 \left(\frac{1}{2}\right)^{2/3} (E_G/kT)^{1/3}$.

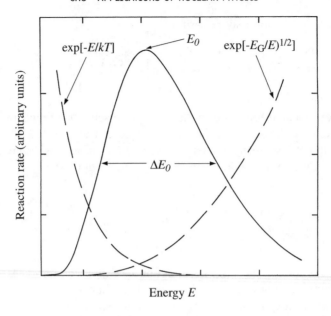

Figure 8.4 The right-hand dashed curve is proportional to the barrier penetration factor and the left-hand dashed curve is proportional to the Maxwell distribution. The solid curve is the combined effect and is proportional to the overall probability of fusion with a peak at E_0 and a width of ΔE_0

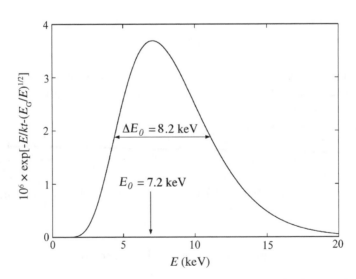

Figure 8.5 The exponential part of the integrand in Equation (8.39) for the case of *pp* fusion at a temperature of 2×10^7 K

If we take the masses to be $A_{a,b}$ in atomic mass units we can evaluate Equation (8.36) using the expression (8.20) for E_G to give

$$R_{ab} = \frac{7.21 \times 10^{-22}}{Z_a Z_b} n_a n_b \frac{(A_a + A_b)}{A_a A_b} \left(\frac{S(E_0)}{1\,\text{MeV b}}\right) \tau^2 \exp[-\tau]\,\text{m}^3\text{s}^{-1}, \qquad (8.43)$$

with

$$\tau = 18.8 (Z_a Z_b)^{2/3} \left(\frac{A_a A_b}{A_a + A_b}\right)^{1/3} \left(\frac{1\,\text{keV}}{kT}\right)^{1/3}. \qquad (8.44)$$

The rate depends very strongly on both the temperature and the nuclear species because of the factor $\tau^2 \exp[-\tau]$. This is illustrated in Figure 8.6 for the pp and $p^{12}C$ reactions, the initial reactions in the pp and CNO cycles.

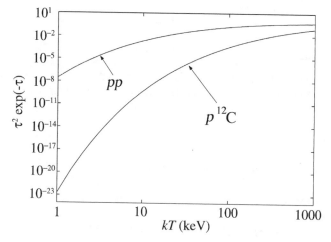

Figure 8.6 The function $\tau^2 \exp(-\tau)$ of Equation (8.43) for the pp and $p^{12}C$ reactions

8.2.4 Fusion reactors

There is currently an international large-scale effort to achieve controlled fusion in the laboratory, with the eventual aim of producing power. For this, the pp reactions are far too slow to be useful. However, the Coulomb barrier for the deuteron ^2_1H is the same as for the proton and the exothermic reactions

$$^2_1\text{H} + {}^2_1\text{H} \rightarrow {}^3_2\text{H} + n + 3.27\,\text{MeV} \qquad (8.45a)$$

and

$$^2_1\text{H} + {}^2_1\text{H} \rightarrow {}^3_1\text{H} + p + 4.03\,\text{MeV} \qquad (8.45b)$$

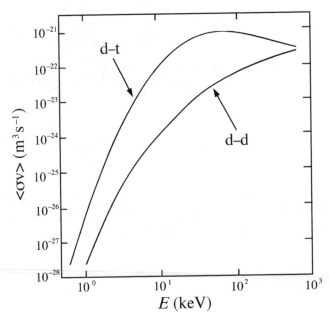

Figure 8.7   Values of the quantity $\langle \sigma v \rangle$ for the d--t reaction of Equation (8.46) and the combined d--d reactions of Equations (8.45) (adapted from Ke82 and reproduced by permission of Annual Reviews)

suggest that deuterium might be a suitable fuel for a fusion reactor. Deuterium is also present in huge quantities in sea water and is easy to separate at low cost.

An even better reaction in terms of energy output is deuterium–tritium fusion:

$$^2_1\text{H} + {}^3_1\text{H} \rightarrow {}^4_2\text{He} + n + 17.62 \text{ MeV}. \tag{8.46}$$

The values of $\langle \sigma v \rangle$ for the d–t reaction of Equation (8.46) and the combined d–d reactions of Equations (8.45) are shown in Figure 8.7. It can be seen that the deuterium–tritium (d–t) reaction has the advantage over the deuterium–deuterium (d–d) reaction of a much higher cross-section. The heat of the reaction is also greater. The principal disadvantage is that tritium does not occur naturally (it has a mean life of only 17.7 years) and has to be manufactured, which increases the overall cost. From Figure 8.7 it can be seen that the rate for the d–t reaction peaks at about $E = kT = 30-40 \text{ keV}$ and a working energy where the cross-section is still considered reasonable is about 20 keV, i.e. 3×10^8 K.

The effective energy produced by the fusion process will be reduced by the heat radiated by the hot plasma. The mechanism for this is predominantly electron bremmstrahlung. The power loss per unit volume due to this process is proportional to $T^{1/2}Z^2$, where Z is the atomic number of the ionized atoms. Thus for a plasma with given constituents and at a fixed ion density, there will be a minimum temperature below which the radiation losses will exceed the power produced by fusion. For example, for the d–t reaction with an ion density $10^{21}\,\mathrm{m}^{-3}$, $kT_{min} \approx 4\,\mathrm{keV}$. It would be 10 times larger for the d–d reaction of Equation (8.45a) because of the form of $\langle \sigma v \rangle$ (see Figure 8.7), which is another reason for using the d–t reaction. In practice, the situation is worse than this because most of the neutrons in Equation (8.46) will escape, so even at the theoretical 'break-even' temperature, external energy would have to be supplied to sustain the fusion process. Only when the energy deposited in the plasma by the α-particles exceeds the radiation losses would the reaction be self-sustaining. This is referred to as the 'ignition point'.

A numerical expression that embodies these ideas is the so-called *Lawson criterion*, which provides a measure of how close to practicality is a particular reactor design. We will assume a d–t reaction. To achieve a temperature T in a deuterium–tritium plasma, there has to be an input of energy $4n_d(3kT/2)$ per unit volume. Here n_d is the number density of deuterium ions and the factor of 4 comes about because n_d is equal to the number density of tritium ions and the electron density is twice this, giving $4n_d$ particles per unit volume. The reaction rate in the plasma is $n_d^2 \langle \sigma_{dt} v \rangle$. If the plasma is confined for time t_c, then per unit volume of plasma,

$$L \equiv \frac{\text{energy output}}{\text{energy input}} = \frac{n_d^2 \langle \sigma_{dt} v \rangle\, t_c Q}{6 n_d\, k\, T} = \frac{n_d \langle \sigma_{dt} v \rangle\, t_c Q}{6\, k\, T}, \qquad (8.47)$$

where Q is the energy released in the fusion reaction. For a useful device, $L > 1$. For example, If we assume $kT = 20\,\mathrm{keV}$ and use the experimental value $\langle \sigma_{dt} v \rangle \approx 10^{-22}\,\mathrm{m}^3\,\mathrm{s}^{-1}$, then the Lawson criterion may be written

$$n_d\, t_c > 7 \times 10^{19}\,\mathrm{m}^{-3}\mathrm{s}. \qquad (8.48)$$

Thus either a very high particle density or a long confinement time, or both, is required.

At the temperatures required for fusion, any material container will vaporize and so the central problem is how to contain the plasma for sufficiently long times for the reaction to take place. The two main methods are *magnetic confinement* and *inertial confinement*. Both techniques present enormous technical challenges. In practice, most work has been done on magnetic confinement and so this method will be discussed in more detail than the inertial confinement method.

In magnetic confinement, the plasma is confined by magnetic fields and heated by electromagnetic fields. Firstly we recall the behaviour of a particle of charge q in a uniform magnetic field $\mathbf{B}$, taking the two extreme cases where the velocity $\mathbf{v}$ of the particle is (a) at right angles to $\mathbf{B}$ and (b) parallel to $\mathbf{B}$. In case (a) the particle traverses a circular orbit of fixed radius (compare the principle of the cyclotron discussed in Chapter 4) and in case (b) the path is a helix of fixed pitch along the direction of the field (compare the motion of electrons in a time projection chamber, also discussed in Chapter 4). Two techniques have been proposed to stop particle losses: magnetic 'mirrors' and a geometry that would ensure a stable indefinite circulation. In the former, it is arranged that the field in a region is greater at the boundaries of the region than in the interior. Then as the particle approaches the boundary, the force it experiences will develop a component that points into the interior where the field is weaker. Thus the particle is trapped and will oscillate between the interior and the boundaries.[11] However, most practical work has been done on case (b) and for that reason we will restrict our discussion to this technique.

The simplest configuration is a toroidal field produced by passing a current through a doughnut-shaped solenoid. In principle, charged particles in such a field would circulate endlessly, following helical paths along the direction of the magnetic field. In practice, the field would be weaker at the outer radius of the torus and the non-uniformity of the field would produce instabilities in the orbits of some particles and hence lead to particle loss. To prevent this a second field is added called a poloidal field. This produces a current around the axis of the torus and under the combined effect of both fields, charged particles in the plasma execute helical orbits about the mean axis of the torus. Most practical realizations of these ideas are devices called *tokamaks*, in which the poloidal field is generated along the axis of the torus through the plasma itself.

One of the largest tokamaks in existence is the Joint European Torus (JET), which is a European collaboration and sited at the Culham Laboratory in Berkshire, UK. A schematic view of the arrangement of the fields in JET is shown in Figure 8.8(a). This shows the external coils that generate the main toroidal field. The poloidal field is generated by transformer action on the plasma. The primary windings of the transformer are shown with the plasma itself forming the single-turn secondary. The current induced in the plasma not only generates the poloidal field, but also supplies several megawatts of resistive heating to the plasma. However, even this is insufficient to ensure a sufficient temperature for fusion and additional energy is input via other means, including rf sources.

In the inertial confinement method, small pellets of the deuterium–tritium 'fuel' mixture are bombarded with intense energy from several directions simultaneously which might, for example, be supplied by pulsed lasers. As material is ejected from the surface, other material interior to the surface is imploded, compressing the core of the pellet to densities and temperatures where fusion can take place. The laser pulses are extremely short, typically $10^{-7} - 10^{-9}$ s, which is many orders of magnitude shorter than the times associated with the pulsed poloidal current in a tokamak (which could be as long as 1s), but this is compensated for by much higher plasma densities.

Considerable progress has been made towards the goal of reaching the ignition point. However, although appropriate values of n_d, t_c, and T have been obtained

---

[11]The Van Allen radiation belts that occur at high altitudes consist of charged particles from space that have become trapped by a magnetic mirror mechanism because the Earth's magnetic field is stronger at the poles than at the equator.

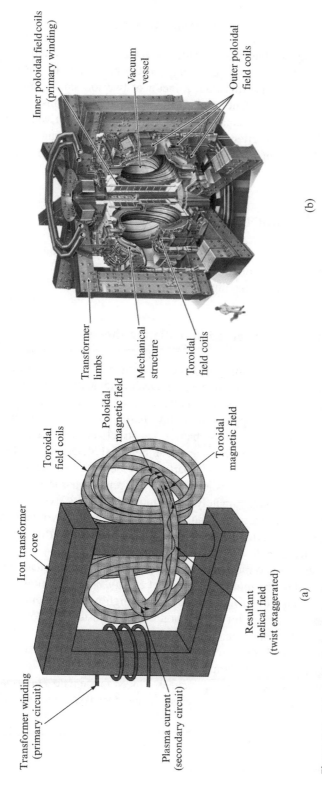

Figure 8.8 Schematic diagrams showing: (a) the main magnetic field components of the JET tokamak; (b) how these elements are incorporated into the JET device (courtesy of EFDA–JET)

separately, to date no device has yet succeeded in achieving the Lawson criterion. Tokamaks have reached the break-even point, but the best value of the Lawson ratio that has been achieved is still about a factor of two too small. Much work remains to be done on this important problem and in recognition of this at least one major new tokamak machine is planned as a global collaboration. Even when the ignition point is achieved, experience with fission power reactors suggests that it will probably take decades of further technical development before fusion power becomes a practical reality.

8.3   Biomedical Applications

The application of nuclear physics to biomedicine is a very large subject and for reasons of space we will therefore concentrate on just two topics: the therapeutic uses of radiation and medical imaging.

8.3.1   Biological effects of radiation: radiation therapy

Radiation therapy is a long-standing treatment for cancer, often combined with chemotherapy and/or surgery. By damaging DNA, the ability of the cell to reproduce is inhibited and so tumour tissue can, in principle, be destroyed. However, the same of course applies to healthy tissue so, when using radiation in a medical environment, a balance has to be struck between the potential diagnostic and/or therapeutic benefits and the potential deleterious effects of damage done by the radiation. This is a particularly delicate balance for cancer treatment because, as we shall see below, highly oxygenated tissue has a greater sensitivity to radiation and unfortunately many tumours are less oxygenated than healthy tissue and therefore more resistant to radiation. We start by reviewing the biological effects of radiation and then describe the use of various types of radiation for cancer treatment.

Exposure of living tissue to radiation is a complex process. Immediate physical damage may be caused by the initial deposition of energy, but in addition there can also be secondary damage due to the production of highly active chemicals. The latter may not be evident in full for several hours after exposure. For low levels of radiation this effect is the only one. High levels of damage may lead to the rapid death of living cells, but cells that survive in a damaged form may still have serious consequences. However caused, damage to the DNA of the nucleus of cells can result in long-term biological effects, such as cancer or genetic abnormalities, which may not reveal themselves for years, even decades, after the original exposure.[12]

---

[12]This has been known for a long time. For example, Hermann Muller was awarded the 1946 Nobel Prize in Physiology and Medicine for his discovery that mutations can be induced by X-rays.

To make descriptions like 'low-level' and 'high-level' used above meaningful needs a more detailed discussion, including the question of how dosages are defined. We will do this only very briefly. Roughly speaking, the average absorbed dose is the total energy deposited per unit mass of tissue. This is measured in 'grays', defined by $1\,\text{Gy} = 1\,\text{J}\,\text{kg}^{-1}$, which has largely replaced the older unit of the 'rad' ($1\,\text{Gy} = 100\,\text{rads}$). However, in practice, biological effects depend not only on the total dose, but also on other factors, including the type of radiation, the rate of deposition and whether the whole organ is uniformly radiated. These considerations lead to the definitions used in medical applications of *equivalent* and *effective doses*, where multiplicative weighting factors are included to take account of different types of radiation and different organs being radiated. To distinguish these latter doses from the simple absorbed dose, the sievert (Sv) unit is used, also defined as $1\,\text{J}\,\text{kg}^{-1}$ because the weighting factors are dimensionless. For example, the dose rate absorbed in tissue at a distance r from an external source of activity $\mathscr{A}$ emitting γ-rays of energy E_γ is given approximately by

$$\frac{\mathrm{d}D}{\mathrm{d}t}(\mu\text{Sv}\,\text{h}^{-1}) \approx \frac{\mathscr{A}(\text{MBq}) \times E_\gamma(\text{MeV})}{6r^2(\text{m}^2)} \tag{8.49a}$$

and for an internal source emitting radiation of energy E_R, the effective dose rate for an organ of mass M is

$$\frac{\mathrm{d}D}{\mathrm{d}t} = \frac{\mathscr{A}E_\text{R}f}{M}, \tag{8.49b}$$

where f is the fraction of the energy deposited in the organ.

To get some idea of scale, the total annual effective dose to the UK population is approximately $2600\,\mu\text{Sv}$, of which 85 per cent is due to naturally occurring background radiation, although much higher doses can occur in specific cases – for example, workers whose occupational activities expose them to radiation on a daily basis, or people who live in areas rich in granite rocks (which emit radon, the source of about half of the background radiation). The recommended limit for additional whole-body exposure of the general population is $1\,\text{mSv}\,\text{y}^{-1}$.[13]

The primary deposition of energy is due, as in non-living matter, to ionization and excitation of atoms and molecules in the path of the radiation. This occurs on a timescale of $10^{-16}\,\text{s}$ or less and was described in Chapter 4. We can draw on that discussion here, bearing in mind that living tissue consists mainly of light elements and in particular has a high proportion (about 80 per cent) of water.

For heavy particles, such as protons and α-particles, the most important process is ionization via interactions with electrons and the energy losses are given by the Bethe–Bloch formula Equation (4.11). The rate of energy loss by a heavy particle is

---

[13]For a discussion of Equations (8.49) and quantitative issues of acceptable doses for various sections of the population and to different organs, see for example, Chapter 7 of Li01 and Chapter 11 of De99.

high, peaking near the end of its range and so the penetrating power is low. For example, a 1 MeV α-particle travels only a few tens of microns and is easily stopped by skin. However, considerable damage can be caused to sensitive internal organs if an α-emitting isotope is ingested. An exception to the above is neutron radiation, which being electrically neutral does not produce primary ionization. Its primary interaction is via the nuclear strong force and it will mainly scatter from protons contained in the high percentage of water present. The scattered protons will, however, produce ionization as discussed above. The overall effect is that neutrons are more penetrating than other heavy particles and at MeV energies can deposit their energy to a depth of several centimetres. Electrons also lose energy by interaction with electrons, but the rate of energy loss is smaller than for heavy particles. Also, as they have small mass, they are subject to greater scatter and so their paths are not straight lines. In addition, electrons can in principle lose energy by bremsstrahlung, but this is not significant in the low Z materials that make up the patient. The overall result is that electrons are more penetrating than heavy particles and deposit their energy over a greater volume. Finally, photons lose energy via a variety of processes (see Section 4.4.4), the relative importance of which depends on the photon energy. Photons are very penetrating and deposition of their energy is not localized.

In addition to the physical damage that may be caused by the primary ionization process, there is also the potential for chemical damage, as mentioned above. This comes about because most of the primary interactions result in the ionization of simple molecules and the creation of neutral atoms and molecules with an unpaired electron. The latter are called *free radicals* (much discussed in advertising material for health supplements). These reactions occur on much longer timescales of about 10^{-6} s. For example, ionization of a water molecule produces a free electron and a positively charged molecule:

$$\mathrm{H_2O} \xrightarrow{\text{radiation}} \mathrm{H_2O^+} + e^- \qquad (8.50a)$$

and the released electron is very likely to be captured by another water molecule producing a negative ion:

$$e^- + \mathrm{H_2O} \rightarrow \mathrm{H_2O^-}. \qquad (8.50b)$$

Both ions are unstable and dissociate to create free radicals (denoted by black circles):

$$\mathrm{H_2O^+} \rightarrow \mathrm{H^+} + \mathrm{OH^\bullet} \qquad (8.51a)$$

and

$$\mathrm{H_2O^-} \rightarrow \mathrm{H^\bullet} + \mathrm{OH^-}. \qquad (8.51b)$$

Free radicals are chemically very active, because there is a strong tendency for their electrons to pair with one in another free radical. Thus the free radicals in

Equations (8.51) will interact with organic molecules (denoted generically by RH) to produce organic free radicals:

$$RH + OH^{\bullet} \rightarrow R^{\bullet} + H_2O \qquad (8.52a)$$

and

$$RH + H^{\bullet} \rightarrow R^{\bullet} + H_2. \qquad (8.52b)$$

The latter may then induce chemical changes in critical biological structures (e.g. chromosomes) some way from the site of the original radiation interaction that produced them. Alternatively, the radiation may interact directly with the molecule RH again releasing a free radical $R^{\bullet}$:

$$RH \xrightarrow{\text{radiation}} RH^{+} + e^{-}; \quad RH^{+} \rightarrow R^{\bullet} + H^{-}. \qquad (8.53)$$

Finally, if the irradiated material is rich in oxygen, yet another set of reactions is possible:

$$R^{\bullet} + O_2 \rightarrow RO_2^{\bullet}, \qquad (8.54a)$$

followed by

$$RO_2^{\bullet} + RH \rightarrow RO_2H + R^{\bullet}, \qquad (8.54b)$$

with the release of another free radical. This is the *oxygen effect* mentioned above that complicates the treatment of tumours.

Fortunately, for low-level radiation, living matter itself has the ability to repair much of the damage caused by radiation and so low-level radiation does not lead to permanent consequences. Indeed, if this were not so, then life may not have evolved in the way it has, because we are all exposed to low levels of naturally-occurring radiation throughout our lives (which may well have been far greater in the distant past) and the modern use of radiation for a wide range of industrial and medical purposes has undoubtedly increased that exposure. However, the repair mechanism is not effective for high levels of exposure.

In the context of radiation therapy an important quantity is the *linear energy transfer* (LET) which measures the energy deposited per unit distance over the path of the radiation. Except for bremsstrahlung, LET is the same as dE/dx discussed in Chapter 4. High-LET particles are heavy ions and α-particles, which lose their energy rapidly and have short ranges. LET values of the order of 100 keV/mm and ranges 0.1–1.0 mm are typical. Low-LET particles are electrons and photons with LET values of the order of 1 keV/mm and ranges of the order of 1 cm. Much cancer therapy work uses low-LET particles. Treatment consists of directing a beam at a cancer site from several directions to reduce the exposure of healthy tissue, while maintaining the total dose to the tumour. Other techniques include giving the dose in several stages so that the outer regions of the tumour, which are relatively oxygen-rich, are successively destroyed as they become re-oxygenated. Other treatments, particularly for localized cancers, involve the introduction of a radionuclide either physically via a needle or by ingestion/ injection of a compound containing the radionuclide. Chemicals that preferentially target specific organs or bones are commonly used.

Neutron therapy, as an example of a high-LET particle, is not widely used because of the problem of producing a strongly collimated beam plus the difficulty of ensuring that the energy is deposited primarily at the tumour site. Neutrons also share with low-LET radiation the drawback that their attenuation in matter is exponential. On the other hand, the rate of energy loss of protons and other charged particles increases with penetration depth, culminating at a maximum, the *Bragg peak*, close to the end of their range. In principle, this means that a greater fraction of the energy would be deposited at the tumour site and less damage would be caused along the path length to the site. There is also an increasing interest in using heavy charged particles. For example, carbon ions at the beginning of their path in tissue have a low rate of energy loss more like an LET particle, but near the end of their range the local ionization increases dramatically as it approaches the Bragg peak. Thus considerable energy can be deposited at a precise depth without the danger of massive destruction of healthy tissue en route to the target. Another potential advantage is that nuclear interactions along the path length will convert a small fraction of the nuclei to radioactive positron-emitting isotopes which could then be used to image the irradiated region (using the PET technique described below) and thus monitor the effectiveness of the treatment programme. Unfortunately, the use of protons and heavy particles requires access to an accelerator and for this reason proton and heavy ion therapy is not commonly used.

8.3.2 Medical imaging using radiation

There are several techniques for producing images useful for diagnostic purposes and in this section we will describe the principles of the main ones, but without technical details.[14]

Imaging using projected images

The use of an *external* source of radiation for medical imaging is of long-standing and well known. Basically, the system consists of a source placed some distance in

---

[14]A readable account of medical imaging at the appropriate level is given in Chapter 7 of Li01 and a short useful review of the whole field is He97.

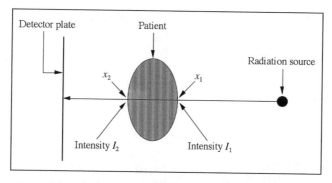

Figure 8.9 Basic layout for imaging using an external source

front the patient and a detector (usually a special type of sensitive film) placed immediately behind the patient. Because the radiation is absorbed according to an exponential law, a measurement of the intensities just before and after the patient yields information on the integrated mean free path (or equivalently the attenuation coefficient $\mu \equiv 1/\lambda$) of the photons in the body.

Thus, referring to Figure 8.9, we have for the ray shown, using Equation (4.17),

$$\ln(I_1/I_2) = \int_{x_1}^{x_2} \mu(x)\, dx. \tag{8.55}$$

The full image reveals variations of this integral only in two dimensions and thus contains no depth information. A three-dimensional effect comes from overlapping shadows in the two-dimensional images and part of the skill of a radiologist is to interpret these effects.

The most commonly used radiation is X-rays. The attenuation coefficient is dependent on the material and is greater for elements with high Z than for elements with low Z. Thus X-rays are good for imaging bone (which contains calcium with $Z = 20$), but far less useful for imaging soft tissue (which contains a high proportion of water).

Images can also be obtained using an *internal* source of radiation. This is done by the patient ingesting, or being injected with, a substance containing a radio-active γ-emitting isotope. As photon detectors are very sensitive, the concentration of the radioisotope can be very low and any risk to the patient is further minimized by choosing an isotope with a short lifetime. If necessary, the radioisotope can be combined in a compound that is known to be concentrated preferentially in a specific organ if that is to be investigated, for example iodine in the thyroid. In practice, more than 90 per cent of routine investigations use the first excited state of $^{99}_{43}\text{Tc}$ as the radioisotope. This has a lifetime of about 6 hours and is easily produced from the β-decay of $^{99}_{42}\text{Mo}$ which has a lifetime of 67 hours. The

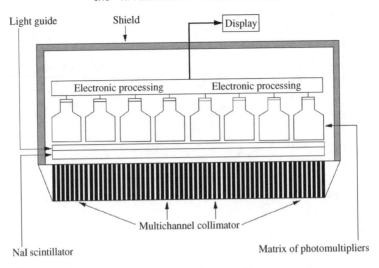

Light guide   Shield   Display

Electronic processing   Electronic processing

Multichannel collimator

NaI scintillator   Matrix of photomultipliers

Figure 8.10   Schematic diagram of a γ-camera

usefulness of this metastable state (written ^{99}Tc^m) is that it emits a single 140 keV photon with negligible β-decay modes, decaying to the very long-lived (2×10^5 years) ground state.

Because the radiation is emitted in all directions, a different technique is used to detect it. The patient is stationary and is scanned by a large-area detector consisting of a collimated single-crystal scintillator, usually NaI, the output from which is viewed by an array of photomultipliers (PMTs) via a light guide (see Section 4.4.2). A schematic diagram of such a γ-camera is shown in Figure 8.10. The output from the scintillator is received in several PMTs and the relative intensities of these signals depend on the point of origin. The signals can be analysed to locate the point to within a few millmetres. The collimator restricts the direction of photons that can be detected and, combined with the information from the PMTs, the overall spatial resolution is typically of the order of 10 mm, provided the region being examined has an attenuation coefficient that differs by at least 10 per cent from its surroundings.

Radioisotope investigations principally demonstrate function rather than anatomy, in contrast to X-ray investigations that show mainly anatomical features. Thus better images of soft tissue, such as tumours, can be obtained than those obtained using external X-rays, because the ability of the tumour to metabolize has been exploited, but the exact location of the tumour with respect to the anatomy is often lost or poorly defined.

Figure 8.11 shows part of a whole body skeletal image of a patient who had been injected with a compound MDP which moves preferentially to sites of bone cancer, labelled with the isotope ^{99}Tc^m. The image clearly shows selective take-up of the isotope in many tumours distributed throughout the body. (The concentration in the bladder is probably not significant.)

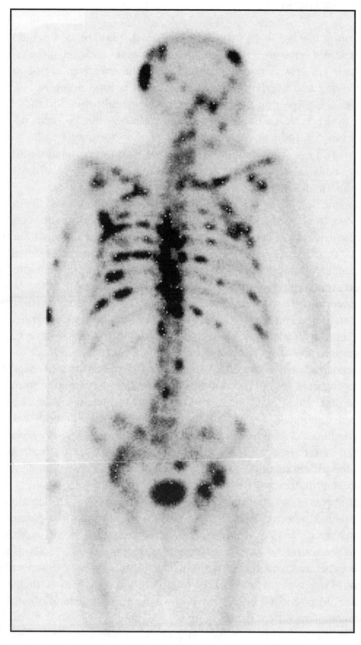

Figure 8.11   Part of a whole-body skeletal image obtained using $^{99}Tc^m$ MDP (image courtesy of Prof. R. J. Ott, Royal Marsden Hospital, London, UK)

Computed tomography

A radiographic image is a two-dimensional display of a three-dimensional structure and although the overlapping images give a useful three-dimensional effect, details are always partially obscured by the superposition of information from underlying and overlying planes. The result is loss of contrast. Thus while images from the projection methods have good spatial resolution they have poor resolution in depth. A major advance which addresses this problem was made in 1971 with the introduction of a new scanning technique called *computed tomography* (CT).[15] This enables a series of two-dimensional sections to be imaged as small as a millimetre across, even when the attenuation coefficient differs by less than 1 per cent from its surroundings.

The principle behind the CT technique is the observation that all the information needed to construct an image of a two-dimensional section of tissue is contained in the one-dimensional projections that cover all possible directions within the plane of the slice. Thus for example, if the slice is in the x–y plane, a projected image of the slice contains information on $\mu(x,y)$ in the form of a set of line integrals of μ taken through the region in a particular direction. As the angle in the plane of the slice is varied, a different representation of $\mu(x, y)$ is obtained in the form of a different set of line integrals. Once a complete set of line integrals has been obtained there are mathematical methods (including some that have been developed by particle physicists to reconstruct events from high-energy collisions) that allow the required two-dimensional function to be reconstructed. Modern high-speed computers are able to perform this construction very rapidly, so that images can now be obtained in real time and motion as fast as heartbeats can be captured.

Computed tomography may be used in conjunction with both external and internal radiation. As an example, the arrangement for a CT X-ray scan is shown schematically in Figure 8.12. In this example (known as a fourth-generation machine), the patient remains stationary within a ring of several hundred detectors (solid-state scintillators are frequently used). Within this ring there is an X-ray source that moves on another ring and provides a fan of X-rays. Each alignment of the source and a detector in the ring defines a line through the patient and the recorded count rate enables a line integral to be computed from Equation (8.55). By moving the source through its full angular range, a complete set of such line integrals is generated, enabling a two-dimensional section to be computed through the patient. This type of scanner is relatively expensive in both capital and maintenance costs and another type (known as a third-generation machine) is more common. This differs from Figure 8.12 in having a single bank of detectors opposite the source and both source and detectors are rotated to cover the full angular range. Although the CT method can produce scans of soft tissue better than conventional X-ray projections, the images are achieved at the expense of the

---

[15]The CT system was devised independently by Godfrey Hounsfield and Allan Cormack who were jointly awarded the 1979 Nobel Prize in Physiology and Medicine for their work.

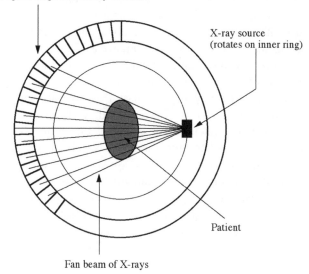

Complete ring of stationary detectors

X-ray source
(rotates on inner ring)

Patient

Fan beam of X-rays

Figure 8.12 Schematic diagram of the arrangement for a CT X-ray scanner

patient receiving a higher dose of potentially harmful radiation. An example of a
CT X-ray scan is shown in Figure 8.13(a).

CT can also be used to construct images obtained from projections from internal
radiation using radioisotopes that emit a single γ-ray. This technique is called
single-photon emission computed tomography (SPECT). The arrangement is in

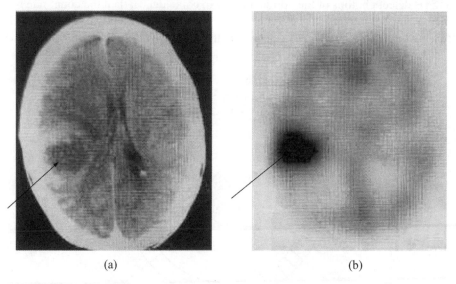

(a)   (b)

Figure 8.13 (a) X-ray CT scan of the brain, and (b) SPECT brain scan using a $^{99}Tc^m$ labelled
blood flow tracer, showing high perfusion in the tumour (indicated by arrows) (image courtesy of
Prof. R. J. Ott, Royal Marsden Hospital, London, UK)

some sense the 'inverse' of that in Figure 8.12. Thus the source is now within the patient and the fixed ring of detectors is replaced by one or more γ-cameras designed so that they can rotate in a circle about the patient. An example of an image obtained using SPECT is shown in Figure 8.13(b).

For a number of technical reasons, including the fact that the emitted radiation is isotropic, there are more stringent requirements on the γ-cameras and SPECT images have a resolution of only about 10 mm. However, although not suitable for accurate quantitative measurement of anatomy, they are of great use for clinical diagnostic work involving function. For example, the technique is used to make quantitative measurements of the functioning of an organ, i.e. clearance rates in kidneys, lung volumes, etc..

Since radionuclide imaging provides functional and physiological information, it would be highly desirable to be able to image the concentrations of elements such as carbon, oxygen and nitrogen that are present in high abundances in the body. The only radioisotopes of these elements that are suitable for imaging are short-lived positron emitters: ^{11}C (half-life ~20 min), ^{13}N (~10 min) and ^{15}O (~2 min). For these emitters, the radiation detected is the two γ-rays emitted when the positron annihilates with an electron. This occurs within a few millimetres from the point of production of the positron, whose initial energy is typically less than 0.5 MeV. The photons each have energies equal to the rest mass of an electron, i.e. 0.511 MeV and emerge 'back-to-back' to conserve momentum. This technique is called positron emission tomography (PET) and was mentioned earlier in connection with radiation treatment using heavy ions.

The arrangement of a PET scanner is shown in Figure 8.14. If the detectors D_1 and D_2 detect photons of the correct energy in coincidence, then the count rate is a measure of the integral of the source activity within the patient along the line AB passing through P. The ring of detectors defines a plane through the patient and the

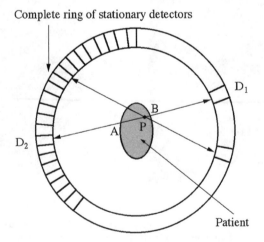

Figure 8.14   Schematic diagram of the arrangement of a PET scanner

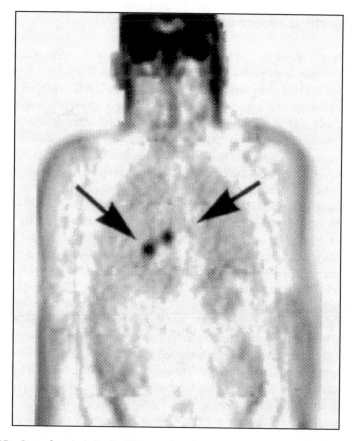

Figure 8.15   Part of a whole-body PET scan showing uptake of the chemical FDG (labelled by $^{99}Tc^m$) in lung cancer (image courtesy of Prof. R. J. Ott, Royal Marsden Hospital, London, UK)

complete set of data from all combinations of detector pairs contains all the information needed to generate the set of line integrals which can be converted into a two-dimensional image of the source using standard CT image reconstruction techniques. An example of an image using the PET technique is shown in Figure 8.15.

This account of medical imaging has ignored many technical points. For example, there are a number of corrections that have to be made to the raw data, particularly in the SPECT technique, and the most useful radioisotopes used in PET are produced in a cyclotron, so the scanner has to be near such a facility, which considerably limits is use. The interested reader is referred to specialized texts for further details.[16]

---

[16]For a more detailed discussion see, for example, De99.

8.3.3   Magnetic resonance imaging

We conclude this brief description of imaging with an account of a remarkable technique which in a relatively short time has become one of the most sophisticated tools for clinical diagnostic work and medical research. It is not only capable of producing images of unprecedented clarity, but it does so in a way that is intrinsically safe and without using potentially harmful ionising radiation.

Magnetic resonance imaging (MRI) is based on the phenomenon of nuclear magnetic resonance that was discovered independently by Bloch and Purcell and used by them to study the structure and diffusion properties of molecules.[17] It is based on the fact that the quantum spin states of nuclei (strictly their associated magnetic moments) can be manipulated by magnetic fields. A brief overview of the method is as follows. First, nuclear spins in tissue are aligned by a powerful static magnetic field, typically in the range 0.2–3 T, usually supplied by a superconducting magnet. As living tissue is predominantly water, the spins in question are mainly those of protons. (Oxygen is an even–even nucleus and so plays no role.) Secondly, oscillating magnetic field pulses at radio frequency are applied in a plane perpendicular to the magnetic field lines of the static field, which causes some of the protons to change from their aligned positions. After each pulse, the nuclei relax back to their original configuration and in so doing they generate signals that can be detected by coils wrapped around the patient. Differences in the relaxation rates and associated signals are the basis of contrast in MRI images. For example, water molecules in blood have different relaxation rates from water molecules in other tissues.

There are many different types of MRI scan, each with their own specialized procedures and the full mathematical analysis of these is complex. We will therefore give only a rather general account concentrating on the basic physics. The interested reader is referred to more detailed texts at an appropriate level.[18]

The proton has spin-$\frac{1}{2}$ and magnetic moment $\mathbf{\mu}_P$. In the absence of an external magnetic field, the two states corresponding to the two values of the magnetic quantum number $m_s = \pm\frac{1}{2}$ are equally populated and the net magnetization $\mathbf{M}$ (i.e. the average magnetic moment per unit volume) is zero. In the presence of a static magnetic field $\mathbf{B}$, taken to be in the z-direction, there is an interaction energy $(-\mathbf{\mu}_P \cdot \mathbf{B})$ and the two states have different energies with different probabilities given by the Boltzmann distribution. The energy difference between the states is $\Delta E = 2\mu_P B = hf$, where f is the Larmor (or nuclear resonance) frequency,[19]

---

[17]Felix Bloch and Edward Purcell were awarded the 1952 Nobel Prize in Physics for their discovery of nuclear magnetic resonance (NMR) and subsequent researches. Although the term NMR is still used in research environments, the term magnetic resonance imaging (MRI) is preferred in clinical environments to prevent patients associating the technique with 'harmful nuclear radiation'.

[18]See, for example, De99, McR03 and Ho97a.

[19]In general, the *nuclear resonance frequency* is defined by $f = |\mathbf{\mu}|B/jh$, where j is the spin of the particle involved and μ is its magnetic dipole moment.

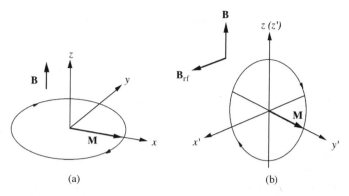

Figure 8.16 (a) Precession of the magnetization **M** in the xy-plane under the action of a torque **M** × **B** resulting from an external field **B**; (b) motion viewed in a frame of reference (x', y', z') rotating at the Larmor frequency about the z-axis — the r.f. pulse **B**$_{\rm rf}$ applied in the x'-direction has rotated **M** so that it points in the y'-direction

which is the frequency of a photon that would correspond to a transition between the two nuclear spin states. The energy difference is small. For example, for a field of 1 T, $\Delta E \approx 1.8 \times 10^{-7}$eV and f is about 43 MHz, i.e. in the radio region of the electromagnetic spectrum. Although there is a net magnetization in the z-direction, the resultant magnetization $\mathbf{M_0}$ is too small to be measured.

The situation changes, however, if **M** no longer points along the z-axis and a signal is generated if the magnetization has a component in the plane orthogonal to **B**. This is illustrated in Figure 8.16(a). In this figure, **M** has been rotated to lie in the xy-plane and since there is an angular momentum associated with the magnetization, **M** will precess about **B** under the action of the torque **M** × **B**.

The rotation can be achieved by applying an alternating r.f. magnetic field **B**$_{\rm rf}$ to the sample at right angles to **B** and at the Larmor frequency. As **M** precesses about **B**, one component of **B**$_{\rm rf}$ rotates in phase with it. The resulting motion is complicated and is best viewed in a frame of reference rotating at the Larmor frequency about the z-axis, which we label by (x', y', z') with z' parallel to z. This is shown in Figure 8.16(b). The full mathematical analysis is given, for example, in the book by Hobbie (see Footnote 18) and we will just quote the result. This is that the magnetization vector can be rotated through an arbitrary angle depending on the strength and duration of the r.f. pulse. In particular, it is possible to rotate it through 90° so that the magnetization vector precesses about the the x'-axis, i.e. rotating with a frequency that depends on the magnitude of the r.f. field. As the r.f. pulse forces all the protons to precess exactly in phase, there will be a component of magnetization along the y-axis in the rotating frame. When the r.f. pulse is turned off, the system returns to equilbrium with **M** aligned along the z-axis by re-emitting the energy absorbed from the r.f. pulse. As it does so, the external field due to **M** will vary with time with the same frequency and can be detected as an induced emf in a coil surrounding the patient. This is the basic MRI signal.

Crucially, the frequency of the external r.f. field must match the Larmor frequency of the protons to be excited.

The induced signal will decay as equilibrium is restored. If **B** were uniform throughout the selected region, all the protons would precess at the same frequency and remain in phase. In that case the interaction of the proton spins with the surrounding lattice, the so-called spin-lattice interactions, would cause **M** to relax to its equilibrium state $\mathbf{M}_0$ parallel to **B**. Under reasonable assumptions the radiated signal is proportional to the difference $(\mathbf{M}_0 - \mathbf{M})$ and decreases exponentially with a characteristic spin-lattice, or longitudinal, relaxation time T_1. Typical spin-lattice relaxation times are of the order of a few hundred milliseconds and are significantly different for different materials, such as muscle, fat and water. However, because there are always small irregularities in the field due to local atomic and nuclear effects, individual protons actually precess at slightly different rates and the signal decays because the component of **M** orthogonal to **B** (i.e. in the xy-plane) decreases as the individual moments loose phase coherence. This decrease is characterized by a second time T_2, called the spin–spin, or transverse, relaxation time. This is normally much shorter than T_1, but again varies with material. Both relaxation times can be measured.

The above assumes that the external field **B** is perfectly uniform, but of course the ideal is not realized in practice. The effects of macroscopic inhomogeneity in the magnetic field can be eliminated by generating so-called *spin echoes*, which may crudely be described as making two 'orthogonal' measurements such that the unwanted effects cancel out exactly in the sum. Many MRI imaging sequences use this technical device and again we refer the interested reader to the literature cited in Footnote 18 for further details.

All the above assumes we are scanning the whole body. The original development of the method as a medical diagnostic technique is due to the realization that gradients in the static magnetic field could be used to encode the signal with precise spatial information and be processed to generate two-dimensional images corresponding to slices through the tissue of the organ being examined.[20] The patient is placed in the fixed field **B** pointing along the z-direction. A second static field $\mathbf{B}_g$ parallel to z, but with a gradient in the z-direction is then applied so that the total static field is a function of z. This means that the Larmor frequency (which is proportional to the magnetic field) will vary as a function of z. Thus when the r.f. field $\mathbf{B}_{rf}$ is applied with a narrow band of frequencies about $f_{r.f.}$, the only protons to be resonantly excited will be those within a narrow slice of thickness dz at the particular value of z corresponding to the narrow band of frequencies. The field $\mathbf{B}_{rf}$ is applied until the magnetization in the slice has been rotated through either 90 or 180° depending on what measurements are to be taken. Both $\mathbf{B}_{rf}$ and $\mathbf{B}_g$ are then turned off.

---

[20]This discovery was first made by Paul Lauterbur and an analysis of the effect was first made by Sir Peter Mansfield. They shared the 2003 Nobel Prize in Physiology and Medicine for their work is establishing MRI as a medical diagnostic technique.

The final step is to obtain a spatial image of the magnetization as a function of x and y. This entails encoding the MRI signal with information linking it to a point of origin in real space. There are many ways this can be done (one utilizes the CT method encountered earlier) and again we refer the interested reader to the specialized texts quoted earlier for the details. The outcome is that **M** and the two relaxation times can both be measured. All three quantities vary spatially within the body and can give valuable biomedical information. For example, relaxation times are usually different for tumour tissue compared with normal tissue.

In some areas MRI scans have considerable advantages over other forms of imaging. For example, the contrast of soft tissue is much better than CT scans, leading to very high quality images, especially of the brain. Examples of such images are shown in Figure 8.17.

As ionizing radiation is not used, MRI is intrinsically safe at the field intensities used. The only exception to this is that because of the presence of high magnetic fields, care must be taken to keep all ferromagnetic objects away from the scanner. This means that patients with heart pacemakers, or metal implants cannot in general be scanned and care has to be taken to screen out people who have had an occupational exposure to microscopic fragments of steel (such as welders) as these may well have lodged in critical organs such as the eyes and the latter could be seriously damaged if the fragments moved rapidly under the action of the very strong magnetic field.

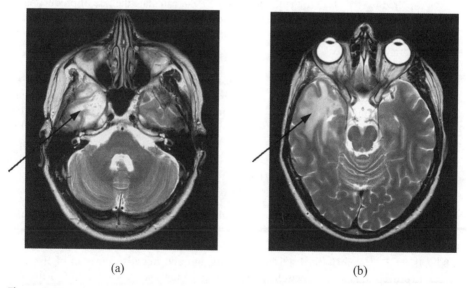

(a)   (b)

Figure 8.17 Two MRI scans of a brain — (a) T1-weighted, and (b) T2-weighted — showing a frontal lobe tumour (images courtesy of the MRI Unit of the Royal Marsden NHS Foundation Trust, London, UK)

Problems

8.1 The fission of ^{235}U is induced by a neutron and the fission fragments are $^{92}_{37}$Rb and $^{140}_{55}$Cs. Use the SEMF to calculate the energy released (in MeV) per fission. Ignore the (negligible) contributions from the pairing term. The reaction is used to power a 100 MW nuclear reactor whose core is a sphere of radius 100 cm. If an average of one neutron per fission escapes the core, what is the neutron flux at the outer surface of the reactor in $m^{-2}\,s^{-1}$? The core is surrounded by $1.3\,m^3$ of ideal gas maintained at a pressure of 1×10^5 Pa and a temperature of 298 K. All neutrons escaping the reactor core pass through the gas. If the interaction cross-section between the neutrons and the gas is 1 mb, calculate the rate of neutron interactions in the gas.

8.2 A neutron with non-relativistic laboratory speed v collides elastically with a nucleus of mass M. If the scattering is isotropic, show that the average kinetic energy of the neutron after the collision is

$$E_{final} = \frac{M^2 + m^2}{(M + m)^2} E_{initial},$$

where $m \equiv m_n$. Use this result to estimate the number of collisions necessary to thermalize neutrons from the fission of ^{235}U using a graphite moderator (assume this is pure ^{12}C).

8.3 A thermal fission reactor uses natural uranium. The energy released from fission is 200 MeV per atom of ^{235}U and the total power output is 500 MW. If all neutrons captured by ^{238}U lead to the production of ^{239}Pu, calculate the rate of production of plutonium in kg/year. The cross-sections at the relevant neutron energy are $\sigma_{capture} = 3\,b$ and $\sigma_{fission} = 600\,b$; and the relative abundance of ^{238}U to ^{235}U in natural uranium is 138:1.

8.4 In a particular thermal reactor, each fission releases 200 MeV of energy with an instantaneous power output $3\,t^{-1.2}$, where t is measured in seconds. After burning with a steady power output $P_0 = 2\,GW$ for a time T, the reactor is shut-down. Show that the mean thermal power P from a fuel rod of the reactor after time t ($> 1\,s$) is approximately

$$P(t) = 0.075P_0 \left[t^{-0.2} - (T + t)^{-0.2} \right]$$

and, taking the mean age of the fuel rods to be 1 year, calculate the power output after 6 months.

8.5 If the Sun were formed 4.6 billion years ago and initially consisted of 9×10^{56} hydrogen atoms and since then has been radiating energy via the PPI chain at a detectable rate of 3.86×10^{26} W, how much longer will it be before the Sun's supply of hydrogen is exhausted (assuming that the nature of the Sun does not change)?

8.6 In the PPI cycle, helium nuclei are produced by the fusion of hydrogen nuclei and 6.55 MeV of electromagnetic energy is produced for every proton consumed. If the electromagnetic radiation energy at the surface of the Earth is $8.4\,\mathrm{J\,cm^{-2}s^{-1}}$ and is due predominantly to the PPI cycle, what is the expected flux of solar neutrinos at the Earth in $\mathrm{cm^{-2}\,s^{-1}}$?

8.7 In a plasma of equal numbers of deuterium and tritium atoms (in practice, deuteron and triton nuclei) at an energy $kT = 10\,\mathrm{keV}$, the Lawson criterion is just satisfied for a total of 5 s. Estimate the number density of deuterons.

8.8 A thermal power station operates using inertial confinement fusion. If the 'fuel' consists of 1 mg pellets of frozen deuterium–tritium mixture, how many would have to be supplied per second to provide an output of 750 MW if the efficiency for converting the material is 25 per cent?

8.9 In some extensions of the standard model (to be discussed in Chapter 9) the proton is unstable and can decay, e.g. via $p \rightarrow \pi^0 + e^+$. If all the energy in such decays is deposited in the body and assuming that an absorbed dose of 5 Gy per year is lethal for humans, what limit does the existence of life place on the proton lifetime?

8.10 The main decay mode of $^{60}_{27}\mathrm{Co}$ is the emission of two photons, one with energy 1.173 MeV and the other with 1.333 MeV. In an experiment, an operator stands 1 m away from an open source of 40 KBq of $^{60}_{27}\mathrm{Co}$ for a total period of 18 h. Estimate the approximate whole-body radiation dose received.

8.11 A bone of thickness b cm is surrounded by tissue with a uniform thickness of t cm. It is irradiated with 140 keV γ-rays. The intensities through the bone (I_b) and through the tissue only (I_t) are measured and their ratio $R \equiv I_b/I_t$ is found to be 0.7. If the attenuation coefficients of bone and tissue at this energy are $\mu_b = 0.29\,\mathrm{cm^{-1}}$ and $\mu_t = 0.15\,\mathrm{cm^{-1}}$, calculate the thickness of the bone.

8.12 The flux of relativistic cosmic ray muons at the surface of the Earth is approximately $250\,\mathrm{m^{-2}s^{-1}}$. Use Figure 4.8 to make a rough estimate of their rate of ionization energy loss as they traverse living matter. Hence estimate in grays the annual human body dose of radiation due to cosmic ray muons.

8.13 Calculate the nuclear magnetic resonance frequency for the nucleus $^{55}_{25}\mathrm{Mn}$ in a field of 2 T if its magnetic dipole moment is 3.46 μ_N.

9

Outstanding Questions and Future Prospects

In this chapter we shall describe a few of the outstanding questions in both nuclear and particle physics and future prospects for their solution. The list is by no means exhaustive (particularly for nuclear physics, which has a very wide range of applications) and concentrates mainly on those areas touched on in earlier chapters. The examples should be sufficient to show that nuclear and particle physics remain exciting and vibrant subjects with many interesting phenomena being discovered and questions awaiting explanations.

9.1 Particle Physics

Unlike nuclear physics, particle physics does have a comprehensive theory, but although the standard model is very successful at explaining a wide range of phenomena, there are still questions that remain to be answered and some hints from experiments of phenomena that lie outside the model, for example neutrino oscillations and the possibility of lepton number violation. In addition, the success of the standard model has spurred physicists to construct theories that incorporate the strong interaction, and even in some cases gravity, in wider unification schemes. A full discussion of these topics is beyond the scope of this book, but in this chapter we will briefly review some of these questions and also look at the rapidly growing field of particle astrophysics.

9.1.1 The Higgs boson

The Higgs boson is an electrically neutral spin-0 boson whose existence is predicted by the unified electroweak theory, but which has not yet been observed.

Nuclear and Particle Physics   B. R. Martin
© 2006 John Wiley & Sons, Ltd

It is required because of a fundamental symmetry associated with theories in which the force carriers are spin-1 bosons. This symmetry is called gauge invariance and has been mentioned in previous chapters. Gauge invariance can be shown to require that the spin-1 'gauge bosons' have zero masses if they are the only bosons in the theory. This is acceptable for QED and QCD, since the gauge bosons are photons and gluons and they do indeed have zero masses. Gauge invariance also plays a crucial role in the unified electroweak theory, where it is needed to ensure the cancellation of the divergences that occur in individual higher-order Feynman diagrams. In this case the result is even stronger and it can be shown that gauge invariance requires that the *all* the fundamental particles – quarks, leptons and gauge bosons – have zero masses if gauge bosons are the only bosons in the theory. This prediction is clearly in contradiction with experiment, because the *W* and *Z* bosons have masses about 80–90 times that of the nucleon. This problem, known as the *origin of mass,* is overcome by assuming that the various particles interact with a new field, called the *Higgs field,* whose existence can be shown to allow the gauge bosons to acquire masses without violating the gauge invariance of the interaction.[1] The 'price' of this is that there must exist electrically neutral quanta associated with the Higgs field, called *Higgs bosons*, in the same way that there are quanta associated with the electromagnetic field, i.e. photons.

We saw in Chapter 3, that there is evidence that neutrinos, originally assumed to have zero masses in the standard model, are in fact not massless. The Higgs mechanism can also, in principle, be invoked to generate masses for neutrinos. However, it would be natural to expect that such masses would then be roughly the same size as the masses generated for the gauge bosons and we have seen that this is clearly not the case. This problem can only be avoided if the coupling of the neutrinos to the Higgs field is at least 12 orders of magnitude smaller than that of the coupling of the top quark. Many physicists reject such an explanation as implausible and alternative mechanisms have been suggested for generating very small neutrino masses. All have problems of their own and at present none is universally accepted. Experiments currently being planned should help resolve the matter.

The existence of the Higgs boson is the most important prediction of the standard model that has not been verified by experiment, and searches for it are of the highest priority. A problem in designing suitable experiments is that its mass is not predicted by the theory. However, its couplings to other particles *are* predicted and are essentially proportional to the masses of the particles to which it couples. The Higgs boson therefore couples very weakly to light particles like neutrinos, electrons, muons and *u, d, s* quarks; and much more strongly to heavy particles like $W^{\pm}$ and Z^0 bosons, and presumably *b* and *t* quarks. Hence attempts to produce Higgs bosons are made more difficult by the need to first produce the very heavy particles to which they couple.

---

[1]This process is called 'spontaneous symmetry breaking' and was mentioned in Chapter 6.

The failure to observe Higgs bosons in present experiments leads to limits on their mass. The best results come from the Large Electron–Positron (LEP) accelerator at CERN. This machine (which is no longer operational) had a maximum energy of 208 GeV, which is enough to produce Higgs bosons with masses up to almost 120 GeV/c^2 in the reaction

$$e^+ + e^- \rightarrow H^0 + Z^0, \tag{9.1}$$

which is expected to occur by the dominant mechanism of Figure 9.1.

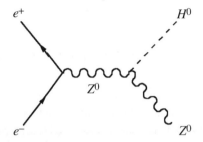

Figure 9.1   Dominant mechanism for Higgs boson production in e^+e^- annihilation

Attempts were made to detect Higgs bosons by their decays to $b\bar{b}$ pairs, where the quarks would be observed as jets containing short-lived hadrons with non-zero beauty. The results were tantalizing. By the time LEP closed down in November 2000 to make way for another project, it had shown that no Higgs bosons existed with a mass less than 113.5 GeV/c^2; and some evidence had been obtained for the existence of a Higgs boson with a mass of 115 GeV/c^2. This is very close to the upper limit of masses that were accessible by LEP, but because the Higgs boson would have a width, its mass distribution would extend down to lower energies and would give a signal. Unfortunately, while this signal was statistically likely to be a genuine result rather than a statistical fluctuation, the latter cannot be completely ruled out.

Future investigations will involve the use of new accelerators currently under construction, particularly the LHC proton–proton collider mentioned in Chapter 4. (One of the detectors at the LHC, ATLAS was shown in Figure 4.19.) This will enable searches to be made for Higgs bosons with masses up to 1 TeV/c^2 via reactions of the type

$$p + p \rightarrow H^0 + X, \tag{9.2}$$

where X is any state allowed by the conservation laws. The mechanism for this reaction is the weak interaction between the constituent quarks of the protons, an example of which is shown in Figure 9.2, where the other quarks in the protons are spectators, as usual.

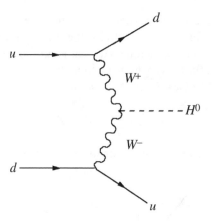

Figure 9.2   An example of a process that can produce Higgs bosons in *pp* collisions

The reaction in Equation (9.2) will take place against a large background of strong interaction processes and the method of detecting it will depend on the actual Higgs boson mass. If $M_H > 2M_W$, then the Higgs boson can decay to a pair of W-mesons or Z^0-mesons, which themselves decay. For example, from the leptonic decay of the Z^0 s, we could have overall the reaction

$$H^0 \rightarrow \ell^+ + \ell^- + \ell^+ + \ell^-, \qquad (\ell = e, \mu). \tag{9.3}$$

This would enable the mass range $200\,\text{GeV}/c^2 \le M_H \le 500\,\text{GeV}/c^2$ to be explored. However, the branching ratios are such that only a few per cent of decays will have such a distinctive signal and other decays modes will also have to be explored. For lower masses such that $M_H < 2M_W$ where these decays are energetically forbidden, one might think of looking for decays to fermion–antifermion pairs. Because the Higgs boson preferentially couples to heavy particles, the dominant decay of this type will be $H^0 \rightarrow b + \bar{b}$ with accompanying jets. This was the method used in the LEP experiments referred to above. Unfortunately, it is very difficult to distinguish these jets from those produced by other means. Rarer decay modes, but with more distinctive signals, will have to be sought, such as $H^0 \rightarrow \gamma + \gamma$, which in the standard model has a branching ratio of about 10^{-3}.

All the above is based on the standard model with a single neutral Higgs boson, but we will see in Section 9.1.3 that realistic extensions of the standard model require several Higgs bosons, not all of which are electrically neutral. Experimental investigations of the Higgs sector will undoubtedly play a central role in the future of particle physics for many years to come.

9.1.2   Grand unification

Whether or not the Higgs boson exists is the most pressing unanswered question of the standard model but, even if it is found with its predicted properties, this is not the end of the story, because one of the goals of particle theory is to have a single universal theory that explains all the phenomena of the subject. Since we already have a unified theory for the weak and electromagnetic interactions, the next logical step is to try to include the strong interaction. Attempts to do this are called *grand unified theories* (GUTs).

We have seen that unification of the weak and electromagnetic interactions does not manifest itself until energies of the order of the W and Z masses. To get some idea of the energy scale of a grand unified theory, we show in Figure 9.3 the couplings[2]

$$g \equiv 2\sqrt{2}g_W, \quad g' \equiv 2\sqrt{2}g_Z \tag{9.4}$$

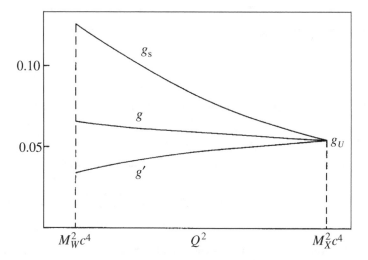

Figure 9.3 Idealized behaviour of the strong and electroweak coupling as functions of the squared energy--momentum transfer Q^2 in a simple grand unified theory; g_U is the unification coupling

and the strong coupling g_s (this is related to α_s by $\alpha_s = g_s^2/4\pi$) as functions of Q^2, the squared energy–momentum transfer in a typical GUT. A naïve extrapolation in Q^2 (using, for example, Equation (5.11)) from the region where these couplings are presently known suggests that they become approximately equal to a single value g_U at the enormous energy $Q^2 = M_X^2 c^4$, where M_X, the so-called *unification mass*, is of the order of 10^{15} GeV/c^2. In practice, which couplings to extrapolate depends on which version of GUT one considers, but if the extrapolation is done accurately the three curves actually fail to meet at a point by an amount that cannot be explained by uncertainties in the models.

---

[2]Recall that the electromagnetic coupling e is related to these couplings by the unification condition Equation (6.71).

There are many potential grand unified theories, but the simplest incorporates the known quarks and leptons into common families. For example, one way is to put the three coloured d-quarks and the doublet $(e^+, \bar{\nu}_e)$ (strictly their right-handed components) into a common family, i.e.

$$(d_r, d_b, d_g, e^+, \bar{\nu}_e). \tag{9.5}$$

The fundamental vertex interactions allowed in this model are shown in Figure 9.4.

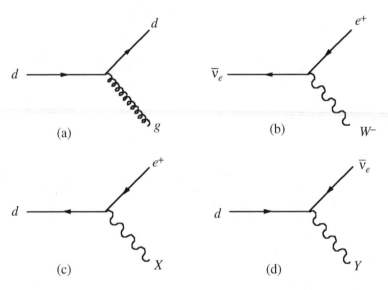

Figure 9.4   Fundamental vertices that can occur for the multiplet of particles in Equation (9.5)

In addition to the known QCD interaction in (a) and the electroweak interaction in (b), there are two new interactions represented by (c) and (d) involving the emission or absorption of two new gauge bosons X and Y with electric charges $-\frac{4}{3}$ and $-\frac{1}{3}$, respectively, and masses of the order of M_X. In this theory all the processes of Figure 9.4 are characterized by a single GUT coupling given by

$$\alpha_U \equiv \frac{g_U^2}{4\pi} \approx \frac{1}{42}, \tag{9.6}$$

which is found by extrapolating the known coupling of the standard model to the energy $M_X c^2$.

This simple model has a number of attractive features. For example, it can be shown that the sum of the electric charges of all the particles in a given multiplet is zero. So, using the multiplet $(d_r, d_b, d_g, e^+, \bar{\nu}_e)$, it follows that

$$3q_d + e = 0, \tag{9.7}$$

where q_d is the charge of the down quark. Thus $q_d = -e/3$ and the fractional charges of the quarks is seen to originate in the fact that they exist in three colour states. By a similar argument, the up quark has charge $q_u = 2e/3$ and so with the usual quark assignment $p = uud$, the proton charge is given by

$$q_p = 2q_u + q_d = e. \tag{9.8}$$

Thus, we also have an explanation of the long-standing puzzle of why the proton and positron have precisely the same electric charge.

GUTs make a number of predictions that can be tested at presently accessible energies. For example, if the three curves of Figure 9.3 really did meet at a point, then the three low-energy couplings of the standard model would be expressible in terms of the two parameters α_U and M_X. This could be used to predict one of the former, or equivalently the weak mixing angle θ_W. The result is $\sin^2 \theta_W = 0.214 \pm 0.004$, which is close to the measured value of 0.2313 ± 0.0003, although not strictly compatible with it. (This is true even if the effect of the Higgs boson is taken into account when evaluating the evolution of the coupling constants.)

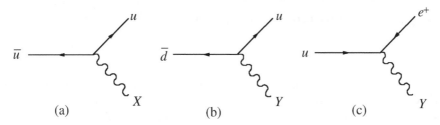

Figure 9.5 The three fundamental vertices predicted by the simplest GUT involving the gauge bosons X and Y (these are in addition to those shown in Figure 9.4)

In addition to the interactions of the X and Y bosons shown in Figure 9.4, there are a number of other possible vertices, which are shown in Figure 9.5. (There is also another set where particles are changed to antiparticles.) A consequence of these interactions and those of Figure 9.4(c) and (d) is the possibility of reactions that conserve neither baryon nor lepton numbers. The most striking prediction of this type is that the proton would be unstable, with decay modes such as $p \rightarrow \pi^0 + e^+$ and $p \rightarrow \pi^+ + \bar{\nu}_e$. Examples of Feynman diagrams for these decays are shown in Figure 9.6 and are constructed by combining the vertices of Figure 9.4 and 9.5. In all such processes, although lepton number L and baryon number B are not conserved, the combination

$$R \equiv B - \sum_\ell L_\ell \quad (\ell = e, \mu, \tau) \tag{9.9}$$

is conserved.

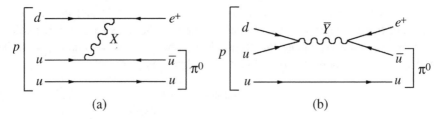

Figure 9.6 Examples of processes that contribute to the proton decay mode $p \rightarrow \pi^0 + e^+$

Since the masses of the X and Y bosons are far larger than the quarks and leptons, we can use the zero-range approximation to estimate the lifetime of proton decay. In this approximation, and by analogy with the lifetime for the muon Equation (7.62), we have for the proton lifetime

$$\tau_p \approx \frac{(M_X c^2)^4}{g_U^4 (M_p c^2)^5}.$$
(9.10)

Taking account of reasonable uncertainties on g_U and M_X, this gives

$$\tau_p \approx 10^{30 \pm 1} \text{ years}.$$
(9.11)

Proton decay via these modes has been looked for experimentally. The most extensive search has been made using the Kamiokande detector described in Chapter 4. To date no events have been observed and this enables a lower limit to be put on the proton lifetime of about 10^{32} years, which rules out the simplest version of a grand unified theory. However, there are other, more complicated, versions that still cannot be completely ruled out by present experiments. Some of these incorporate the idea of *supersymmetry* which is described below.

Finally, GUTs may offer an explanation for the very small neutrino masses observed in the oscillation experiments discussed in Chapter 3. In Section 6.3 we discussed the possibility that the neutrino was its own antiparticle (a so-called Majorana neutrino). In GUTs the right-handed neutrino states are predicted to be very massive (of order 10^{17} GeV/c^2) and mix with the massless left-handed neutrinos of the standard model to give physical neutrinos with masses

$$m_\nu \sim m_L^2 / M_X,$$
(9.12)

where m_L is the typical mass of a charged lepton or quark.

9.1.3 Supersymmetry

One of the problems with GUTs is that if there are new particles associated with the unification energy scale, then they would have to be included as additional contributions in the higher-order calculations in the electroweak theory, for example for the mass of the W-boson. These contributions would upset the delicate cancellations that ensure finite results from higher-order diagrams in the standard model, unless there were some way of cancelling these new contributions. Supersymmetry (SUSY) does exactly this.

Supersymmetry is the proposal that every known elementary particle has a partner, called a *superpartner*, which is identical to it all respects except its spin. Spin-$\frac{1}{2}$ particles have spin-0 superpartners and spin-1 particles have spin-$\frac{1}{2}$ superpartners. To distinguish between a spin-$\frac{1}{2}$ particle and its superpartner, an 's' is attached to the front of its name in the latter case. Thus, for example, a spin-$\frac{1}{2}$ electron has a spin-0 *selectron* as its superpartner. The full set of elementary particles and their superpartners in the simplest SUSY model (the so-called Minimal Supersymmetric Standard Model – MSSM) is shown in Table 9.1. (This is actually a simplification because even the simplest SUSY requires a number of different Higgs bosons, not all electrically neutral.)

Table 9.1 The particles of the MSSM and their superpartners

Particle	Symbol	Spin	Superparticle	Symbol	Spin
Quark	q	$\frac{1}{2}$	Squark	$\tilde{q}$	0
Electron	e	$\frac{1}{2}$	Selectron	$\tilde{e}$	0
Muon	μ	$\frac{1}{2}$	Smuon	$\tilde{\mu}$	0
Tauon	τ	$\frac{1}{2}$	Stauon	$\tilde{\tau}$	0
W-boson	W	1	Wino	$\tilde{W}$	$\frac{1}{2}$
Z-boson	Z	1	Zino	$\tilde{Z}$	$\frac{1}{2}$
Photon	γ	1	Photino	$\tilde{\gamma}$	$\frac{1}{2}$
Gluon	g	1	Gluino	$\tilde{g}$	$\frac{1}{2}$
Higgs boson	H	0	Higgsino	$\tilde{H}$	$\frac{1}{2}$

If the symmetry were exact then a particle and its superparticle would have equal masses. This is clearly not the case or such states would have already been found. So supersymmetry is at best an approximate symmetry of nature. Nevertheless, even in an approximate symmetry, the couplings of the two states are equal and opposite, thereby ensuring the required cancellation, providing their masses are not too large. In practice, it is usually assumed in GUTs that incorporate supersymmetry that the masses of the superparticles are of the same order as the masses of the W and Z bosons. With the inclusion of superparticles, the evolution of the coupling constants of the standard model as functions of Q^2 changes slightly and when extrapolated they meet very close to a single point. The unification mass is increased somewhat to about 10^{16} GeV/c^2, while the value of g_U remains roughly constant. Thus the predicted lifetime of the proton is increased to about $10^{32} - 10^{33}$ years, conveniently beyond the 'reach' of current experiments. At the same time, the value of the weak mixing angle is brought into almost exact agreement with the measured value. Whether this is simply a coincidence or not is unclear.

To verify supersymmetry it will of course be essential to detect the superparticles and that will not be easy. For example, the virtual exchange of superparticles could contribute to the deviation of the muon magnetic dipole moment from its Dirac value, although it would be difficult to separate these contributions from other corrections. To date, activity has concentrated on the direct detection of superparticles in reactions. In the simplest version of a SUSY theory, super-particles are produced in pairs (like leptons or strange particles in strong interactions, i.e. associated production) so that the decay of a superparticle must have at least one superparticle in the final state and the lightest such particle will necessarily be stable. Most versions of SUSY theories assume that the lightest particle will be a *neutralino* $\tilde{\chi}_0$, which is the name given to a mixture of the photino, the higgsino and the zino, the three spin-$\frac{1}{2}$ superparticles that interact purely by the electroweak interaction. If this is the case, a possible reaction that could be studied is

$$e^+ + e^- \rightarrow \tilde{e}^+ + \tilde{e}^-, \tag{9.13}$$

followed by the decays

$$\tilde{e}^\pm \rightarrow e^\pm + \tilde{\chi}^0, \tag{9.14}$$

giving overall

$$e^+ + e^- \rightarrow e^+ + e^- + \tilde{\chi}_0 + \tilde{\chi}_0. \tag{9.15}$$

The cross-section for Equation (9.13) is predicted to be comparable to that for producing pairs of ordinary charged particles. As the neutralinos only have weak interactions they will be undetectable in practice and so the reaction would be characterized by e^+e^- pairs in the final state with only a fraction (typically 50 per cent) of the initial energy and not emerging 'back-to-back' (because it is not a two-body reaction). This and many other reactions have been studied, mainly in experiments at LEP, but to date no evidence for the existence of superparticles has been found. The null results have enabled lower limits to be set on the masses of neutralinos and sleptons of various flavours in the range, $40 - 100$ GeV/c^2. This is not very useful in practice, as the masses are believed to be of the order of the W and Z masses. Much larger lower limits for the masses of gluinos and squarks have been obtained in experiments using the CDF detector that was described in Chapter 4 (see Figure 4.18). The search for supersymmetric particles will be a major activity of detectors at the LHC accelerator currently under construction at CERN.[3]

Undeterred by the lack of immediate success of supersymmetry, some bold physicists have attempted to incorporate gravity into even larger unified schemes.

---

[3]For a review of the current state of experimental searches for superparticles see, for example, Ei04.

The problems here are formidable, not least of which is that the divergences encountered in trying to quantize gravity are far more severe than those in either QCD or the electroweak theory and there is at present no successful 'stand-alone' quantum theory of gravity analogous to the former two. The theories that have been proposed that include gravity invariably replace the idea of point-like elementary particles with tiny quantized *strings* as a device to reduce these technical problems and are formulated in many more dimensions (usually 10 or 11) than we observe in nature. More recently, even strings have been superceded by theories based on mathematical objects called membranes, or simply *branes*. The problem with these theories, leaving aside their formidable mathematical complexity, is that they apply at an energy scale where gravitational effects are comparable to those of the gauge interactions, i.e. at energies defined by the so-called *Planck* mass M_P, which is given by

$$M_P = \left(\frac{\hbar c}{G}\right)^{1/2} = 1.2 \times 10^{19} \, \text{GeV}/c^2, \qquad (9.16)$$

where G is the gravitational constant.[4] This energy is so large that it is difficult to think of a way that the theories could be tested at currently accessible energies, or even indeed at energies accessible in the conceivable future. Their appeal at present is, therefore, the mathematical beauty and 'naturalness' that their sponsors claim for them. Needless to say, experimentalists will remain sceptical until definite experimental tests can be suggested and carried out.

9.1.4 Particle astrophysics

Particle physics and astrophysics interact in an increasing number of areas and the resulting field of particle astrophysics is a rapidly expanding one. The interactions are particularly important in the field of cosmology where, for example, the detection of neutrinos can provide unique cosmological information. Another reason is because the conditions in the early Universe implied by standard cosmological theories (the big bang model) can only be approached, however remotely, in high-energy particle collisions. At the same time, these conditions occurred at energies that are relevant to the grand unified and SUSY theories of particle physics and so offer a possibility of testing the predictions of such theories. This is important because, as mentioned above, it is difficult to see other ways of testing such predictions. For reasons of space, we will discuss just three examples of particle astrophysics. We will return to the question of conditions in the early universe in Section 9.2.2.

---

[4]This implies that strings have dimensions of order $\ell_P \sim \hbar/M_P c = 1.6 \times 10^{-35}$ m.

Neutrino astrophysics

We have seen in Chapter 3 that cosmic rays and emissions from the Sun are important sources of information about neutrinos and have led us to revise the view that neutrinos are strictly massless, as is assumed in the standard model. At the same time, there is considerable interest in studying ultra high-energy neutrinos as a potential source of information about galactic and extra-galactic objects and hence cosmology in general.

One of the first neutrino astrophysics experiments was the observation of neutrinos from a supernova. Supernovas are very rare events where a star literally explodes with a massive output of energy over a very short timescale measured in seconds. The mechanism for this (briefly) is as follows. If a star has a mass greater than about 11 solar masses, it can evolve through all stages of fusion, ending in a core of iron surrounded by shells of lighter elements. Because energy cannot be released by the thermonuclear fusion of iron, the core will start to contract under gravity. Initially this is resisted by the pressure of the dense gas of degenerate electrons in the core (*electron degeneracy pressure*), but as more of the outer core is burned and more iron deposited in the core, the resulting rise in temperature makes the electrons become increasingly relativistic. When the core mass reaches about 1.4 solar masses (the so-called *Chandrasekhar limit*), the electrons become ultra relativistic and they can no longer support the core. At this point the star is on the brink of a catastrophic collapse.

The physical reactions that lead to this are as follows. Firstly, photodisintegration of iron (and other nuclei) takes place,

$$\gamma + {}^{56}\mathrm{Fe} \rightarrow 13\,{}^{4}\mathrm{He} + 4n, \qquad (9.17)$$

which further heats the core and enables the photodisintegration of the helium produced, i.e.

$$\gamma + {}^{4}\mathrm{He} \rightarrow 2p + 2n. \qquad (9.18)$$

As the core continues to collapse, the energy of the electrons present increases to a point where the weak interaction

$$e^{-} + p \rightarrow n + \nu_e \qquad (9.19)$$

becomes possible and eventually the hadronic matter of the star is predominantly neutrons. This stage is therefore called a *neutron star*. The collapse ceases when the gravitational pressure is balanced by the neutron degeneracy pressure. At this point the radius of the star is typically just a few kilometres. The termination of the collapse is very sudden and as a result the core material produces a shock wave that travels outwards through the collapsing outer material, leading to a supernova (actually a so-called Type II supernova). Initially there is an intense burst of ν_e

with energies of a few MeV from the reaction of Equation (9.19). This lasts for a few milliseconds because the core rapidly becomes opaque even to neutrinos and after this the core material enters a phase where all its constituents (nucleons, electrons, positrons and neutrinos) are in thermal equilibrium. In particular, all flavours of neutrino are present via the reactions

$$\gamma \rightleftharpoons e^+e^- \rightleftharpoons \nu_\ell\bar{\nu}_\ell, \qquad (\ell = e, \mu, \tau) \tag{9.20}$$

and these will eventually diffuse out of the collapsed core and escape. Neutrinos of all flavours, with average energies of about 15 MeV, will be emitted in all directions over a period of 0.1–10 s. Taken together, the neutrinos account for about 99 per cent of the total energy released in a supernova. Despite this, the output in the optical region is sufficient to produce a spectacular visual effect.

The first experiments to detect neutrinos from a supernova were an earlier version of the Kamiokande experiment described in Chapter 3 and the IMB collaboration, which also used a water Čerenkov detector. Both had been constructed to search for proton decay as predicted by GUTs, but by good fortune both detectors were 'live' in 1987 at the time of a spectacular supernova explosion (now named SN1987A) and both detected a small number of antineutrino events. The data are shown in Figure 9.7. The Kamiokande experiment detected 12 $\bar{\nu}_e$ events and the IMB experiment eight events, both over a time interval of approximately 10 s and with energies in the range 0–40 MeV. These values are consistent with the estimates for the neutrinos that would have been produce by the reaction in Equation (9.20) and then diffused from the supernova after the initial pulse.

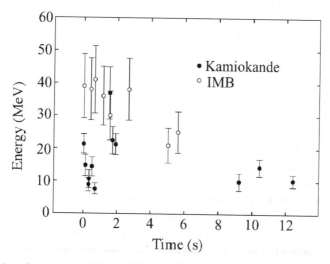

Figure 9.7 Data for neutrinos from SN1987A detected in the Kamiokande and IMB experiments: the threshold for detecting neutrinos in the experiments are 6 MeV (Kamiokande) and 20 MeV (IMB) -- in each case the first neutrino detected is assigned the time zero

The data can be used to make an estimate of the neutrino mass as follows. The time of arrival on Earth of a neutrino i is given by

$$t_i = t_0 + \left(\frac{L}{c}\right)\left[1 + \frac{m^2 c^4}{2E_i^2}\right], \tag{9.21}$$

where t_0 is the time of emission from the supernova and (m, E_i) are the mass and total energy of the neutrino. Thus

$$(\Delta t)_{ij} \equiv t_i - t_j = \frac{L m^2 c^4}{2c}\left[\frac{1}{E_i^2} - \frac{1}{E_j^2}\right]. \tag{9.22}$$

Using data for pairs of neutrinos, Equation (9.22) leads to the result

$$m_{\bar{\nu}_e} \leq 20\,\mathrm{eV}, \tag{9.23}$$

which, although larger than the value from tritium decay, is still a remarkable measurement.

The neutrinos from SN1987A were of low energy, but there is also a great interest in detecting ultra high-energy neutrinos. For example, it is known that there exist point sources of γ-rays with energies in the TeV range, many of which have their origin within so-called *active galactic nuclei*. It is an open question whether this implies the existence of point sources of neutrinos with similar energies. The neutrinos to be detected would be those travelling upwards through the Earth, as the signal from downward travelling particles would be swamped by neutrinos produced via pion decay in the atmosphere above the detector. Like all weak interactions the intrinsic rate would be very low, especially so for such high-energy events, but this is partially compensated by the fact that the ν–nucleon cross-section increases with energy, as we showed in Chapter 6.

To detect neutrinos in the TeV energy range using the Čerenkov effect in water requires huge volumes, orders-of-magnitude larger than used in the Super-Kamiokande detector. An ingenious solution to this problem is to use the vast quantities of water available in liquid form in the oceans, or frozen in the form of ice at the South Pole, and several experiments have been built, or are being built, using these sources. The largest so far is the Antartic Muon and Neutrino Detector Array (AMANDA) which is sited at the geographical South Pole. A schematic diagram of this detector is shown in Figure 9.8.

The detector consists of strings of optical modules containing photomultiplier tubes that convert the Čerenkov radiation to electrical signals. The enlarged inset in Figure 9.8 shows the details of an optical module. They are located in the ice at great depths by using a novel hot-water boring device. The ice then refreezes around them. In the first phase of the experiment in 1993/94 (AMANDA-A) four detector strings were located at depths of between 800 and 1000 m. The ice at

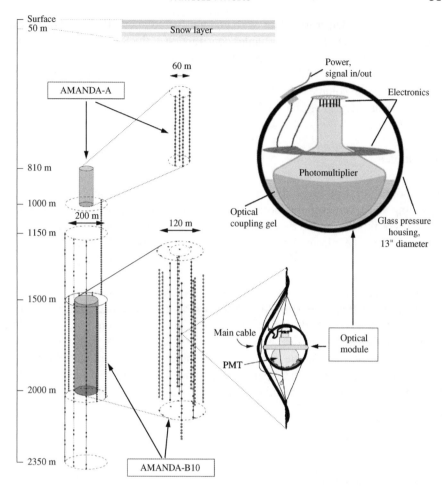

Figure 9.8   A schematic diagram of the AMANDA neutrino detector

these depths is filled with air bubbles and so the detectors are not capable of precision measurements, but they proved the validity of the technique. In the next phase a few years later (AMANDA-B10), 10 more strings containing 320 optical modules were located at depths between 1.5 and 2.0 km, where the properties of ice are suitable for muon detection. Finally, the current version of the detector (AMANDA-II) has an additional nine strings extending to a depth of 2.35 km. In total there are 680 optical modules covering a cylindrical volume with a cross-sectional diameter of 120 m.

The AMANDA detector has successfully detected atmospheric neutrinos and has produced the most detailed map of the high-energy neutrino sky to date. However, no source of continuous emission has yet been observed that would be a candidate for a point source.

AMANDA can detect neutrinos with energies up to about 10^{15} eV, but an even bigger detector, called IceCube, is under construction at the South Pole. This uses 80 strings each containing 60 optical modules regularly spaced over an area of 1 km² at depths of between 1.4 and 2.4 km (the volume covered by AMANDA is only 1.5 per cent of the volume to be covered by IceCube) and will be capable of detecting neutrinos with energies as high as 10^{18} eV. IceCube is due for completion in 2010.

Dark matter

The modern description of the universe is based on the observation that it is expanding and assumes that the origin of this is a sudden explosion at some time in the past. For this reason the description is called the *big bang model*. However, this does not mean an explosion from a singular space–time point. Because the universe appears isotropic at large distance scales, there can be no preferred points in space and so the big bang must have occurred everywhere at once, thus ensuring that the expansion appears the same to all observers irrespective of their locations. Two pieces of evidence for this model are the existence of a cosmic background radiation consistent with a black-body spectrum at an effective temperature of 2.7 K, and the cosmic abundance of light elements.[5] Whether the expansion will continue indefinitely depends on the average density of the universe ρ. The critical density ρ_c at present times, below which the expansion will continue indefinitely, and above which it will eventually halt and the universe start to contract, can be written

$$\rho_c = \frac{3H_0^2}{8\pi G} \sim 10^{-26}\, \mathrm{kg\ m^{-3}} \approx 5.1\,(\mathrm{GeV}/c^2)\mathrm{m}^{-3}, \tag{9.24}$$

where G is the gravitational constant and we have used the best current value for Hubble's constant H_0 to evaluate Equation (9.24). In the most popular version of the model, called the *inflationary big bang model*, the relative density

$$\Omega \equiv \rho/\rho_c = 1. \tag{9.25}$$

The relative density is conveniently written as the sum of three components,

$$\Omega = \Omega_r + \Omega_m + \Omega_\Lambda, \tag{9.26}$$

where Ω_r is the contribution due to radiation, Ω_m is that due to matter and Ω_Λ is related to a term in the equation governing the evolution of the universe that

---

[5]For an accessible discussion of the big bang model and other matters discussed in this section see, for example, Pe03.

contains a so-called *cosmological constant* Λ. The latter contribution can be estimated from various cosmological observations, including recently measured temperature fluctuations in the microwave background radiation. Its value is about 0.7 and is the largest contribution to Ω. The value of the radiation term is of the order of 10^{-5} and so makes a negligible contribution to Ω. Finally, the total matter density contribution can be deduced from the gravitational energy needed for consistency with observations on the rotation of galaxies and the kinematics of large-scale structures in the universe. Its value is about 0.3. Thus we see that the value of Ω is consistent with Equation (9.25), although the uncertainties are considerable. An unsatisfactory feature is that the origin of the largest term, also referred to as *dark energy*, is totally unknown.

The contribution of baryons to the mass term may be inferred from knowledge of how nuclei are formed in the universe (*nucleosynthesis*) and its value is about 0.05, of which only about 20 per cent is accounted for in the form of stars, gas and dust, i.e. in the form of visible luminous baryonic matter. There could be other sources of non-luminous baryonic matter, for example in the form of brown dwarfs and small black holes the size of planets, and there is experimental evidence that such 'massive, compact halo objects' (MACHOs) do indeed exist, but in unknown quantities. However, it is not thought that they alone can account for the 'missing' matter. Thus we are forced to conclude that the bulk of matter, as much as 85 per cent, is non-baryonic. It is referred to collectively as *dark matter*.

There are several dark matter candidates. Massive neutrinos might be one possibility. Such particles would have to be heavy enough to have been non-relativistic in the early stages of the universe (so-called *cold dark matter*), because if they were relativistic (*hot dark matter*) they would have rapidly dispersed, giving rise to a uniform energy distribution in space. Calculations suggest that in this case there would have been insufficient time for the observed galaxies to have formed. Although neutrinos may still play a minor role in contributing to the matter deficit, it is now believed that the bulk of the contribution comes from cold dark matter in the form of 'weakly interacting massive particles' (WIMPs). Although there are no known particles that have the required properties, for various reasons the most likely candidates are SUSY particles and in particular the lightest such state, usually taken to be the neutralino.

Experiments such as AMANDA can search for WIMPs, but they were not designed to do so as a priority. However, several dedicated experiments have been mounted to detect WIMPs by detecting the recoil energy of interacting nuclei, which is about 50 keV. Such recoils can, in principle, be detected in a number of ways. For example, in semiconductors such as GaAs, free charge will be produced that can be detected electronically; in a scintillator such as NaI the emission of photons can be detected using photomultipliers; and in crystals at low temperatures the energy can be converted to phonons that can be detected by a very small rise in temperature. In practice, the problems are formidable because of the very low expected event rate. This can be calculated from the expected WIMP velocities and assumed masses. For example, if WIMPs are identified with neutralinos, then expectations range from 1–10 events per kg of detector per week. This is very small compared with the event rate from naturally occurring radioactivity, including that in the materials of the detectors themselves. The former is minimized by working deep underground to shield the detector from cosmic rays and in areas with geological structures where radioactive rocks are absent; and the latter is minimized by building detectors of extreme purity. Finally, WIMP recoils should exhibit a small seasonal time variation due to the motion of the Earth around the Sun and the motion of the Sun within the galaxy. One experiment claims to have seen this variation. Present experiments are at an early stage, but some versions of SUSY theories with low-mass neutralinos can probably already be ruled out.[6]

Matter–antimatter asymmetry

One of the most striking facts about the universe is the paucity of antimatter compared with matter. There is ample evidence for this. For example, cosmic rays are overwhelmingly composed of matter and what little antimatter is present is compatible with its production in intergalactic collisions of matter with photons. Neither do we see intense outbursts of electromagnetic radiation that would accompany the annihilation of clouds of matter with similar clouds of antimatter. The absence of antimatter is completely unexpected because, in the original big bang, it would be natural to assume a total baryon number $B = 0$.[7] Then during the period when kT was large compared with hadron energies, baryons and antibaryons would be in equilibrium with photons via reversible reactions such as

$$p + \bar{p} \rightleftharpoons \gamma + \gamma \qquad (9.27)$$

and this situation would continue until the temperature fell to a point where the photons no longer had sufficient energy to produce $p\bar{p}$ pairs and the expansion had proceeded to a point where the density of protons and antiprotons was such that their mutual annihilation became increasingly unlikely. The critical temperature is $kT \approx 20$ MeV and at this point the ratios of baryons and antibaryons to photons 'freezes' to values that can be calculated to be

$$N_B/N_\gamma = N_{\bar{B}}/N_\gamma \sim 10^{-18}, \qquad (9.28)$$

---

[6]An up-to-date review of the status of dark matter searches is given in Pe03 and Ei04.

[7]One could of course simply bypass the problem by arbitrarily assigning an initial non-zero baryon number to the universe, but it would have to be exceedingly large to accommodate the observed asymmetry, as well as being an unaesthetic solution.

with of course $N_{\bar{B}}/N_B = 1$. These ratios would then be maintained in time, whereas the actual observed ratios are

$$N_B/N_\gamma \approx 10^{-9}, \quad N_{\bar{B}}/N_\gamma \sim 10^{-13}, \tag{9.29}$$

with $N_{\bar{B}}/N_B \sim 10^{-4}$. The simple big bang model fails spectacularly.

The conditions whereby a baryon–antibaryon asymmetry could arise were first stated by Sakharov. It is necessary to have: (a) an interaction that violates baryon number conservation, (b) an interaction that violates charge conjugation, and (c) a non-equilibrium situation must exist at some point to 'seed' the process. We have seen in Chapter 6 that there is evidence that CP is violated in the decays of some neutral mesons, but its source and size are not compatible with that required for the observed baryon–antibaryon asymmetry and we must conclude that there is another, as yet unknown, source of CP violation. Likewise a method for generating a non-equilibrium situation is also unknown, although it may be that the baryon-violating interactions of GUTs, which are necessary for condition (a), may provide one. Clearly, matter–antimatter asymmetry remains a serious unsolved problem.

9.2 Nuclear Physics

Despite more than a century of research, nuclear physics is by no means a 'closed' subject. Even the basic strong nucleon–nucleon force is not fully understood at a phenomenological level, let alone in terms of the fundamental quark–gluon strong interaction. Indeed one of the outstanding problems of nuclear physics is to understand how models of interacting nucleons and mesons arise as approximations to the quark–gluon picture of QCD and where these two descriptions merge. A related question is whether the nuclear environment modifies the quark–gluon structure of nucleons and mesons. It follows from our lack of knowledge in these areas that the properties of nuclei cannot at present be calculated from first principles, although some progress has been made in this direction. Meanwhile, in the absence of a fundamental theory to describe the nuclear force, we have seen in earlier chapters that specific models and theories are used to interpret the phenomena in different areas of nuclear physics. Current nuclear physics models must break down at very high energy-densities and at sufficiently high temperatures the distinction between individual nucleons in a nucleus should disappear. This is the regime that is believed to have existed in the very early times of the universe and is of great interest to astrophysicists.

Nuclear physics is a mature subject and has implications in many other areas of physics and wide applications in industry, biology and medicine that are at the core of the subject. Examples include: the nuclear physics input required to understand many processes that occur in cosmology and astrophysics, such as supernovae and the production of chemical elements; the many applications of NMR, such as studies of protein structure and its use in medical diagnostics; and industrial applications such as the production of power. In Chapter 8 we touched on just three applications and the 'applied' problems to be solved in those – safe disposal of nuclear waste, better medical imaging diagnostics and therapeutics, controlled nuclear fusion, etc. – are as demanding as the 'fundamental' ones, simply different. They are also vitally important for the future well-being of everyone. In the sections that follow we will take a brief look at a few of these pure and applied problems.[8]

9.2.1   The structure of hadrons and nuclei

In the standard model, the structure of nucleons is specified in terms of quarks and gluons, but questions remain. One concerns the spin of the proton. This must be formed from combining the spins and the relative orbital angular momenta of its constituent quarks and gluons. Measuring these various contributions can be done in deep inelastic scattering experiments of the type described in Chapter 5, but using spin-polarized targets, sometimes with spin-polarized beams. Experiments to date have shown the surprising result that the spins of all the quarks and antiquarks together contribute only about 20–30 per cent to the total spin of the proton (the so-called 'proton spin crisis'). There is some information that the angular momentum contributions of the quarks play an important role, but very little is known about the contribution of the total angular momentum of the gluon. This is an area where the type of experiment that can be pursued at the CEBAF accelerator described in Section 4.2.2 will be vital in unravelling the details of each contribution and thus further testing QCD.

Nucleons and mesons are the building blocks of nuclear matter, but there is no guarantee that the properties of these particles in nuclei are identical to those exhibited as free particles. According to QCD the properties of hadrons are strongly influenced by the sea of quark–antiquark pairs and gluons that we have seen in Chapter 5 are always present around confined quarks due to quantum fluctuations. However, these influences could well be different in the case of closely spaced nucleons in nuclear matter from those for a free nucleon. Indeed there are theoretical predictions that the probability of finding a $q\bar{q}$ pair decreases as the density of the surrounding nuclear matter increases. If such effects could be established they would have a profound influence on our understanding of quark confinement.

---

[8]A comprehensive overview of the field as at 1999 is a report of the Board on Physics and Astronomy of the National Research Council, USA: *"Nuclear Physics: The Core of Matter, The Fuel of Stars"*, National Academy Press, Washington, D.C. (1999) – NRC99. Other useful sources are the publications of the Nuclear Physics European Collaboration Committee (NuPecc) and in particular its *"Report on Impact, Applications, Interactions of Nuclear Science"* (2002) and the NuPecc Long-Range Plan 2004.

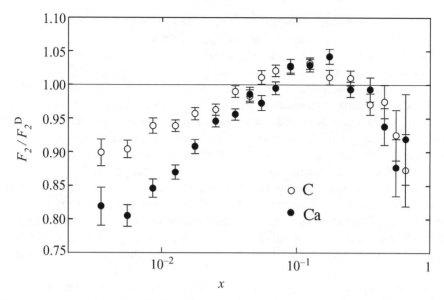

Figure 9.9 The ratios of the F_2 structure function found from nuclear targets to that found from deuterium, as a function of the scaling variable x (Carbon data from Ar95, calcium data from Am95)

Another consequence of these predictions is that the effective masses of hadrons will in general change in nuclear matter, as will their sizes and interactions. There is already some evidence in favour of this suggestion from deep inelastic scattering from nucleons (see Section 5.7) where the structure functions obtained using targets of light and heavier nuclei differ slightly, even after allowing for calculable effects such as nuclear binding energies and the internal Fermi motion of the nucleons. (This is the so-called 'EMC effect', named after the group that first discovered it.) It is illustrated in Figure 9.9, which shows the ratios F_2^{Ca}/F_2^D and F_2^C/F_2^D, i.e. the F_2 nucleon structure function deduced from calcium and carbon targets, divided by the structure function deduced using a deuterium target.

A number of other experiments have been performed to detect the effect of the nuclear environment on effective masses (for example, by determining the mass of mesons produced in nuclear matter) but nothing significant has been found elsewhere. This will be a continuing field of study.

It is also important to study how the interactions of lower-energy hadrons change when they are embedded in nuclear matter. For example, there is considerable interest in the interactions of hadrons containing a strange valence quark. (One reason is that they may play an important role in the high-density matter present in neutron stars.) The lightest mesons that contain a strange valence quark or antiquark are the kaons and these can be implanted in nuclei by nuclear reactions that substitute a strange quark for an up or down quark. (This is an example of a so-called 'hypernucleus'.) Experiments at CEBAF and other laboratories will provide information on the interaction of implanted, negatively charged kaons with the surrounding nucleons in a nucleus.

The facilities at CEBAF and RHIC (the relativistic heavy ion accelerator described in Section 4.2.2) will enable a range of new experimental possibilities to be explored, in addition to those above. One is the intriguing question of the existence of glueballs (mesons made of gluons alone) and hybrid quark–gluon mesons, mentioned in Section 5.2 and vital for the theory of confinement via QCD. The results may well help to find a solution to one of the central questions posed at the start of this section: how are the properties of the strong nuclear force related to the standard model formulation in terms of quarks and gluons?

There are also questions to be answered in the realm of nuclear structure, many with implications elsewhere. For example, can the properties of nuclei be related to those of an underlying nucleon–nucleon interaction and can they be derived from many-body theory? At present we have a good knowledge from scattering experiments of the long-range part of the nucleon–nucleon force in terms of meson exchanges (see Section 7.1), but models that fit data differ about the short-range part. This is not surprising because at separations of less than 1 fm a description in terms of quarks and gluons is necessary and the interface with QCD is critical. Experiments on meson production in nucleon–nucleon collisions are sensitive to the short-range part of the forces and should provide information on this region. On the theoretical side, advances in computer power and calculational techniques have enabled the binding energies of all light nuclei to be successfully calculated using the best available parameterization of the nucleon–nucleon force. However, this is only possible by including an explicit weaker three-nucleon force, which has to be adjusted to obtain the correct binding energies. A satisfactory theory of the three-body force between nucleons is lacking. This work also needs to be extended to heavier nuclei, but present computer power is inadequate to the task using current computational techniques.

One approach to the latter problem is to work within the framework of the shell model, where each nucleon moves in the average potential (the mean field) generated by its interactions with all the other nucleons in the nucleus. We have seen the successes of this approach in simple applications in Section 7.3. When combined with further computational improvements, it has enabled nuclear structure calculations to be extended to $A = 56$. This is an important point for astrophysics, because the details of the nuclear reactions of iron control the critical process occurring in the collapse of a supernova, as we have seen above in Section 9.1.4.

Very often in science new insights are achieved by pushing experiments to their limits. Nuclear physics is no exception. One such limit is the quest for super-heavy elements. Discovery of elements beyond those currently known could explore questions about possible limits on nuclear charges and masses. According to nuclear models there should exist a new group of super-heavy elements with charges Z in the range approximately 114 to 126 that are stabilized by shell effects.

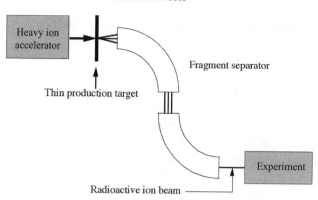

Figure 9.10 An energetic particle (typically several tens of MeV/u to GeV/u) is fragmented in a nuclear reaction in a thin target, and radioactive reaction products are separated in-flight and transported as a secondary beam to the experiment

The heaviest element made to date has $Z = 116$ and was produced by fusion in the reaction $^{48}_{20}\text{Ca} + ^{248}_{96}\text{Cm} \rightarrow ^{292}_{116}\text{Uuh} + 4n$ (the symbol Uuh is used as the element has yet to be named). Strenuous efforts are being made to reach the predicted new island of relative stability. This will require facilities to produce exotic short-lived nuclear beams and there is much development work going on in this area. One example of how such a beam can be formed is shown in Figure 9.10. The other main method employs two independent accelerators: a high-power driver accelerator for production of the short-lived nuclei in a thick target that is directly connected to an ion-source, and a second post-accelerator. Radioactive atoms diffuse out of a hot target into an ion source where they are ionized for acceleration in the post-accelerator.

Fewer than 300 stable nuclei occur naturally (see Figure 2.7) and outside the stability region nuclei decay by the mechanisms discussed in Chapters 2 and 7. In the uncharted regions there are many fundamental questions to be answered, such as what are the limiting conditions under which nuclei can remain bound and do new structures emerge near these limits? The answers to these questions are important because theoretical descriptions of nuclei far from the line of stability suggest that their structures are different from what has been seen in stable nuclei. Nuclei far from stability also play an important role in astrophysics, for example in understanding the processes in supernovae and how elements are synthesized in stars. Another limiting region that is expected to yield interesting information is that of angular momentum. Super-deformed nuclei have been discovered with highly elongated shapes and very rapid rotational motion. The states associated with these shapes are extremely stable. Further investigation of these is expected to yield important information about nuclear structure.

9.2.2　Quark–gluon plasma, astrophysics and cosmology

We have touched on the implications of nuclear physics for astronomy at various places above. Here we look at other areas where improvements in our nuclear physics knowledge would help astrophysics and cosmology.

In QCD, quarks and gluons are confined within hadrons, although the nature of this confinement is still not fully understood. At extremely high energy-densities the quarks and gluons are expected to become deconfined across a volume that is large compared with that of a hadron. They would then exist in a new state of matter, called a *quark–gluon plasma*, which is the state of nuclear matter believed to have existed in the first few microseconds after the big bang (see Figure 9.11).

It is possible to probe this state of matter using the RHIC facility (and also in a few years at the LHC when its construction is complete). RHIC typically collides two counter-circulating beams of fully-stripped gold ions at a maximum energy of 200 GeV per nucleon. If the ions collide centrally (i.e. head-on) several thousand final-state particles are produced. An example of an event seen in the STAR detector (which was shown in Figure 4.20) is illustrated in Figure 9.12. A key

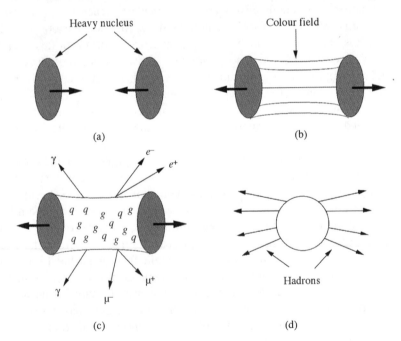

Figure 9.11　Stages in the formation of a quark--gluon plasma and subsequent hadron emission: two heavy nuclei collide at high energies (a) and interact via the colour field (b); the very high energy-density produced causes the quarks and gluons to deconfine and form a plasma that can radiate photons and lepton pairs (c); finally, as the plasma cools, hadrons condense and are emitted (d) (after NRC99, with permission of the National Academics Press)

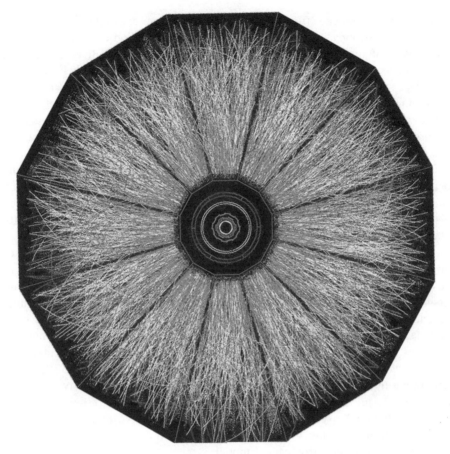

Figure 9.12 View of a 200 GeV gold--gold interaction in the STAR detector at the RHIC accelerator (Courtesy of Brookhaven National Laboratory)

question is whether the energy-density in the collisions is sufficient to have created a quark–gluon plasma and its subsequent cooling phases. There are many signatures for this, including the relative abundances of different final-state particle types (for example, production of the $c\bar{c}$ meson J/Ψ would be suppressed) and measurements are all consistent with the expected temperature at which hadrons would be formed (about 176 MeV, corresponding to about 10^{12} K, close to that predicted by QCD) and that the temperature of the initial fireball is considerably higher.

Future experiments at RHIC will play a crucial role in understanding the basic nature of deconfinement. Questions to be addressed include: what is the nature of matter at the highest densities (very recent experiments at RHIC suggest that the plasma behaves more like a liquid than a gas); under what conditions can a quark–gluon plasma be made; and what are the rules governing the evolution and the transition to and from this kind of matter?

Information gathered from high-energy heavy-ion collisions is potentially important in astrophysics. It will help constrain the equation of state that relates the density of matter in neutron stars and supernovae (as well as in the first microseconds of the early universe) to pressure and temperature. This information will place stronger theoretical constraints on the maximum mass of a neutron star, improving the ability to distinguish neutron stars and black holes.

The synthesis of nuclei in the very early universe is one of the cornerstones of modern astrophysics, but even here there are still surprises. For example, in the discussion of stellar fusion in Chapter 8, we saw that the production of heavy elements involves the rare reaction $3\,(^4\text{He}) \rightarrow {}^{12}\text{C}$ (Equation (8.31)), the occurrence of which depends critically on the existence of a particular excited state of ^{12}C. We also noted that very recently another excited state has been discovered at a somewhat higher energy which has the effect of significantly altering the energy dependence (or equivalently the temperature dependence) of this reaction from the values usually assumed. This could have major consequences for models and theories of stellar evolution. Another recent experiment has measured for the first time the lifetime of the doubly-magic nucleus ^{78}Ni and finds it to be shorter than expected, implying that supernova explosions may produce gold and other heavier elements much faster than had previously been thought. This is important because ^{78}Ni is believed to produce more than half the elements heavier than iron in the universe. A reaction of great current interest is the synthesis of ^{16}O from the reaction of ^4He with ^{12}C (Equation (8.32)), which determines the relative sizes of the carbon and oxygen shells of massive stars that later explode in supernovae. The sizes of these shells are a crucial factor in predicting the nucleosynthesis that occurs during the explosion. Nuclear physicists are currently trying to measure the rate of this reaction with sufficient accuracy to constrain astrophysical models.

One of the outstanding theoretical challenges in nuclear astrophysics is to understand the process by which a massive, fully-evolved star ejects its mantle while its core collapses to a neutron star or black hole. In Section 9.1.4 above we gave a simple description of this process involving the collapse of the iron core to several times the density of nuclear-matter, thereby producing a powerful shock wave that travels outward through the mantle of the star. This shock wave, aided by the heating of the matter by neutrinos emitted by the newly formed neutron star, is responsible for the ejection of the mantle. This is the broad-brush picture, but there is still no satisfactory theory that can account for the observed frequency of supernovae. Efforts to understand dense nuclear matter and to predict the properties of neutron stars depend on knowledge of nuclear interactions gained in the laboratory. Heavy-ion collisions will help us better understand the interactions of mesons in hot, dense nuclear matter, which is crucial to the issue of meson condensation in neutron stars. Future studies of neutron-rich nuclei, near the limit of stability, in radioactive ion beam facilities, as mentioned in Section 9.2.1 above, will allow more accurate modelling of nuclear forces in neutron star crusts.

9.2.3   Symmetries and the standard model

An important symmetry that can be tested in nuclear physics is time reversal. We have seen in Section 6.6 that *CP* invariance is violated in the weak decays of *K* and *B* mesons and by inference so is *T* invariance, provided *CPT* invariance holds. However, we have also seen in Section 9.1.4 above that the mechanism of violation that can explain meson decays is unable to explain the observed matter/antimatter asymmetry in the universe. Thus it is likely there exists another *CP*-violating mechanism and hence another source of *T* violation.

There are several ways in principle of exploring *CP* violation in the context of atomic and nuclear physics. One way involves antihydrogen. This was first produced in a controlled experiment in 2002 by mixing cold antiprotons with a dense positron plasma confined by electromagnetic fields in a so-called 'Penning trap'. If atoms of antihydrogen could be trapped for extended periods their properties could be compared with those of hydrogen and this might shed light on matter–antimatter asymmetry. *CP* violation can also be probed by searching for electric dipole moments (EDMs) of the neutron, atoms or the electron. In the case of atoms, an EDM could arise if the electron had an electric dipole moment or if there were a *T*-violating interaction within the nucleus. Static EDMs are forbidden if *T* invariance is exact and so a non-zero value would imply *CP* violation, assuming *CPT* invariance holds. The present 90 per cent confidence limit on the EDM d_n of the neutron is $d_n < 6.3 \times 10^{-26}$ e cm and that for the electron is $d_e < 1.6 \times 10^{-27}$ e cm. Improving these presents formidable experimental challenges. Nevertheless, several experiments are planned or are underway to measure EDMs, with the aim of reducing the bounds to regions where they could test the predictions of current theories. For the standard model these are $d_n \sim 10^{-31}$ e cm and $d_e \sim 10^{-38}$ e cm, although some extensions of the standard model discussed in Section 9.1.3 above predict considerably larger values. Limits on the existence of atomic and neutron EDMs already provide constraints on some of the most plausible extensions to the standard model. It is also possible that *T*-violation might show up in the decay of an unstable system. Modern experiments are searching for *T*-violating correlations in the β-decay of neutrons, mesons and particular nuclei.

Atomic/nuclear physics can also provide information on the standard model in other areas of the weak interactions. For example, a recent (2005) study of the β-decay of a metastable state of ^{38}K in an atomic trap has enabled severe limits to be placed on a possible spin-0 particle to augment the spin-1 *W*-meson exchange. The mixing between the weak and electromagnetic interactions can also be studied. This is characterized by the Weinberg angle, which can be measured in the parity-violating interaction between electrons and the nuclei of particular atoms. This was mentioned at the end of Section 6.7. Parity mixing has been seen in several atomic systems. The best measurement at present has been made using ^{133}Cs atoms, although the limits on the Weinberg angle do not yet compete with those obtained from particle physics experiments. Other experiments plan to study this effect in atomic francium, where the parity-mixing effect should be about 18 times larger. (The effect of an electric dipole moment of the electron is also expected to be greatly enhanced in francium.) Unfortunately, francium is an extremely rare element with no stable isotopes and so experiments will be carried out with a small number of radioactive atoms collected in a magneto-optic trap.

9.2.4   Nuclear medicine

In Section 8.3.1, we reviewed the use of radiation techniques for cancer therapy. We also briefly mentioned that in principle heavier particles had advantages over photons. For example, because of the form of the Bragg curve, protons deposit more of their energy where they stop, not where they enter the body. Also their depth of penetration can be precisely controlled so that they stop within the tumour, thus allowing radiologists to increase the radiation dose to the tumour while reducing the dose to healthy tissues.

This is illustrated in Figure 9.13, which compares the treatment plans (i.e. simulations of the pattern of radiation that the patient would receive) for treating a case of advanced pancreatic cancer. Figure. 9.13(a) shows an X-ray plan using a 'state-of-the-art' nine-beam X-ray system. The amount of radiation received by nearby organs and other critical areas (kidneys, liver and spinal chord) is seen to be a substantial fraction of the dose received by the region of the cancer. This is contrasted with the results of Figure 9.13(b), which is for treatment using a single proton beam. Although there is some unwanted exposure at the input site

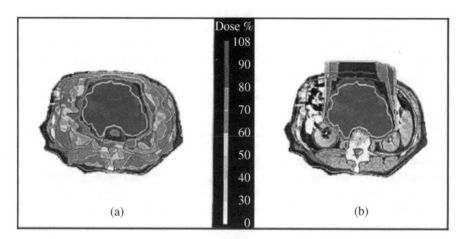

Figure 9.13 Treatment plans for a large pancreatic tumour: (a) using a nine-beam X-ray system; (b) using a single proton beam. The diffuse grey areas in (a) indicate the spread of energy deposition outside the region of the tumour (adapted from Zu00, copyright Elsevier, with permission)

(which could be lessened by a system of multiple beams or a rotating beam), the radiation energy is concentrated much more within the area of the tumour.

Although they have great potential, the problem with using particle beams is the practical one of access to suitable accelerators. There is considerable effort being made to design proton accelerators for cancer therapy and more than 20 centres now exist worldwide specifically for proton therapy. Research is also continuing with other forms of radiation therapy using neutrons and heavy ions. Neutrons produce a high linear energy transfer (LET) and this overcomes a cancer cell's resistance to radiation damage more effectively than low-LET photon, electron or proton radiation. Thus neutrons appear to be more biologically effective in killing cancers than many other forms of radiation, especially in oxygen-poor cells. Beams of heavy ions, such as carbon or neon, with energies of 400–800 MeV per nucleon, are nearly ideal dose delivery vehicles for radiation therapy. Limited studies with carbon and neon beams have been conducted and doubtless these studies will increase in the future.

Progress in the NMR technique in medicine continues. For example, recent advances have enabled a variation known as functional MRI (fMRI) to be developed that exploits the paramagnetic behavior of deoxyhaemoglobin in red blood cells. When in a magnetic field, a blood vessel containing deoxyhaemoglobin distorts the field in its immediate environs, with the degree of distortion increasing with the concentration of deoxyhaemoglobin. This distortion affects the behaviour of water protons in the environs and, consequently, the magnetic-resonance signal arising from these protons. Neural activation of a region of the brain stimulates increased arterial flow of oxygenated blood, thereby decreasing the concentration of deoxyhemoglobin in the region. Changes in the magnetic-resonance signal can be detected and displayed as functional-MRI images. These so-called BOLD (blood-oxygen-level dependent) images enable studies to be made of the way the brain works by taking MRI images in real time while the patient is performing specific tasks. In this way areas of the brain can be studied that are associated with particular activities or sensations.

As another example, the gases ^{3}He and ^{129}Xe have the magnetic properties needed for MRI and the atomic structure needed to retain their polarization for hours at a time. They can be introduced into lungs, allowing MRI studies of lung function. Because of the strong signal provided by the polarized nuclei in the gas atoms, the MRI scans are short and can be synchronized with breathing. Developments are also being made towards general high-speed imaging, which would be useful for claustrophobic patients and children who are unable to be in the confined environment of a conventional MRI magnet for sometimes up to an hour.

Perhaps the greatest potential of all lies in the imaging of nuclei other than hydrogen, particularly the phosphorus nucleus. Phosphorus is a major constituent of the molecules adenosine triphosphate (ATP) and phosphocreatine, which mediate the transfer of energy in living cells. From knowledge of such concentrations it is possible to infer the metabolic status of internal organs, and it may eventually be possible to add this capability to an imaging instrument. The future will undoubtedly see both an improvement in the quality of NMR images and a growing diversity of applications for nuclear magnetic resonance in clinical practice.

An area that was not mentioned in Chapter 8 is the use of radioactive nuclear isotopes produced by accelerators or nuclear reactors in many areas of biological and biomedical research. For example, by inserting such radioisotopes as ^{14}C and tritium, it is possible to obtain information on how molecules move through the body, what types of cells contain receptors, and what kinds of compounds bind to these receptors. Radioisotopes are also used directly to treat disease and radioactive tracers are indispensable tools for the new forensic technique of DNA fingerprinting, as well as for the Human Genome Project.

9.2.5   Power production and nuclear waste

Nuclear fusion still holds the promise of unlimited power without the problem of radioactive waste, but the road to realization of this goal is long and we are far from the end. In Section 8.2 we introduced the Lawson criterion as a measure of how close a design was to the ignition point, i.e. the point at which a fusion reaction becomes self-sustaining. To date no device has yet succeeded in achieving the Lawson criterion and much work remains to be done on this important problem. In recognition of this, at least one major new tokamak machine (to be built in France) is planned as a global collaboration, but even when the ignition point is attained, based on experience with fission reactors, it could be many decades before that achievement is translated into a practical power plant.

In the shorter term and assuming that renewable sources of energy are insufficient to fulfil the world's increasing energy needs, it does seem as if power plants based on fission reactions are the only hope of replacing fossil fuels in the future. The problems of reactor safety and the safe disposal of radioactive waste are therefore paramount.

The waste from light-water reactors, the most common type of power reactor, has two major components: the actinides, i.e. any of the series of radioactive elements with atomic numbers between 89 and 103 (mainly uranium but also smaller amounts of heavier elements, the transuranic elements like plutonium and the minor actinides such as neptunium, americium and curium) and fission products, which are medium-weight elements from fission processes in the nuclear fuel. While it is generally agreed that radioactive nuclei with relatively short lifetimes can be safely stored in deep geological disposal facilities, the same is not true of waste with very long lifetimes, some of which are water-soluble and so have the potential to contaminate ground water. An additional problem is the disposal of material that could be used for nuclear weapons, i.e. ^{239}Pu and ^{235}U. One option for handling waste with very long lifetimes, which was mentioned as a theoretical possibility in Section 8.1.3, is to transmute it by neutron reactions into shorter-lived, or even stable, isotopes that can be dealt with by conventional storage.

The idea of using an accelerator to produce materials that can only be made artificially has been around for more than 40 years, but more recently there has been considerable interest and research in this idea to 'incinerate' nuclear waste with the aim of reducing the waste lifetimes to less than 100 years. This is referred to as ADS – Accelerator Driven System. In one proposed scheme, uranium and most of the plutonium would be separated prior to proton irradiation and used again as reactor fuel. The most important long-lived components of the remaining waste would be isotopes of neptunium, americium, curium and iodine, some with half-lives of 10 000 years or more. The approach would be to irradiate this material with a new source of fast neutrons produced by spallation reactions (cf. the discussion of producing neutron beams by this process in Section 4.2.3) initiated using protons from a high-current accelerator. In this way the capacity to 'burn' long-lived fission products and actnides is greatly increased, leaving waste with much shorter lifetimes which can be disposed of by conventional means. The accelerator would deliver a high-power (10–20 mA) proton beam of about 1 GeV energy to a heavy metal (spallation) target surrounded by the nuclear waste to be incinerated. The accelerator–waste combination would be operated at a subcritical level – by itself it could not sustain a chain reaction – so that no reactor-core meltdown accident could occur.

It has been suggested that this concept might be carried one step further, and a particle beam might be used to produce additional neutrons directly in a nuclear-reactor-like core. Versions of this concept have been studied in America and by a European group. The latter is based on a proposal by Rubbia[9] and is called the Energy Amplifier. In this scheme, the core of the reactor would again be sub-critical, and the accelerator beams would provide sufficient additional neutrons via the spallation reaction to run the reactor. An idealized possible set-up is shown in Figure 9.14.

Because the spallation neutrons would have high energy, a less enriched element, such as natural thorium, could serve as the fuel. Thorium has the great advantage over uranium in being an abundant element that does not require costly isotope separation. Although the thorium fuel would not require enrichment, it would need to be recharged every 5 years or so. The proposal has a number of other advantages over a conventional power reactor, including: it is sub-critical without the spallation neutrons and so is inherently safe – a meltdown or explosion is not possible; radioactive waste is consumed in the reactor and no long-lived waste is produced; there is no overlap with the nuclear weapons fuel cycle and so the energy amplifier cannot be used as the basis for producing materials for nuclear weapons, making installations politically acceptable worldwide.

---

[9]The same man who shared the 1984 Nobel Prize in Physics for the discovery of the W and Z bosons.

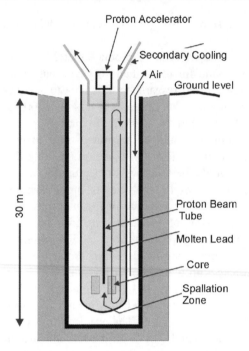

Figure 9.14   Schematic diagram of a possible configuration of an energy amplifier; in this design the coolant and spallation metal is molten lead (from Sc01, copyright Cavendish Press Ann Arbor 2001, with permission)

The possible energy flow in a commercial system is shown in Figure 9.15. This assumes a 1 GeV, 20 ma proton beam requiring about 20 MW of input power. The latter is taken from the output of the reactor leaving a net electrical output of 580 MW, i.e. a gain factor of about 30.

The current situation on the energy amplifier is that a European collaboration has shown that initial partitioning at the level of 95–99 per cent is possible

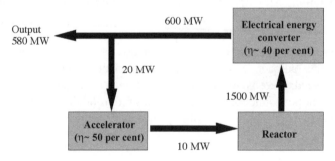

Figure 9.15   Possible energy flows in an energy amplifier system; the conversion efficiencies are denoted by η

depending on the actinide species. They have also carried out a number of successful reactor transmutation and spallation studies and the first full ADS experiment (TRADE) is currently under construction. This consists of coupling a cyclotron delivering a 140 MeV, 0.5–1.0 ma proton beam to an existing 1MW water-cooled reactor sited in Italy and uses a spallation target of tantalum. The operation date is planned for 2007/08. Additional work is being carried out in Belgium on coupling a 350 MeV, 5 ma proton beam to a 100 MW subcritical reactor (the Myrrha experiment) and has already shown that some long-lived isotopes can be successfully incinerated. Although ADS has enormous potential, there are still a great many problems to be overcome and questions to be answered. The estimated time for completion of research and development work and commencement of an industrial plant based on ADS could be as long as 50 years.

Appendix A
Some Results in Quantum Mechanics

A.1 Barrier Penetration

Consider the one-dimensional potential shown in Figure A.1(a). Free particles of mass m and energy E represented by plane waves are incident from the left and encounter the constant rectangular barrier of height V, where $V > E$.

In region I $(x < 0)$, there is an incoming wave e^{ikx}, where the wave number k is given by

$$\hbar^2 k^2 = 2mE, \tag{A.1}$$

and also a wave reflected at the barrier travelling from right to left of the form e^{-ikx}. Thus the total wavefunction in region I is

$$\psi_1(x) = Ae^{ikx} + Be^{-ikx}, \tag{A.2}$$

where A and B are complex constants. Within the barrier, region II $(0 < x < a)$, the solution of the Schrödinger equation is a decaying exponential plus an exponential wave reflected from the boundary at $x = a$, i.e. the total wavefunction is

$$\psi_2(x) = Ce^{-\kappa x} + De^{\kappa x}, \tag{A.3}$$

where C and D are complex constants and κ is given by

$$\hbar^2 \kappa^2 = 2m(V - E). \tag{A.4}$$

Finally, in region III $(x > a)$ to the right of the barrier, there is only an outgoing wave of the form

$$\psi_3(x) = Fe^{ikx}, \tag{A.5}$$

where again F is a complex constant.

Nuclear and Particle Physics   B. R. Martin
© 2006 John Wiley & Sons, Ltd

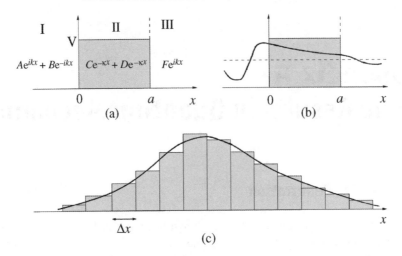

Figure A.1 Rectangular barrier with (a) wavefunction solutions, and (b) form of the incoming and outgoing waves; (c) modelling an arbitrary smooth barrier as a series of rectangular barriers

We are interested in the *transmission coefficient T*, defined by

$$T \equiv |F/A|^2. \tag{A.6}$$

The values of F and A are found by imposing continuity of the wavefunction and its first derivative, i.e. matching the values of these quantities at the two boundaries $x = 0$ and $x = a$. The algebra may be found in any introductory book on quantum mechanics.[1] The result is

$$T = \left| \frac{2k\kappa e^{-ika}}{2k\kappa \cosh(\kappa a) - i(k^2 - \kappa^2) \sinh(\kappa a)} \right|^2. \tag{A.7}$$

The corresponding incident and transmitted waves are shown in Figure A.1(b) (the reflected waves are not shown).

For large κa, which corresponds to small penetrations, we can make the replacement

$$\sinh(\kappa a) \approx \cosh(\kappa a) \approx \frac{1}{2} e^{\kappa a} \tag{A.8}$$

and hence

$$T \approx \left(\frac{4k\kappa}{k^2 + \kappa^2} \right)^2 e^{-2\kappa a}. \tag{A.9}$$

---

[1]See, for example, Chapter 6 of Me61.

The first factor is due to the reflection losses at the two boundaries $x = 0$ and $x = a$ and the decreasing exponential describes the amplitude decay within the barrier. The first factor is slowly varying with energy and is usually neglected.

The result of Equation (A.9), ignoring the first factor, may be used to find the transmission coefficient for an arbitrary smoothly-varying barrier by modelling it as a series of thin rectangular barriers. This is illustrated in Figure A.1(c). Thus by replace $2\kappa a$ by $2 \sum \kappa(x)\Delta x$ and taking the limit of small Δx, the summation goes over to an integral, i.e.

$$2\kappa a \rightarrow 2 \int dx \left\{ \frac{2m}{\hbar^2} [V(x) - E] \right\}^{\frac{1}{2}} \tag{A.10}$$

and

$$T \approx \exp\left[-2 \int dx \left\{ \frac{2m}{\hbar^2} [V(x) - E] \right\}^{\frac{1}{2}} \right]. \tag{A.11}$$

This is the essence of what is known as the WKB approximation in quantum mechanics. Equation (A.11) was used in Section 7.6 to discuss α-decay and in Section 8.2.1 to discuss nuclear fusion.

A.2 Density of States

Consider a spinless particle of mass m confined within a cube of sides L and volume $V = L^3$, oriented so that one corner is at the origin $(0,0,0)$ and the edges are parallel to the x, y and z axes. If the potential is zero within the box, then the walls represent infinite potential barriers and the solutions of the Schrödinger equation must therefore vanish on all faces of the cube. It is straightforward to show that the solutions of the Schrödinger equation satisfying these boundary conditions are standing waves of the form

$$\psi(x, y, z) = C \sin(k_x x) \sin(k_y y) \sin(k_z z), \tag{A.12}$$

where C is a constant and the components of the wave number $\mathbf{k} = (k_x, k_y, k_z)$ take the values

$$k_x = \frac{n_x \pi}{2}, \quad k_y = \frac{n_y \pi}{2}, \quad k_z = \frac{n_z \pi}{2}, \quad (n_x, n_y, n_z) = 1, 2, 3.... \tag{A.13}$$

The energy of the particle is given by

$$E = \frac{\hbar^2}{2m}(k_x^2 + k_y^2 + k_z^2) = \frac{\hbar^2 k^2}{2m} = \frac{(\hbar\pi)^2}{8m}(n_x^2 + n_y^2 + n_z^2), \tag{A.14}$$

where $k \equiv |\mathbf{k}| = p/\hbar$ and p is the particle's momentum. Negative values of the integers do not lead to new states since they merely change the sign of the wave function Equation (A.12) and phase factors have no physical significance.

The allowed values of $\mathbf{k}$ form a cubic lattice in the quadrant of 'k-space' where all the values of (n_x, n_y, n_z) are positive. Since each state corresponds to one combination of (n_x, n_y, n_z), the number of allowed states is equal to the number of lattice points. The spacing between the lattice points is (L/π), so the density of points per unit volume in $\mathbf{k}$-space is $(L/\pi)^3$. The number of lattice points $n(k_0)$ with k less than some fixed value k_0 is the number contained within a volume that for large values of k_0 may be well approximated by the quadrant of a sphere of radius k_0, i.e.

$$n(k_0) = \frac{1}{8}\frac{4}{3}\pi k_0^3 \left(\frac{L}{\pi}\right)^3 = \frac{V}{(2\pi)^3}\frac{4\pi k_0^3}{3}. \tag{A.15}$$

Hence the number of points with k in the range $k_0 < k < (k_0 + dk_0)$ is

$$dn(k_0) = \frac{V}{(2\pi)^3} 4\pi k_0^2 dk_0. \tag{A.16}$$

The *density of states* is defined as $\rho(k_0) \equiv dn(k_0)/dk_0$ and so is given by

$$\rho(k_0) = \frac{V}{(2\pi)^3} 4\pi k_0^2. \tag{A.17}$$

Thus $\rho(k_0)dk_0$ is the number of states with k between k_0 and $k_0 + dk_0$, or equivalently

$$\rho(p)dp = \frac{4\pi V}{(2\pi\hbar)^3} p^2 dp \tag{A.18}$$

is the number of states with momentum between p and $p + dp$. This is the form used in Equation (7.1) when discussing the Fermi energy in the Fermi gas model. Equation (A.18) can also be written in terms of energy using $E = p^2/2m$, when it becomes

$$\rho(E)dE = \frac{4\pi V}{(2\pi\hbar)^3} m p\, dE \tag{A.19}$$

and this was the form used in discussing β-decay in Section 7.7.2.

Although the above derivation is for a particle confined in a box, the same technique can be used for scattering problems. In this case we can consider a large volume $V = L^3$ and impose 'periodic' boundary conditions

$$\psi(x+L, y, z) = \psi(x, y+L, z) = \psi(x, y, z+L) = \psi(x, y, z). \tag{A.20}$$

Instead of standing waves, the solutions of the Schrödinger equation consistent with Equation (A.20) are the travelling waves

$$e^{i\mathbf{k}\cdot\mathbf{r}} = e^{ik_xx}e^{ik_yy}e^{ik_zz} \tag{A.21}$$

where

$$k_x = \frac{2n_x\pi}{L}, \quad k_y = \frac{2n_y\pi}{L}, \quad k_z = \frac{2n_z\pi}{L}, \quad n_x, n_y, n_z = 0, \pm 1, \pm 2 \ldots \tag{A.22}$$

The density of lattice points in **k**-space now becomes $(L/2\pi)^3$, but unlike the standing wave case, permutations of signs in Equation (A.22) *do* produce new states and the whole quadrant of lattice points has to be considered. Thus these two effects 'cancel out' and we arrive at the same result for the density of states in Equations (A.18) and (A.19). This approach was used in discussing the formal definitions of cross sections in Chapter 1.

All the above is for spinless particles. If the particle has spin then the density of states must be multiplied by the appropriate spin multiplicity factor, taking account of the Pauli principle as necessary. Thus, for example, for spin-$\frac{1}{2}$ particles, with two spin states, Equation (A.19) becomes

$$\rho(E)dE = \frac{8\pi V}{(2\pi\hbar)^3} mp\, dE. \tag{A.23}$$

A.3 Perturbation Theory and the Second Golden Rule

Without detailed proof, we will outline the derivation from perturbation theory of the important relationship between the transition probability per unit time for a process and its matrix element.[2]

In perturbation theory, the Hamiltonian at time t may be written in general as

$$H(t) = H_0 + V(t), \tag{A.24}$$

where H_0 is the unperturbed Hamiltonian and $V(t)$ is the perturbation, which we will assume is small. The solution for the eigenfunctions of H starts by expanding in terms of the complete set of energy eigenfunctions $|u_n\rangle$ of H_0, i.e.

$$|\psi(t)\rangle = \sum_n c_n(t)|u_n\rangle e^{-iE_nt/\hbar}, \tag{A.25}$$

---

[2]We follow the derivation given in Chapter 9 of Ma92.

where E_n are the corresponding energies. If $|\psi(t)\rangle$ is normalized to unity, then the squared coefficient $|c_n(t)|^2$ is the probability that at time t the system is in a state $|u_n\rangle$. Substituting Equation (A.25) into the Schrödinger equation leads to a differential equation for the transition coefficients:

$$i\hbar \frac{dc_f(t)}{dt} = \sum_n V_{fn}(t)e^{i\omega_{fn}t}c_n(t), \qquad (A.26)$$

where the matrix element $V_{fn} \equiv \langle u_f |V(t)|u_n\rangle$ and the angular frequency $\omega_{fn} \equiv (E_f - E_n)/\hbar$. If we assume initially $(t = 0)$ that the system is in a state $|u_i\rangle$, then $c_n(0) = \delta_{ni}$ and the solutions for $c_f(t)$ are found by substituting this result into the right-hand side of Equation (A.22) giving, *to first-order in V,*

$$c_i(t) = 1 + \frac{1}{i\hbar} \int_0^t V_{ii}(t')dt' \qquad (A.27a)$$

and

$$c_f(t) = \frac{1}{i\hbar} \int_0^t V_{fi}(t')e^{i\omega_{fi}(t')}dt' \qquad (f \neq i). \qquad (A.27b)$$

For $f \neq i$, the quantity $|c_f(t)|^2$ is the probability, in first-order perturbation theory, that the system has made a transition from state i to state f.

The above is for a general time-dependent perturbation $V(t)$, but the results can also be used to describe other situations, for example where the perturbation is zero up to some time t_0 and a constant thereafter. In this case, the integrals in Equations (A.27) can be evaluated and, in particular, Equation (A.27b) gives, again to first-order in V,

$$c_f(t) = \frac{V_{fi}}{\hbar\omega_{fi}}\left[1 - e^{i\omega_{fi}t}\right] \qquad (A.28)$$

and hence the probability of the transition $i \to f$ is

$$P_{fi}(t) = |c_f(t)|^2 = \frac{4|V_{fi}|^2}{\hbar^2}\left[\frac{\sin^2(\frac{1}{2}\omega_{fi}t)}{\omega_{fi}^2}\right]. \qquad (A.29)$$

The function in the square brackets in Equation (A.29) is shown in Figure A.2.

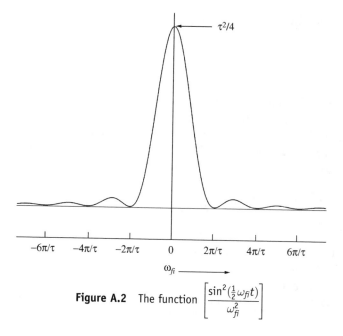

Figure A.2   The function $\left[\dfrac{\sin^2\left(\frac{1}{2}\omega_{fi}t\right)}{\omega_{fi}^2}\right]$

For sufficiently large values of t, it has the form of a large central peak with much smaller side oscillations. In this case P_{fi} is only appreciable if

$$\hbar|\omega_{fi}| = |E_f - E_i| \leq 2\pi\hbar/t \tag{A.30}$$

and then the square bracket can be replaced by a Dirac delta function[3], i.e.

$$\lim_{t\to\infty} \frac{\sin^2\left(\frac{1}{2}\omega_{fi}t\right)}{\omega_{fi}^2} = \frac{1}{2}\pi\hbar t\delta(E_f - E_i), \tag{A.31}$$

where the external factors are to preserve the normalization. Then

$$P_{fi}(t) = t\frac{2\pi}{\hbar}|V_{fi}|^2\delta(E_f - E_i) \tag{A.32}$$

---

[3]The Dirac delta function was the first so-called 'generalized function'. It is defined by the two conditions: (i) $\delta(x' - x) = 0$ if $x \neq x'$ and (ii) $\int_{-\infty}^{+\infty} \delta(x' - x)dx' = 1$. It follows that if $f(x)$ is a function continuous in the interval $x_1 < x < x_2$, then $\int_{x_1}^{x_2} f(x')\delta(x' - x)dx' = f(x)$ if $x_1 < x < x_2$ or $= 0$ if $x < x_1$ or $x > x_2$.

and the transition probability per unit time is

$$\frac{\mathrm{d}P_{fi}(t)}{\mathrm{d}t} = \frac{2\pi}{\hbar}|V_{fi}|^2\delta(E_f - E_i).$$

(A.33)

The above assumes that the final state is discrete, but it is more common for the final states to form a continuum defined by the density of states $\rho(E)$ derived in Section A.2 above. In this case, since $\rho(E)\mathrm{d}E$ is the number of states with energy between E and $E + \mathrm{d}E$, we can write the transition rate per unit time $\mathrm{d}T_{fi}/\mathrm{d}t$ to a group of states f with energies in this range as

$$\frac{\mathrm{d}T_{fi}}{\mathrm{d}t} = \int \frac{\mathrm{d}P_{fi}(t)}{\mathrm{d}t}\rho(E_f)\mathrm{d}E_f = \frac{2\pi}{\hbar}\left[|V_{fi}|^2\rho(E_f)\right]_{E_f=E_i},$$

(A.34)

where the integral has been evaluated using the properties of the delta function. Equation (A.34) is called the Second Golden Rule (sometimes Fermi's Second Golden Rule, although strictly the result is not due to Fermi) and has been used in several places in this book, for example in Chapter 7 when discussing nuclear β-decay.

Appendix B
Relativistic Kinematics

In particle physics, most scattering interactions take place between particles whose speeds are comparable with the speed of light c. This is often true even in decays, particularly if light particles are emitted. The requirements of special relativity therefore cannot be ignored. In nuclear physics accurate predictions can also often only be obtained if relativistic effects are taken into account. In this appendix we review (usually without proof) some relativistic kinematical results and the use of invariants to simplify calculations.

B.1   Lorentz Transformations and Four-Vectors

Consider a particle of mass m in an inertial frame of reference S. Its co-ordinates are $(t, \mathbf{r}) \equiv (t, x, y, z)$ and its speed is $u = |\mathbf{u}|$, where $\mathbf{u}$ is its velocity. In a second inertial frame S' its co-ordinates are $(t', \mathbf{r}') \equiv (t', x', y', z')$ and its speed is $u' = |\mathbf{u}'|$ where $\mathbf{u}'$ is its velocity. If S and S' coincide at $t = 0$ and S' is moving with uniform speed v in the positive z-direction with respect to S, then the two sets of coordinates are related by the *Lorentz transformation*

$$
\begin{aligned}
x' &= x \\
y' &= y \\
z' &= \gamma(v)(z - vt) \\
t' &= \gamma(v)(t - vz/c^2)
\end{aligned}
\tag{B.1}
$$

where $\gamma(v) = (1 - \beta^2)^{-\frac{1}{2}}$ is the *Lorentz factor* and $\beta \equiv v/c$. From the definition of velocity and using these transformations, the particle's speed in S' is related to its speed in S by

$$
u' = \frac{u - v}{1 - uv/c^2}
\tag{B.2}
$$

Nuclear and Particle Physics   B. R. Martin
© 2006 John Wiley & Sons, Ltd

and hence

$$\gamma(u') \equiv [1 - (u'/c)^2]^{-\frac{1}{2}} = \gamma(u)\gamma(v)(1 - uv/c^2). \tag{B.3}$$

As $v \to 0$, the transformations in Equations (B.1) approach the Galilean transformations.

The most general Lorentz transformation has its simplest form in terms of *four-vectors*, whose general form is $a = (a_0, a_1, a_2, a_3) = (a_0, \mathbf{a})$. Then Equations (B.1) become

$$a'_0 = \gamma(a_0 - va_3/c); \quad a'_1 = a_1; \quad a'_2 = a_2; \quad a'_3 = \gamma(a_3 - va_0/c). \tag{B.4}$$

For example, the space-time four-vector is $x = (ct, \mathbf{x})$ and when used in Equations (B.4) reproduces Equations (B.1). The scalar product of two four-vectors a and b is defined as

$$ab \equiv a_0 b_0 - \mathbf{a} \cdot \mathbf{b} \tag{B.5}$$

and is an *invariant*, i.e. is the same in all inertial frames of references.

The basic four-vector in particle kinematics is the *four-momentum*, defined by

$$P \equiv mu, \tag{B.6}$$

where m is the *rest mass* and u is the *four-velocity*, whose components are $u = \gamma(v)(c, \mathbf{v})$, where $\mathbf{v}$ is the three-velocity and $v \equiv |\mathbf{v}|$. In terms of the *total energy* E (i.e. including the rest mass) and the three-momentum $\mathbf{p}$,

$$P = (E/c, \mathbf{p}). \tag{B.7}$$

Thus for two four-momenta P_1 and P_2 the invariant scalar product is

$$P_1 P_2 = E_1 E_2/c^2 - \mathbf{p}_1 \cdot \mathbf{p}_2 \tag{B.8}$$

and for $P_1 = P_2 = P$,

$$P^2 = E^2/c^2 - \mathbf{p}^2. \tag{B.9}$$

However, from Equations (B.5) and (B.6) we have $u^2 = c^2$ and hence $P^2 = m^2 c^2$, so combining this with Equation (B.9) gives

$$E^2 = \mathbf{p}^2 c^2 + m^2 c^4. \tag{B.10}$$

It follows that

$$E = \gamma(v)mc^2, \quad \mathbf{p} = \gamma(v)m\mathbf{v}, \quad \mathbf{v} = c^2 \mathbf{p}/E. \tag{B.11}$$

The Lorentz transformations for energy and momentum follow from these definitions and Equations (B.4). Thus, in S' we have

$$E' = m c^2 \gamma(u') = \gamma(v)(E - vp) \tag{B.12a}$$

and

$$p' = mu'\gamma(u') = \gamma(v)(p - vE/c^2), \tag{B.12b}$$

where $p = |\mathbf{p}|$ and $p' = |\mathbf{p}'|$. For a set of N non-interacting particles,

$$p'_z = \gamma(v)(p_z - vE/c^2); \quad p'_x = p_x; \quad p'_y = p_y; \tag{B.13a}$$

and

$$E' = \gamma(v)(E - vp_z), \tag{B.13b}$$

where

$$E = \sum_{i=1}^{N} E_i \quad \text{and} \quad \mathbf{p} = \sum_{i=1}^{N} \mathbf{p}_i. \tag{B.13c}$$

In the general case where the relative velocity $\mathbf{v}$ of the two frames is in an arbitrary direction, the transformations in Equations (B.12) become

$$\mathbf{p}' = \mathbf{p} + \gamma \mathbf{v} \left(\frac{\gamma \mathbf{v} \cdot \mathbf{p}}{\gamma + 1} - E \right) \frac{1}{c^2}, \quad E' = \gamma(E - \mathbf{v} \cdot \mathbf{p}). \tag{B.14}$$

B.2 Frames of Reference

The two most commonly used frames of reference for particle kinematics are the *laboratory system* (LS) and the *centre-of-mass system* (CMS). We will start by discussing these in the context of two-particle scattering. In the LS, a moving projectile a in a beam strikes a target particle b at rest, i.e.

$$P_a = (E_a/c, \mathbf{p}_a), \quad P_b = (m_b c, \mathbf{0}). \tag{B.15}$$

In the CMS, the three-momenta of the two particles a and b are equal and opposite, so that the total momentum is zero,[1] i.e.

$$P_a = (E_a/c, \mathbf{p}_a), \quad P_b = (E_b/c, \mathbf{p}_b), \tag{B.16a}$$

---

[1]Although 'centre-of-mass' system is the most frequently used name, some authors refer to this as the 'centre-of-momentum' system. Logically, a better name would be 'zero-momentum' frame.

with

$$\mathbf{p}_a + \mathbf{p}_b = \mathbf{0}. \tag{B.16b}$$

In a colliding beam accelerator, these two views become mixed. The colliding particles are both moving, but only if they have equal momenta and collide at zero crossing angle is the system identical to the centre-of-mass system.

The four-vectors of the initial-state particles in the two systems may be written (L = laboratory, T = target)

$$P_a = (E_L/c, 0, 0, p_L), \quad P_T = (m_T c, 0, 0, 0) \quad \text{LS} \tag{B.17a}$$

with $E_L^2 = m_B^2 c^4 + p_L^2 c^2$ (B = beam), and

$$P_a = (E_a/c, 0, 0, p), \quad P_b = (E_b/c, 0, 0, -p) \quad \text{CMS} \tag{B.17b}$$

with $E_a^2 = m_B^2 c^4 + p^2 c^2$ and $E_b^2 = m_T^2 c^4 + p^2 c^2$.
The Lorentz transformations between them are

$$p = \gamma(p_L - vE_L/c^2), \quad E_a = \gamma(E_L - vp_L), \tag{B.18}$$

where

$$v = \frac{c^2 p_L}{E_L + m_T c^2}, \quad \gamma = \frac{E_L + m_T c^2}{c^2 \sqrt{s}}, \quad v\gamma = \frac{p_L}{\sqrt{s}} \tag{B.19}$$

and s is the *invariant mass squared* of the system defined by

$$s \equiv (p_a + p_b)^2/c^2 = [(E_a + E_b)^2 - (\mathbf{p}_a c + \mathbf{p}_b c)^2]/c^4. \tag{B.20}$$

In particular, in the LS,

$$s = m_T^2 + m_B^2 + 2m_T E_L/c^2. \tag{B.21}$$

This result was used in Chapter 4 when discussing the relative merits of fixed-target and colliding beam accelerators.

Substituting Equations (B.19) into Equations (B.18) gives

$$p = \frac{p_L m_T}{\sqrt{s}}, \quad E_a = \frac{m_B^2 c^2 + m_T E_L}{\sqrt{s}} \tag{B.22a}$$

and similarly for particle b:

$$p = \frac{p_L m_T}{\sqrt{s}}, \quad E_b = \frac{m_T^2 c^2 + m_T E_L}{\sqrt{s}}. \tag{B.22b}$$

Finally we state, without proof, the transformation of scattering angles for the specific case of laboratory and centre-of-mass systems. Consider the general scattering reaction

$$B(E_L, \mathbf{p}_L) + T(m_T^2, \mathbf{0}) \rightarrow P(E, \mathbf{q}) + \cdots\cdots, \tag{B.23}$$

where B is a beam particle incident on a target particle T at rest in the laboratory system and P is one of a number of possible particles in the final state. If $\mathbf{p}_L$ is taken along the z-direction, then

$$\mathbf{p}_L = (0, 0, p_L) \quad \text{and} \quad \mathbf{q} = (0, q \sin\theta_L, q \cos\theta_L), \tag{B.24}$$

where θ_L is the scattering angle in the laboratory system, i.e. the angle between the beam direction and $\mathbf{q}$. In the CMS,

$$\mathbf{p}'_B + \mathbf{p}'_T = \mathbf{0}, \tag{B.25}$$

where $\mathbf{p}'_B$ and $\mathbf{p}'_T$ are the CMS momenta of the beam and target, respectively. The relation between the scattering angle θ_C in this system and θ_L is

$$\tan\theta_L = \frac{1}{\gamma(v)} \frac{q' \sin\theta_C}{q' \cos\theta_C + vE'/c^2}, \tag{B.26}$$

where

$$E' = m_P c^2 \gamma(u) \quad \text{and} \quad q' = m_P u \gamma(u) \tag{B.27}$$

and u is the magnitude of the velocity of P in the centre-of-mass frame.

It is instructive to consider the form of Equation (B.26) at high energies. From Equation (B.19) the velocity of the transformation is

$$v = p_L c^2 [E_L + m_T c^2]^{-1}, \tag{B.28}$$

so at high energies where $E_L^2 \approx p_L c \gg m_B c^2, m_T c^2$, $v \approx c(1 - m_T c/p_L) \approx c$ and

$$\gamma(v) \approx \left(\frac{p_L}{2m_T c}\right)^{1/2}. \tag{B.29}$$

Substituting Equations (B.27), (B.28) and (B.29) into Equation (B.26) gives

$$\tan\theta_L \approx \left(\frac{2m_T c}{p_L}\right)^{1/2} \cdot \frac{u \sin\theta_C}{u \cos\theta_C + c}. \tag{B.30}$$

Thus, unless $u \approx c$ and $\cos\theta_C \approx -1$, the final-state particles will lie in a narrow cone about the beam direction in the laboratory system. Similarly, when a high-energy particle decays, its decay products will emerge predominantly at small angles to the initial beam direction.

B.3 Invariants

The transformations between laboratory and centre-of-mass systems for energy and momentum have been worked out explicitly above, but a more efficient way is to work with quantities that are invariants, i.e. have the same values in all inertial frames. We have already met one of these: s the invariant mass squared, defined in Equation (B.20). We will now find expressions for the energy and momentum in terms of invariants for both the LS and the CMS.

First, in the LS, from Equations (B.15), we have

$$\mathbf{p}_B = \mathbf{0}, \quad E_B = m_B c. \tag{B.31}$$

However, from Equation (B.23),

$$s = m_B^2 + m_T^2 + 2m_T E_L / c^2 \tag{B.32}$$

i.e.

$$E_L = \frac{(s - m_T^2 - m_B^2)c^2}{2m_T} \tag{B.33}$$

and so

$$p_L^2 = \frac{E_L^2}{c^2} - m_B^2 c^2 = \frac{(s - m_B^2 - m_T^2)^2 c^2 - 4m_B^2 m_T^2 c^2}{4m_T^2}. \tag{B.34}$$

This can be written in the useful compact form

$$p_L = \frac{c}{2m_T} \lambda^{\frac{1}{2}}(s, m_B^2, m_T^2), \tag{B.35a}$$

where the *triangle function* λ is defined by

$$\lambda(x, y, z) \equiv (x - y - z)^2 - 4yz. \tag{B.35b}$$

This function is invariant under all permutations of its arguments and in particular Equation (B.35a) can be written in the form

$$p_L = \frac{c}{2m_T} \left\{ \left[s - (m_T + m_B)^2 \right] \left[s - (m_T - m_B)^2 \right] \right\}^{\frac{1}{2}}. \tag{B.36}$$

In a similar way it is straightforward to show that, in the CMS,

$$p = \frac{c}{2\sqrt{s}} \left\{ \left[s - (m_T + m_B)^2 \right] \left[s - (m_T - m_B)^2 \right] \right\}^{\frac{1}{2}} \tag{B.37}$$

from which it follows that

$$E_a = \frac{(s + m_B^2 - m_T^2)c^2}{2\sqrt{s}}, \quad E_b = \frac{(s - m_B^2 + m_T^2)c^2}{2\sqrt{s}}. \tag{B.38}$$

The above formulae have many applications. For example, if we wish to produce particles with a certain mass M, the minimum laboratory energy of the beam particles is, from Equation (B.33),

$$E_L(\text{min}) = \frac{M^2 c^2 - m_B^2 c^2 - m_T^2 c^2}{2 m_T}. \tag{B.39}$$

In the case of the decay of a particle A to a set of final-state particles $i = 1, 2, 3, \ldots, N$, i.e.

$$A \rightarrow 1 + 2 + 3 + \cdots + N, \tag{B.40}$$

the invariant mass W of the final-state particles is given by

$$W^2 c^4 = \left(\sum_i E_i \right)^2 - \left(\sum_i \mathbf{p}_i c \right)^2 = E_A^2 - (\mathbf{p}_A c)^2 = M_A^2 c^4. \tag{B.41}$$

Hence the mass of the decaying particle is equal to the invariant mass of its decay products. The latter can be measured if the particle is too short-lived for its mass to be measured directly.

Problems

B.1 The *Mandelstam variables s, t and u* are defined for the reaction $A + B \rightarrow C + D$ by

$$s = (p_A + p_B)^2/c^2, \quad t = (p_A - p_C)^2/c^2, \quad u = (p_A - p_D)^2/c^2,$$

where p_A etc. are the relevant energy-momentum four-vectors.

(a) Show that

$$s + t + u = \sum_{j=A,B,C,D} m_j^2.$$

(b)   In the case of elastic scattering show that $t = -2p^2(1 - \cos\theta)/c^2$, where $p \equiv |\mathbf{p}|$, $\mathbf{p}$ is the centre-of-mass momentum of particle A and θ is its scattering angle in the CMS.

B.2   A pion travelling with speed $v \equiv |\mathbf{v}|$ in the laboratory decays via $\pi \rightarrow \mu + \nu$. If the neutrino emerges at right angles to $\mathbf{v}$, find an expression for the angle θ at which the muon emerges.

B.3   A pion at rest decays via $\pi \rightarrow \mu + \nu$. Find the speed of the muon in terms of the masses involved.

B.4   A neutral particle X^0 decays via $X^0 \rightarrow A^+ + B^-$. The momentum components of the final-state particles are measured to be (in GeV/c):

	p_x	p_y	p_z
A^+	−0.488	−0.018	2.109
B^-	−0.255	−0.050	0.486

Test the hypotheses that the decay is (a) $D^0 \rightarrow \pi^+ + K^-$ and (b) $\Lambda \rightarrow p + \pi^-$.

B.5   In a fixed-target $e^- p$ scattering experiment, show that the squared four-momentum transfer is given by $Q^2 \approx 2E^2(1 - \cos\theta)/c^2$, where E is the total laboratory energy of the initial electron and θ is the laboratory scattering angle.

B.6   Calculate the minimum laboratory energy $E_{\min}$ of the initial proton for the production of antiprotons in a fixed-target experiment using the reaction $pp \rightarrow ppp\bar{p}$. If the protons are bound in nuclei, show that taking the internal motion of the nucleons into account leads to a smaller minimum energy given by

$$E'_{\min} \approx (1 - p/m_p c)E_{\min},$$

where p is the modulus of the average internal longitudinal momentum of a nucleon. Use a typical value of p to calculate $E'_{\min}$.

B.7   A particle A decays at rest via $A \rightarrow B + C$. Find the total energy of B in terms of the three masses.

B.8   A meson M decays via $M \rightarrow \gamma\gamma$. Find an expression for the angle in the laboratory between the two momentum vectors of the photons in terms of the photon energies and the mass of M.

B.9   Pions and protons, both with momentum 2 GV/c, travel between two scintillation counters distance L m apart. What is the minimum value of L necessary to differentiate between the particles if the time-of-flight can be measured with an accuracy of 200 ps?

B.10 A photon is Compton scattered off a stationary electron through a scattering angle of 60° and its final energy is half its initial energy. Calculate the value of the initial energy in MeV.

Appendix C
Rutherford Scattering

C.1 Classical Physics

In Chapter 1 we commented on the experiments of Geiger and Marsden that provided evidence for the existence of the nucleus. They scattered low-energy α-particles from thin gold foils and observed that sometimes the projectiles were scattered through large angles, in extreme cases close to $180°$. If we start for the moment by ignoring the fact that there is a Coulomb interaction present, then it is easy to show that this behaviour is incompatible with scattering from light particles such as electrons.

Consider the non-relativistic elastic scattering of an α-particle of mass m_α and initial velocity $\mathbf{v}_i$ from a target of mass m_t stationary in the laboratory. If the final velocities are $\mathbf{v}_f$ and $\mathbf{v}_t$, respectively, then we have the situation as shown in Figure C.1.

Conservation of linear momentum and kinetic energy are:

$$m_\alpha \mathbf{v}_i = m_\alpha \mathbf{v}_f + m_t \mathbf{v}_t \tag{C.1}$$

and

$$m_\alpha v_i^2 = m_\alpha v_f^2 + m_t v_t^2, \tag{C.2}$$

where $v_i = |\mathbf{v}_i|$ etc.. Squaring Equation (C.1) we obtain

$$m_\alpha v_i^2 = m_\alpha v_f^2 + \frac{m_t^2}{m_\alpha} v_t^2 + 2m_t(\mathbf{v}_f \cdot \mathbf{v}_t) \tag{C.3}$$

and hence from Equation (C.2),

$$v_t^2 \left(1 - \frac{m_t}{m_\alpha}\right) = 2\mathbf{v}_f \cdot \mathbf{v}_t. \tag{C.4}$$

Nuclear and Particle Physics   B. R. Martin
© 2006 John Wiley & Sons, Ltd

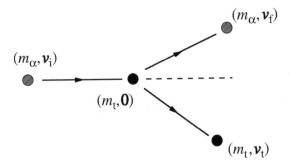

Figure C.1 Kinematics of the Geiger and Marsden experiment

Thus, if the target is an electron, with $m_t = m_e \ll m_\alpha$, the directions of motion of the outgoing α-particle and the recoiling target are essentially along the direction of the initial α-particle and no large-angle scatterings are possible. Such events could, in principle, be due to multiple small-angle scattering, but the thinness of the gold foil target rules this out.[1] If, however, $m_t = m_{Au} \gg m_\alpha$, then the left-hand side of Equation (C.4) will be negative and large scattering angles are possible.

The above only makes plausible the existence of a heavy nucleus, because it has ignored the existence of the Coulomb force, so we now have to take this into account. We will do this first using non-relativistic classical mechanics.

Consider the non-relativistic Coulomb scattering of a particle (the projectile) of mass m and electric charge ze from a target particle of mass M and electric charge Ze. The kinematics of this are shown in Figure C.2. The target mass is assumed to be sufficiently large that its recoil may be neglected. The initial velocity of the projectile is $\mathbf{v}$ and it is assumed that in the absence of any interaction it would travel in a straight line and pass the target at a distance b (called the *impact*

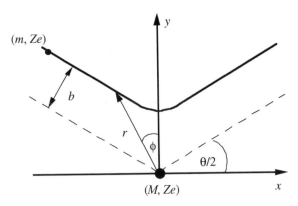

Figure C.2 Kinematics of Rutherford scattering

---

[1]For completeness one should also show that the observations cannot be due to scattering from the diffuse positive charge present. This was done by the authors of the original experiment.

parameter). The derivation follows from considering the implications of linear and angular momentum conservation.

Angular momentum conservation implies that

$$mvb = mr^2 \frac{d\phi}{dt}, \tag{C.5}$$

where $v = |\mathbf{v}|$. Since the scattering is symmetric about the y-axis, the component of linear momentum in the y-direction is initially $p = -mv\sin(\theta/2)$ and changes to $+mv\sin(\theta/2)$ after the interaction, i.e. the total change in momentum in the y-direction is

$$\Delta p = 2mv\sin(\theta/2). \tag{C.6}$$

The change in momentum may also be calculated by integrating the impulse in the y-direction due to the Coulomb force on the projectile. This gives

$$\Delta p = \int\limits_{-\infty}^{+\infty} \frac{zZe^2}{4\pi\varepsilon_0 r^2} \cos\phi \, dt, \tag{C.7}$$

where we have taken $t = 0$ to coincide with the origin of the x-axis. Using Equation (C.5) to change variables, Equation (C.7) may be written

$$2mv\sin(\theta/2) = \frac{zZe^2}{4\pi\varepsilon_0} \left(\frac{1}{bv}\right) \int\limits_{-\phi}^{+\phi} \cos\phi \, d\phi, \tag{C.8}$$

which, using $\phi = (\pi - \theta)/2$, yields

$$b = \frac{zZe^2}{8\pi\varepsilon_0} \cdot \frac{1}{E_{kin}} \cot(\theta/2), \tag{C.9}$$

where $E_{kin} = \frac{1}{2}mv^2$ is the kinetic energy of the projectile.

Finally, we need to calculate the differential cross-section. If the initial flux of projectile particles crossing a plane perpendicular to the beam direction is J, then the intensity of particles having impact parameters between b and $b + db$ is $2\pi b \, J db$ and this is equal to the rate dW at which particles are scattered into a solid angle $d\Omega = 2\pi\sin\theta d\theta$ between θ and $\theta + d\theta$. Thus

$$dW = 2\pi bJ \, db. \tag{C.10}$$

However, from Equation (1.47) and considering a single target particle,

$$dW = J\frac{d\sigma}{d\Omega}d\Omega = 2\pi J \sin\theta \, d\theta \frac{d\sigma}{d\Omega}, \tag{C.11}$$

i.e.

$$\frac{d\sigma}{d\Omega} = \frac{b}{\sin\theta} \cdot \frac{db}{d\theta}. \tag{C.12}$$

The right-hand side of Equation (C.12) may be evaluated from Equation (C.9) and gives

$$\frac{d\sigma}{d\Omega} = \left(\frac{zZe^2}{16\pi\,\varepsilon_0\,E_{\text{kin}}}\right)^2 \operatorname{cosec}^4\,(\theta/2). \tag{C.13}$$

This is the final form of the Rutherford differential cross-section for non-relativistic scattering.

C.2 Quantum Mechanics

While Equation (C.13) is adequate to describe the Geiger and Marsden experiments, in the case of electron scattering we need to take account of both relativity and quantum mechanics. This may be done using the general formalism for the differential cross-section in terms of the scattering potential that was derived in Chapter 1.

The starting point is Equation (1.55), which in the present notation is

$$\frac{d\sigma}{d\Omega} = \frac{1}{4\pi^2\hbar^4}\frac{p'^2}{vv'}\left|\mathcal{M}(\mathbf{q}^2)\right|^2, \tag{C.14}$$

where $\mathbf{v}$ and $\mathbf{p}$ are the velocity and momentum respectively of the projectile (which for convenience we take to have a unit negative charge) with $v = |\mathbf{v}|$, $p = |\mathbf{p}|$ and the primes refer to the final-state values. The matrix element is given by

$$\mathcal{M}(\mathbf{q}) = \int V(\mathbf{x})e^{i\mathbf{q}\cdot\mathbf{x}/\hbar}d\mathbf{x}, \tag{C.15}$$

where $\mathbf{q} = \mathbf{p} - \mathbf{p}'$ is the momentum transfer. $V(\mathbf{x})$ is the Coulomb potential

$$V(\mathbf{x}) = V_{\text{C}}(\mathbf{x}) = -\frac{\alpha Z(\hbar c)}{r}, \tag{C.16}$$

where $r = |\mathbf{x}|$ and Ze is the charge of the target nucleus. Inspection of the integral in Equation (C.15) shows that it diverges at large r. However, in practice, charges are always screened at large distances by intervening matter and so we will interpret the integral as

$$\mathscr{M}_C(q) = \underset{\lambda \to 0}{\text{Lt}} \int \left(-\frac{Z\alpha(\hbar c)e^{-\lambda r}}{r} \right) e^{i\mathbf{q}\cdot\mathbf{x}/\hbar} \, d^3\mathbf{x}. \tag{C.17}$$

To evaluate this, take $\mathbf{q}$ along the x-axis, so that in spherical polar coordinates $\mathbf{q} \cdot \mathbf{x} = qr\cos\theta$. The angular integration may then be done and yields

$$\mathscr{M}_C(q) = -\frac{4\pi(\hbar c)Z\alpha\hbar}{q} \underset{\lambda \to 0}{\text{Lt}} \int_0^\infty e^{-\lambda r} \sin(qr/\hbar) dr. \tag{C.18}$$

The remaining integral may be done by parts (twice) and taking the limit $\lambda \to 0$ gives

$$\mathscr{M}_C(\mathbf{q}) = -\frac{4\pi(\hbar c)Z\alpha\hbar^2}{q^2}. \tag{C.19}$$

Finally, substituting Equation (C.19) into Equation (C.14) gives

$$\frac{d\sigma}{d\Omega} = 4Z^2\alpha^2(\hbar c)^2 \frac{p'^2}{vv'q^4}, \tag{C.20}$$

which is the general form of the Rutherford differential cross-section. To see that this is the same as Equation (C.13) in the non-relativistic limit, we may substitute the non-relativistic approximations

$$p^2 = p'^2 = 2mE_{kin}, \quad \text{and} \quad v = v' = \sqrt{2E_{kin}/m}, \tag{C.21}$$

together with the kinematic relation for the scattering angle

$$q = 2p\sin(\theta/2), \tag{C.22}$$

into Equation (C.20). The result in Equation (C.13) follows immediately.

Because we are assuming that the target mass is heavy so that its recoil may be neglected, to a good approximation $p = p'$ and $E = E'$, where E is the total energy of the electron. Also for relativistic electrons $v = v' \approx c$ and $E \approx pc$. Using these conditions together with Equation (C.22) in Equation (C.20), gives the relativistic result for the Rutherford differential cross-section in the convenient form:

$$\frac{d\sigma}{d\Omega} = \frac{Z^2\alpha^2(\hbar c)^2}{4E^2 \sin^4(\theta/2)}, \tag{C.23}$$

which is the form used in Chapter 2 and elsewhere.

Problems

C.1 Calculate the differential cross-section in mb/sr for the scattering of a 20 MeV α-particle through an angle $20°$ by a nucleus $^{209}_{83}$Bi, stating any assumptions made. Ignore spin and form factor effects.

C.2 Show that in Rutherford scattering at a fixed impact parameter b, the distance of closest approach d to the nucleus is given by $d = b[1 + \text{cosec}\,(\theta/2)]/\text{cosec}\,(\theta/2)$, where θ is the scattering angle.

C.3 Find an expression for the impact parameter b in the case of small-angle Rutherford scattering. A beam of protons with speed $v = 4 \times 10^7$ ms^{-1} is incident normally on a thin foil of $^{194}_{78}$Pt, thickness 10^{-5} m (density $= 2.145 \times 10^4$ kg m^{-3}). Estimate the proportion of protons that experience double scattering, where each scattering angle is at least $5°$.

Appendix D
Solutions to Problems

Chapter 1

1.1 Substituting the operators $\mathbf{p} = -i\hbar\partial/\partial\mathbf{x}$ and $E = i\hbar\partial/\partial t$ into the mass–energy relation $E^2 = p^2 c^2 + M^2 c^4$ and allowing the operators to act on the function $\phi(\mathbf{x}, t)$, leads immediately to the Klein–Gordon equation. To verify that the Yukawa potential $V(r)$ is a static solution of the equation, set $V(r) = \phi(\mathbf{x})$, where $r = |\mathbf{x}|$, and use

$$\nabla^2 = \frac{\partial^2}{\partial r^2} + \frac{2}{r}\frac{\partial}{\partial r}$$

together with the expression for the range, $R = \hbar/Mc$.

1.2 Using Equation (1.11), gives

$$\hat{P}Y_1^1 = \sqrt{\frac{3}{8}}\sin(\pi - \theta)e^{i(\pi + \phi)} = -\sqrt{\frac{3}{8}}\sin(\theta)e^{i\phi} = -Y_1^1,$$

and hence Y_1^1 is an eigenfunction of parity with eigenvalue -1.

1.3 Because the initial state is at rest, it has $L = 0$ and thus its parity is $P_i = P_p P_{\bar{p}}(-1)^L = -1$, where we have used the fact that the fermion–antifermion pair has overall negative intrinsic parity. In the final state, the neutral pions are identical bosons and so their wavefunction must be totally symmetric under their interchange. This implies even orbital angular momentum L' between them and hence $P_f = P_\pi^2(-1)^{L'} = 1 \neq P_i$. The reaction violates parity conservation and is thus forbidden as a strong interaction.

1.4 Since $\hat{C}^2 = 1$, we must have $\hat{C}^2|b, \psi_b\rangle = C_b \hat{C}|\bar{b}, \psi_{\bar{b}}\rangle = |b, \psi_b\rangle$, implying that $\hat{C}|\bar{b}, \psi_{\bar{b}}\rangle = C_{\bar{b}}|b, \psi_b\rangle$ with $C_b C_{\bar{b}} = 1$ independent of C_b. The result follows because an eigenstate of $\hat{C}$ must contain only particle–antiparticle pairs $b\bar{b}$, leading to the intrinsic parity factor $C_b C_{\bar{b}} = 1$, independent of C_b.

Nuclear and Particle Physics B. R. Martin
© 2006 John Wiley & Sons, Ltd

1.5 The parity of the deuteron is $P_d = P_p P_n (-1)^{L_{pn}}$. Since the deuteron is an S-wave bound state, $L_{pn} = 0$ and so, using $P_p = P_n = 1$, gives $P_d = 1$. The parity of the initial state is therefore $P_i = P_{\pi^-} P_d (-1)^{L_{\pi d}} = P_{\pi^-}$, because the pion is at rest and so $L_{\pi d} = 0$. The parity of the final state is $P_f = P_n P_n (-1)^{L_{nn}} = (-1)^{L_{nn}}$ and therefore $P_{\pi^-} = (-1)^{L_{nn}}$. To find L_{nn} impose the condition that $\psi_{nn} = \psi_{\text{space}} \psi_{\text{spin}}$ must be antisymmetric. Examining the spin, Equation (1.17) shows that there are two possibilities for ψ_{spin}: either the symmetric $S = 1$ state or the $S = 0$ antisymmetric state. If $S = 0$, then ψ_{space} would have to be symmetric, implying L_{nn} would be even, but the total angular momentum would not then be conserved. Thus $S = 1$ is implied and ψ_{space} is antisymmetric, i.e. $L_{nn} = 1, 3, \cdots$. The only way to combine L_{nn} and S to give $J = 1$ is with $L_{nn} = 1$ and hence $P_{\pi^-} = -1$.

1.6 (a) $\nu_e + e^+ \rightarrow \nu_e + e^+$;

(b) $p + p \rightarrow p + p + \pi^0 + \pi^0$;

(c) $\bar{p} + n \rightarrow \pi^- + \pi^0 + \pi^0, \quad \pi^- + \pi^+ + \pi^-$.

1.7 (a) $\nu_e + \nu_\mu \rightarrow \nu_e + \nu_\mu$.

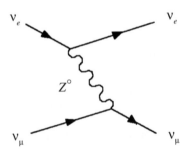

(b) $n \rightarrow p + e^- + \bar{\nu}_e$.

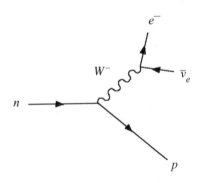

(c) $e^+ + e^- \rightarrow e^+ + e^-$.

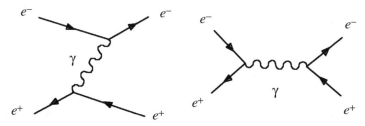

(d) $\gamma + \gamma \rightarrow e^+ + e^-$.

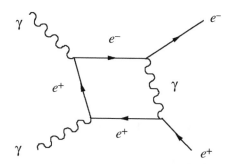

1.8 If an exchanged particle approaches to within a distance d fm, this is equivalent to a momentum transfer $q = \hbar/d = (0.2/d)$ GeV/c. Thus, $q = 0.2$ GeV/c for $d = 1$ fm and $q = 200$ GeV/c for $d = 10^{-3}$ fm. The scattering amplitude is given by $f(q^2) = -g^2\hbar^2 \left[q^2 + m_x^2 c^2\right]^{-1}$, where m_x is the mass of the exchanged particle. Thus,

$$R(q^2) \equiv \frac{f_{\text{EM}}(q^2)}{f_{\text{Weak}}(q^2)} = \frac{q^2 c^2 + m_W^2 c^4}{q^2 c^2 + m_\gamma^2 c^4},$$

since $g_{\text{EM}} \approx g_{\text{Weak}}$. Using $m_\gamma = 0$ and $m_W = 80$ GeV/c^2, gives

$$R(0.2 \text{ GeV/c}) \approx 1.6 \times 10^5 \text{ fm} \quad \text{but} \quad R(200 \text{ GeV/c}) \approx 1.2 \text{ fm}.$$

1.9 Using spherical polar coordinates, we have $\mathbf{q} \cdot \mathbf{x} = qr\cos\theta$ and $d^3\mathbf{x} = r^2\,dr\,d\cos\theta\,d\phi$, where $q = |\mathbf{q}|$. Thus, from Equation (1.38),

$$f(q^2) = \frac{-g^2}{4\pi} \int\limits_0^{2\pi} d\phi \int\limits_0^\infty dr\, r^2 \frac{e^{-r/R}}{r} \int\limits_{-1}^{+1} d\cos\theta\, \exp(iqr\cos\theta/\hbar)$$

$$= \frac{-g^2\hbar}{2iq} \int\limits_0^\infty dr\, e^{-r/R} [\exp(iqr\cos\theta/\hbar)]_{-1}^{+1} = \frac{-g^2\hbar}{2iq} \int\limits_0^\infty dr\, e^{-r/R} \left[e^{iqr/\hbar} - e^{-iqr/\hbar}\right]$$

$$= \frac{-g^2\hbar^2}{q^2 + m^2 c^2}$$

1.10   Let one of the beams (labelled by 1) refer to the 'beam' and let the other beam (labelled by 2) refer to the 'target'. Then in Equation (1.43), $n_b = nN_1/2\pi RA$ and $v_i = 2\pi R/T$, where R is the radius of the circular path. Thus the flux is $J = n_b v_i = nN_1 f/A$, where f is the frequency. Also $N = N_2$, so finally the luminosity is $L = JN = nN_1N_2 f/A$.

1.11   From Equation (1.44c), $\sigma = WM_A/I(\rho t)N_A$. Since the scattering is isotropic, the total number of protons emitted from the target is $W = 20 \times (4\pi/2 \times 10^{-3})$ $= 1.25 \times 10^5 \, \text{s}^{-1}$. I can be calculated from the current, noting that the α-particles carry two units of charge, and is $I = 3.13 \times 10^{10} \, \text{s}^{-1}$. The density of the target is $\rho t = 1 \, \text{mg cm}^{-2} = 10^{-32} \, \text{kg fm}^{-2}$. Putting everything together gives $\sigma = 161 \, \text{mb}$.

Chapter 2

2.1   From Equation (2.21),

$$F(q^2) = \frac{4\pi\hbar}{q} \int_0^r \rho r \sin b(r) dr \left[4\pi \int_0^r r^2 \, dr\right]^{-1} = 3[\sin b(a) - b(a)\cos b(a)]b^{-3},$$

where $b(r) = qr/\hbar$. To evaluate this we need to find a and q. For the latter, we have

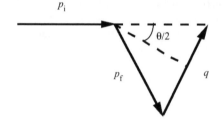

from which $q = 2p\sin(\vartheta/2) = 57.5 \, \text{MeV/c}$. Also, we know that $a = 1.21A^{\frac{1}{3}}$ fm and so for $A = 56$, $a = 4.63$ fm and $qa/\hbar = 1.35$ radians. Finally, using this in the integral, gives $F = 0.829$ and hence the reduction is $F^2 = 0.69$.

2.2   Setting $q = |\mathbf{q}|$ in Equation (2.26), we have

$$F(q^2) = \frac{1}{Ze} \int f(\mathbf{x}) \sum_{n=0}^{\infty} \frac{1}{n!} \left(\frac{iqr\cos\theta}{\hbar}\right)^n d^3\mathbf{x}.$$

Using $d^3\mathbf{x} = r^2 d\cos\theta \, d\phi$ and doing the ϕ integral, gives

$$F(q^2) = \frac{2\pi}{Ze} \int\int f(r)r^2\left[1 + \frac{iqr\cos\theta}{\hbar} - \frac{q^2r^2\cos^2\theta}{\hbar^2} + \dots\right] dr \, d\cos\theta$$

$$= \frac{4\pi}{Ze} \int_0^\infty f(r)r^2 dr - \frac{4\pi q^2}{6Ze\hbar^2} \int_0^\infty f(r)r^4 dr + \dots$$

However, from Equation (2.17), $Z e = 4\pi \int_0^\infty f(r)\, r^2 dr$ and from Equation (2.25),
$Z e \langle r^2 \rangle = 4\pi \int_0^\infty f(r)\, r^4 dr$ so $F(\mathbf{q}^2) = 1 - \frac{q^2}{6\hbar^2}\langle r^2 \rangle + \cdots$

2.3 From Equation (2.28), $\langle r^2 \rangle = 6\hbar^2[1 - F(q^2)]/q^2$, where $q = 2E\sin(\theta/2)$. Thus, $q = 43.6\,\mathrm{MeV/c}$. Also, $F^2 = 0.65$ and so $\sqrt{\langle r^2 \rangle} = 6.56\,\mathrm{fm}$.

2.4 The charge distribution is spherical, so the angular integrations in the general result of Equation (2.17) may be done, giving

$$F(\mathbf{q}^2) = \left[\int_0^\infty \rho(r)[\sin(qr/\hbar)/(qr/\hbar)]4\pi r^2 dr\right]\left[\int_0^\infty \rho(r)4\pi r^2 dr\right]^{-1}.$$

Substituting for $\rho(r)$, setting $x = r/a$ and using $\int_0^\infty x\exp(-x)\,dx = 1$, gives, after integrating by parts (twice),

$$F(\mathbf{q}^2) = \left(\frac{\hbar}{qa}\right)\int_0^\infty e^{-x}\sin\left(\frac{qax}{\hbar}\right)dx = \frac{1}{[1 + q^2 a^2/\hbar^2]}.$$

2.5 In 1 g of the isotope there are initially $N_0 = \left(1\,\mathrm{g}/208 \times 1.66 \times 10^{-24}\,\mathrm{g}\right)$. Thus $N_0 = 2.9 \times 10^{21}$ atoms. At time t there are $N(t) = N_0 e^{-t/\tau}$ atoms, where τ is the mean life of the isotope. Thus, provided $t \ll \tau$, the average decay rate is

$$\frac{N_0 - N(t)}{t} \approx \frac{N_0}{\tau} = \frac{75}{0.1 \times 24}\,\mathrm{h}^{-1}.$$

Thus, $\tau = 2.4N_0/75\,\mathrm{h} \approx 10^{16}$ years.

2.6 The count rate is proportional to the number of ^{14}C atoms present in the sample. If we assume that the abundance of ^{14}C has not changed with time, the artefact was made from living material and is predominantly carbon, then at the time it was made $(t = 0)$, 1 g would have contained 5×10^{22} carbon atoms of which $N_0 = 6 \times 10^{10}$ would have been ^{14}C. Thus the average count rate would have been $N_0/\tau = 13.8\,\mathrm{m}^{-1}$. At time t, the number of ^{14}C atoms would be $N(t) = N_0 \exp(-t/\tau)$ and $N(t)/N_0 = e^{-t/\tau} = 2.1/13.8$, from which $t = \tau \ln 6.57 = 1.56 \times 10^4$ years. The artefact is approximately 16 000 years old.

2.7 If the transition rate for $^{212}_{86}\mathrm{Rn}$ decay is ω_1 and that for $^{208}_{84}\mathrm{Po}$ is ω_2 and if the numbers of each of these atoms at time t is $N_1(t)$ and $N_2(t)$, respectively, then the decays are governed by Equation (2.43), i.e. $N_2(t) = \omega_1 N_1(0)\left[\exp(-\omega_1 t) - \exp(-\omega_2 t)\right][\omega_2 - \omega_1]^{-1}$. The latter is a maximum when $dN_2(t)/dt = 0$, i.e. when $\omega_2 \exp(-\omega_2 t) = \omega_1 \exp(-\omega_1 t)$, with $t_{max} = \ln(\omega_1/\omega_2)(\omega_1 - \omega_2)^{-1}$. Using $\omega_1 = 4.12 \times 10^{-2}\,\mathrm{min}^{-1}$ and $\omega_2 = 6.58 \times 10^{-7}\,\mathrm{min}^{-1}$, gives $t_{max} = 265\,\mathrm{min}$.

2.8 The total decay rate of both modes of $^{138}_{57}$La is

$$(1 + 0.5) \times (7.8 \times 10^2)\,\mathrm{kg^{-1}\,s^{-1}} = 1.17 \times 10^3\,\mathrm{kg^{-1}\,s^{-1}}.$$

Also, since this isotope is only 0.09 per cent of natural lanthanum, the number of $^{138}_{57}$La atoms per kg is $N = (9 \times 10^{-4}) \times (1000/138.91) \times (6.022 \times 10^{23})$, i.e. $N = 3.90 \times 10^{21}\,\mathrm{kg^{-1}}$. The rate of decays is $-dN/dt = \omega N$, where ω is the transition rate, and in terms of this the mean lifetime $\tau = 1/\omega$. Thus,

$$\tau = \frac{N}{-dN/dt} = \frac{3.90 \times 10^{21}}{1.17 \times 10^3}\,\mathrm{s} = 3.33 \times 10^{18}\mathrm{s} = 1.06 \times 10^{11}\ \text{years}.$$

2.9 The energy released is the increase in binding energy. Now from the SEMF, Equations (2.46)–(2.52),

$$BE(35, 87) = a_v(87) - a_s(87)^{2/3} - a_c\frac{(35)^2}{(87)^{1/3}} - a_a\frac{(87 - 70)^2}{348},$$

$$BE(57, 145) = a_v(145) - a_s(145)^{2/3} - a_c\frac{(57)^2}{(145)^{1/3}} - a_a\frac{(145 - 114)^2}{580},$$

$$BE(92, 235) = a_v(235) - a_s(235)^{2/3} - a_c\frac{(92)^2}{(235)^{1/3}} - a_a\frac{(235 - 184)^2}{940}.$$

The energy released is thus

$$E = BE(35, 87) + BE(57, 145) - BE(92, 235)$$
$$= -3\,a_v - 9.153\,a_s + 476.7 a_c + 0.280\,a_a$$

which using the values given in Equation (2.54) gives $E = 154\,\mathrm{MeV}$.

2.10 The most stable nucleus for fixed A has a Z-value given by $Z = \beta/2\gamma$, where from Equation (2.58), $\beta = a_a + (M_n - M_p - m_e)$ and $\gamma = a_a/A + a_c/(A)^{1/3}$. Changing α would not change a_a, but would effect the Coulomb coefficient because a_c is proportional to α. For $A = 111$, using the value of a_a from Equation (2.54) gives $\beta = 93.93\,\mathrm{MeV}/c^2$ and $\gamma = 0.839 + 0.208\,a_c\,\mathrm{MeV}/c^2$. For $Z = 47$, $a_c = 0.770\,\mathrm{MeV}/c^2$. This is a change of about 10 per cent from the value given in Equation (2.54) and so α would have to change by the same percentage.

2.11 In the rest frame of the $^{269}_{108}$Hs nucleus, $m_\alpha v_\alpha = m_{Sg} v_{Sg}$. The ratio of the kinetic energies is $E_{Sg}/E_\alpha = m_\alpha/m_{Sg}$ and the total kinetic energy is $E_\alpha(1 + m_\alpha/m_{Sg}) = 9.370\,\mathrm{MeV}$. Thus, $m_{Hs}c^2 = (m_{Sg} + m_\alpha)c^2 + 9.370\,\mathrm{MeV} = 269.154\,\mathrm{u}$.

2.12 If there are N_0 atoms of $^{238}_{94}$Pu at launch, then after t years the activity of the source will be $A(t) = N_0 \exp(-t/\tau)/\tau$, where τ is the lifetime. The instantaneous power is then $P(t) = A(t) \times 0.05 \times 5.49 \times 1.602 \times 10^{-13}\,\mathrm{W} > 200\,\mathrm{W}$. Substituting the value given for τ, gives $N_0 = 1.88 \times 10^{25}$ and hence the weight of $^{238}_{94}$Pu at launch would have to be at least $\left(\dfrac{1.88 \times 10^{25}}{6.02 \times 10^{23}}\right)\left(\dfrac{238}{1000}\right)\mathrm{kg} = 7.43\,\mathrm{kg}$.

2.13 If there were N_0 atoms of each isotope at the formation of the planet ($t = 0$), then after time t the numbers of atoms are $N_{205}(t) = N_0 \exp(-t/\tau_{205})$ and $N_{204}(t) = N_0 \exp(-t/\tau_{204})$, with

$$\frac{N_{205}(t)}{N_{204}(t)} = \exp\left[-t\left(\frac{1}{\tau_{205}} - \frac{1}{\tau_{204}}\right)\right] = \frac{n_{205}}{n_{204}} = 2 \times 10^{-7}.$$

Now $\tau_{204} \gg \tau_{205}$, so $t = \tau_{205}\ln(2 \times 10^7) = 2.6 \times 10^8$ years.

2.14 We first calculate the mass difference between $[p + {}^{46}_{21}\text{Sc}]$ and $[n + {}^{46}_{22}\text{Ti}]$. Using the information given, we have

$$M(21,46) - [M(22,46) + m_e] = 2.37\,\text{MeV}/c^2 \quad \text{and} \quad M_n - (M_p + m_e) = 0.78\,\text{MeV}/c^2$$

and hence $[M_p + M(21,46)] - [M_n + M(22,46)] = 1.59\,\text{MeV}/c^2$. We also need the mass differences $[M_\alpha + M(20,43)] - [M_n + M(22,46)] = 0.07\,\text{MeV}/c^2$. We can now draw the energy level diagram where the centre-of-mass energy of the resonance is at (see Equation (2.10)) $2.76 \times (45/47) = 2.64\,\text{MeV}$.

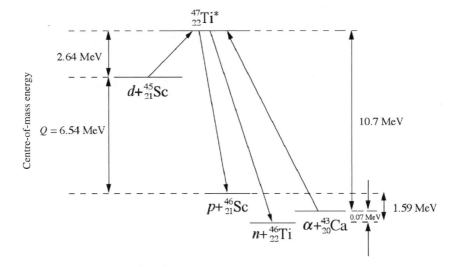

Thus the resonance could be excited in the ${}^{43}_{20}\text{Ca}(\alpha, n){}^{46}_{22}\text{Ti}$ reaction at an α-particle laboratory energy of $10.7 \times (47/43) = 11.7\,\text{MeV}$.

2.15 We have $dN(t)/dt = P - \lambda N$, from which

$$Pe^{\lambda t} = e^{\lambda t}\left(\lambda N + \frac{dN(t)}{dt}\right) = \frac{d}{dt}\left(Ne^{\lambda t}\right).$$

Integrating and using the fact that $N = 0$ at $t = 0$ to determine the constant of integration, gives the required result.

2.16 The number of ^{35}Cl atoms in 1 g of the natural chloride is

$$N = 2 \times 0.758 \times N_A/\text{molecular weight} = 7.04 \times 10^{21}.$$

The activity $\mathscr{A}(t) = \lambda N = P(1 - e^{-\lambda t}) \approx P\lambda t$, since $\lambda t \ll 1$. So

$$t = \frac{\mathscr{A}(t)}{P\lambda} = \frac{\mathscr{A}(t)t_{1/2}}{\ln 2 \times \sigma \times F \times N}.$$

Substituting $\mathscr{A}(t) = 3 \times 10^5$ Bq and using the other constants given, yields $t = 1.55$ days.

2.17 At very low energies we may assume the scattering has $\ell = 0$ and so in Equation (1.63) we have $j = \frac{1}{2}, s_n = \frac{1}{2}$ and $s_u = 0$. Thus,

$$\sigma_{\max} = \frac{\pi\hbar^2}{q_n^2} \frac{(\Gamma_n\Gamma_n + \Gamma_n\Gamma_\gamma)}{\Gamma^2/4} = \frac{4\pi\hbar^2\Gamma_n}{q_n^2\Gamma},$$

Therefore, $\Gamma_n = q_n^2\Gamma\sigma_{\max}/4\pi\hbar^2 = 0.35 \times 10^{-3}$ eV and $\Gamma_\gamma = \Gamma - \Gamma_n = 9.65 \times 10^{-3}$ eV.

Chapter 3

3.1 (a) Forbidden: violates L_μ conservation, because $L_\mu(\nu_\mu) = 1$, but $L_\mu(\mu^+) = -1$.

   (b) Forbidden: violates electric charge conservation, because Q (left-hand side) $= 1$, but Q (right-hand side) $= 0$.

   (c) Forbidden: violates baryon number conservation because B (left-hand side) $= 1$, but B (right-hand side) $= 0$.

   (d) Allowed: conserves L_μ, B, Q etc. (violates S, but this is allowed because it is a weak interaction).

3.2 (a) The quark compositions are: $D^- = d\bar{c}$; $K^0 = d\bar{s}$; $\pi^- = d\bar{u}$ and since the dominant decay of a c-quark is $c \to s$, we have

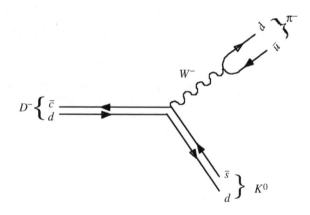

(b) The quark compositions are: $\Lambda = sud$; $p = uud$ and since the dominant decay of an s-quark is $s \rightarrow u$, we have

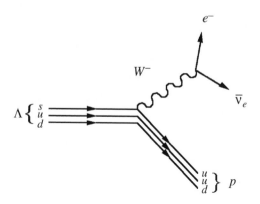

3.3 (a) This would be a baryon because $B = 1$ and the quark composition would be ssb which is allowed in the quark model.

(b) This would be a meson because $B = 0$, but would have to have both an $\bar{s}$- and a $\bar{b}$-quark. However, $Q(\bar{s} + \bar{b}) = 2/3$, which is incompatible with the quark model and anyway combinations of two antiquarks are not allowed. Thus this combination is forbidden.

3.4 'Low-lying' implies that the internal orbital angular momentum between the quarks is zero. Hence the parity is $P = +$ and ψ_{space} is symmetric. Since the Pauli principle requires the overall wavefunction to be antisymmetric under the interchange of any pair of like quarks, it follows that ψ_{spin} is antisymmetric. Thus, any pair of like quarks must have antiparallel spins, i.e. be in a spin-0 state.

Consider all possible baryon states qqq, where $q = u$, d, s. There are six combinations with a single like pair: uud, uus, ddu, dds, ssu, ssd, with the spin of (uu) etc. equal to zero. Adding the spin of the third quark leads to six states with $J^P = \frac{1}{2}^+$. In principle, there could be six combinations with all three quarks the same – uuu, ddd, sss – but in practice these do not occur because it is impossible to arrange all three spins in an antisymmetric way. Finally, there is one combination where all three quarks are different: uds. Here there are no restrictions from the Pauli principle, so for example, the ud pair could have spin-0 or spin-1. Adding the spin of the s-quark leads to two states with $J^P = \frac{1}{2}^+$ and 1 with $J^P = \frac{3}{2}^+$.

Collecting the results, gives an octet of $J^P = \frac{1}{2}^+$ states and a singlet $J^P = \frac{3}{2}^+$ state. This is **not** what is observed in nature. In Chapter 5 we will see what additional assumptions have to be made to reproduce the observed spectrum.

3.5 (a)

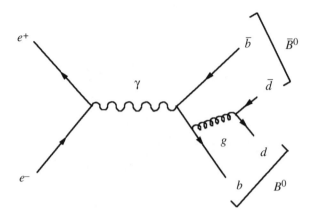

(b)

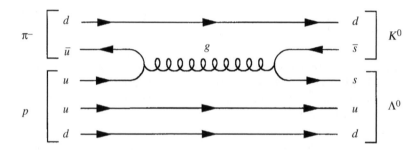

3.6 The ground state mesons all have $L = 0$ and $S = 0$. Therefore they all have $P = -1$. Only in the case of the neutral pion is their constituent quark and antiquark also particle and antiparticle. Thus C is only defined for the π^0 and is $C = 1$. For the excited states, $L = 0$ still and thus $P = -1$ as for the ground states. However, the total spin of the constituent quarks is $S = 1$ and so for the ρ^0, the only state for which C is defined, $C = -1$.

For the excited states, by definition there is a lower mass configuration with the same quark flavours. As the mass differences between the excited states and their ground states is greater than the mass of a pion, they can all decay by the strong interaction. In the case of the charged pions and kaons and the neutral kaon ground states, there are no lower mass configurations with the same flavour structure and so the only possibility is to decay via the weak interaction, with much longer lifetimes.

In the case of ρ^0 decay, the initial state has a total angular momentum of 1 and since the pions have zero spin, the $\pi\pi$ final state must have $L = 1$. While this is possible for $\pi^+\pi^-$, for the case of $\pi^0\pi^0$ it violates the Pauli Principle and so is forbidden.

3.7 In the initial state, $S = -1$ and $B = 1$. To balance strangeness (conserved in strong interactions), in the final state $S(Y^-) = -2$ and to balance baryon number,

$B(Y^-) = 1$. As charm and beauty for the initial state are both zero, these quantum numbers are zero for the Y. The quark content is therefore dss. In the decay, the strangeness of the Λ is -1 and so strangeness is not conserved. This is therefore a weak interaction and its lifetime will be in the range 10^{-7}–10^{-13} s.

3.8 The quark composition is $\Sigma = uds$, then $(\mathbf{S}_u + \mathbf{S}_d)^2 = S_u^2 + S_d^2 + 2\mathbf{S}_u \cdot \mathbf{S}_d = 2\hbar^2$ and hence $\mathbf{S}_u \cdot \mathbf{S}_d = \hbar^2/4$. Then, from the general formula given in Equation (3.84), setting $m_u = m_d = m$, we have

$$M_\Sigma = 2m + m_s + b\left[\frac{\mathbf{S}_u \cdot \mathbf{S}_d}{m^2} + \frac{\mathbf{S}_d \cdot \mathbf{S}_s + \mathbf{S}_u \cdot \mathbf{S}_s}{mm_s}\right]$$

$$= 2m + m_s + b\left[\frac{\mathbf{S}_u \cdot \mathbf{S}_d}{m^2} + \frac{\mathbf{S}_1 \cdot \mathbf{S}_2 + \mathbf{S}_1 \cdot \mathbf{S}_3 + \mathbf{S}_2 \cdot \mathbf{S}_3 - \mathbf{S}_u \cdot \mathbf{S}_d}{mm_s}\right]$$

which, using $\mathbf{S}_1 \cdot \mathbf{S}_2 + \mathbf{S}_1 \cdot \mathbf{S}_3 + \mathbf{S}_2 \cdot \mathbf{S}_3 = -3\hbar^2/4$ from Equation (3.89), gives

$$M_\Sigma = 2m + m_s + \frac{b}{4}\left[\frac{1}{m^2} - \frac{4}{mm_s}\right].$$

3.9 The initial reacton is strong because it conserves all individual quark numbers. The Ω^- decay is weak because strangeness changes by one unit and the same is true for the decays of the Ξ^0, K^+ and K^0. The decay of the π^+ is also weak because it involves neutrinos and finally the decay of the π^0 is electromagnetic because only photons are involved.

3.10 The Feynman diagram is:

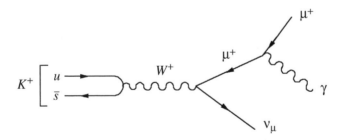

The two vertices where the W-boson couples are weak interactions and have strengths $\sqrt{\alpha_W}$. The remaining vertex is electromagnetic and has strength $\sqrt{\alpha_{EM}}$. So the overall strength of the diagram is $\alpha_W\sqrt{\alpha_{EM}}$.

3.11 From Equation (3.27a), we have $P(\bar{\nu}_e \to \nu_x) = \sin^2(2\alpha)\sin^2[\Delta(m^2c^4)L/(4\hbar cE)]$, which for maximal mixing ($\alpha = \pi/4$) gives $P(\bar{\nu}_e \to \nu_x) = \sin^2[1.27\Delta(m^2c^4)L/E]$ where L is measured in m, E in MeV and $\Delta(m^2c^4)$ in $(eV)^2$. If $P(\bar{\nu}_e \to \bar{\nu}_e) = 0.90 \pm 0.10$, then at 95 per cent confidence level, $1.0 \le P(\bar{\nu}_e \to \nu_x) \ge 0.70$ and hence $0.012 \le \Delta(m^2c^4) \le 0.019 (eV)^2$.

3.12 Reactions (a), (d) and (f) conserve all quark numbers individually and hence are strong interactions. Reaction (e) violates strangeness and is a weak interaction. Reaction (c) conserves strangeness and involves photons and hence is an electromagnetic interaction. Reaction (b) violates both baryon number and electron lepton number and is therefore forbidden.

3.13 The doublet of $S = +1$ mesons (K^+, K^0) has isospin $I = \frac{1}{2}$, with $I_3(K^+) = \frac{1}{2}$ and $I_3(K^0) = -\frac{1}{2}$. The triplet of $S = -1$ baryons $(\Sigma^+, \Sigma^0, \Sigma^-)$ has $I = 1$, with $I_3 = 1, 0, -1$ for Σ^+, Σ^0 and Σ^-, respectively. Thus (K^+, K^0) is analogous to the (p, n) isospin doublet and $(\Sigma^+, \Sigma^0, \Sigma^-)$ is analogous to the (π^+, π^0, π^-) isospin triplet. Hence, by analogy with Equations (3.54a) and (3.54b),

$$M(\pi^- p \to \Sigma^- K^+) = \frac{1}{3}M_3 + \frac{2}{3}M_1; \quad M(\pi^- p \to \Sigma^0 K^0) = \frac{\sqrt{2}}{3}M_3 - \frac{\sqrt{2}}{3}M_1$$

and

$$M(\pi^+ p \to \Sigma^+ K^+) = M_3,$$

where $M_{1,3}$ are the amplitudes for scattering in a pure isospin state $I = \frac{1}{2}, \frac{3}{2}$, respectively. Thus,

$$\sigma(\pi^+ p \to \Sigma^+ K^+) : \sigma(\pi^- p \to \Sigma^- K^+) : \sigma(\pi^- p \to \Sigma^0 K^o)$$
$$= |M_3|^2 : \frac{1}{9}|M_3 + 2M_1|^2 : \frac{2}{9}|M_3 - M_1|^2.$$

3.14 Under charge symmetry, $n(udd) \rightleftharpoons p(duu)$ and $\pi^+(u\bar{d}) \rightleftharpoons \pi^-(d\bar{u})$ and since the strong interaction is approximately charge symmetry, we would expect $\sigma(\pi^+ n) \approx \sigma(\pi^- p)$ at the same energy, with small violations due to electromagnetic effects and quark mass differences. However, $K^+(u\bar{s})$ and $K^-(s\bar{u})$ are not charge symmetric and so there is no reason why $\sigma(K^+ n)$ and $\sigma(K^- p)$ should be equal.

Chapter 4

4.1 In an obvious notation,

$$E_{CM}^2 = (E_e + E_p)^2 - (\mathbf{p}_e c + \mathbf{p}_p c)^2 = (E_e^2 - \mathbf{p}_e^2 c^2) - (E_p^2 - \mathbf{p}_p^2 c^2) + 2E_e E_p - 2\mathbf{p}_e \cdot \mathbf{p}_p c^2$$
$$= m_e^2 c^4 + m_p^2 c^4 + 2E_e E_p - 2\mathbf{p}_e \cdot \mathbf{p}_p c^2$$

At the energies of the beams, masses may be neglected and so with $p = |\mathbf{p}|$,

$$E_{CM}^2 = 2E_e E_p - 2p_e p_p c^2 \cos(\pi - \theta) = 2E_e E_p [1 - \cos(\pi - \theta)],$$

where θ is the crossing angle. Using the values given, gives $E_{CM} = 154\,\text{GeV}$. In a fixed-target experiment, and again neglecting masses, $E_{CM}^2 = 2E_e E_p - 2\mathbf{p}_e \cdot \mathbf{p}_p c^2$, where $E_e = E_L$, $E_p = m_p c^2$, $\mathbf{p}_p = 0$. Thus, $E_{CM} = \left[2m_p c^2 E_L\right]^{1/2}$ and for $E_{CM} = 154\,\text{GeV}$, this gives $E_L = 1.26 \times 10^4\,\text{GeV}$.

4.2 For constant acceleration, the ions must travel the length of the drift tube in half a cycle of the rf field. Thus, $L = v/2f$, where v is the velocity of the ion. Since the energy is far less than the rest mass of the ion, we can use non-relativistic kinematics to find v, i.e. $v = c\sqrt{200/(12 \times 931.5)} = 4.01 \times 10^7\,\text{m\,s}^{-1}$ and finally $L = 1\,\text{m}$.

4.3 A particle with mass m, charge q and speed v moving in a plane perpendicular to a constant magnetic field of magnitude B will traverse a circular path with radius of curvature $r = mv/qB$ and hence the cyclotron frequency is $f = v/2\pi r = qB/2\pi m$. At each traversal the particle will receive energy from the rf field, so if f is kept fixed, r will increase (i.e. the trajectory will be a spiral). Thus if the final energy is E, the extraction radius will be $R = \sqrt{2mE}/qB$. To evaluate these expressions we use $q = 2e = 3.2 \times 10^{-19}\,\text{C}$, together with $B = 0.8\,\text{T} = 0.45 \times 10^{30}(\text{MeV}/c^2)\text{s}^{-1}\,\text{C}^{-1}$ and thus $f = 6.15\,\text{MHz}$ and $R = 62.3\,\text{cm}$.

4.4 A particle with unit charge e and momentum p in the uniform magnetic field B of the bending magnet will traverse a circular trajectory of radius R, given by $p = BR$. If B is in T, R in m and p in GeV/c, then $p = 0.3BR$. Referring to the figure below, we have $\theta \approx L/R = 0.3\,LB/p$ and $\Delta\theta = s/d = 0.3BL\Delta p/p^2$. Solving for d using the data given, gives $d = 9.3\,\text{m}$.

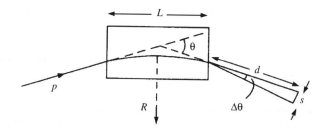

4.5 The Čerenkov condition is $\beta n \geq 1$. So, for the pion to give a signal, but not the kaon, we have $\beta_\pi n \geq 1 \geq \beta_K n$. The momentum is given by $p = mv\gamma$ where $\gamma = (1 - v^2/c^2)^{-1/2}$, so eliminating γ gives $\beta = v/c = (1 + m^2 c^2/p^2)^{-1/2}$. For $p = 20\,\text{GeV}/c$, $m_\pi = 0.14\,\text{GeV}/c^2$ and $m_K = 0.49\,\text{GeV}/c^2$, $\beta_\pi = 0.99997$ and $\beta_K = 0.99970$, so the condition on the refractive index is $3 \times 10^{-4} \geq (n - 1)/n \geq 3 \times 10^{-5}$. Using the largest value of $n = 1.0003$, we have

$$N = 2\pi\alpha \left(1 - \frac{1}{\beta_\pi^2 n^2}\right)\left(\frac{1}{\lambda_1} - \frac{1}{\lambda_2}\right)$$

as the number of photons radiated per metre, where $\lambda_1 = 400\,\text{nm}$ and $\lambda_2 = 700\,\text{nm}$. Numerically, $N = 26.5$ photons/m and hence to obtain 200 photons requires a detector of length 7.5 m. (You could also use

$$N = 2\pi\alpha \left(1 - \frac{1}{\beta_\pi^2 n^2}\right)\left(\frac{\lambda_2 - \lambda_1}{\lambda^2}\right)$$

where λ is the mean of λ_1 and λ_2, which would give 24.5 photons/m and a length of 8.2 m.)

4.6 Luminosity may be calculated from the formula for colliders, $L = n N_1 N_2 f / A$, where n is the number of bunches, N_1 and N_2 are the numbers of particles in each bunch, A is the cross-sectional area of the beam and f is its frequency. We have, $n = 12$, $N_1 = N_2 = 3 \times 10^{11}$, $A = (0.02 \times 10^{-2})\,cm^2$ and $f = (3 \times 10^{10}/8\pi \times 10^5)\,s^{-1}$, so finally $L = 6.44 \times 10^{31}\,cm^{-2}\,s^{-1}$.

4.7 (a) The b quarks are not seen directly but, instead, they fragment (hadronize) to B-hadrons, i.e. hadrons containing b quarks. So one characteristic is the presence of hadrons with non-zero beauty quantum numbers. As these hadrons are unstable and the dominant decay of b-quarks is to c-quarks, a second characteristic is the presence of hadrons with non-zero values of the charm quantum number.

We need to observe the point where the e^+e^- collision occurred and the point of origin of the decay products of the B-hadrons. The difference between these two is due to the lifetime of the B-hadrons. As the difference will be very small, precise position measurements are required. The daughter particles may be detected using a silicon micro-vertex detector and an MWPC. In addition, any electrons from the decays could be detected by an MWPC or an electromagnetic calorimeter. The same is true for muons in the decay products, except they are not readily detected in the calorimeter as they are very penetrating. However, if one places an MWPC behind a hadron calorimeter then one can be fairly confident that any particle detected is a muon, as everything else (except neutrinos) will have been stopped in the calorimeter.

(b) In the electronic decay mode, the electron can be measured in both a MWPC and an EM calorimeter. For high energies the better measurement is made in the calorimeter. The neutrino does not interact unless there is a very large mass of material (thousands of tons) and so its presence must be inferred by imposing conservation of energy and momentum. In a colliding beam machine, the original colliding particles have zero transverse momentum and a fixed energy. If one adds up all the energy and momentum of all the final-state particles, then any imbalance compared to the initial system can be attributed to the neutrino.

For the muonic mode, the muon can be measured in the MWPC but cannot be measured well in the calorimeter because it only ionizes to a very small extent. Since the muons only interact to a small extent they (along with neutrinos) are generally the only particles that emerge from a hadronic calorimeter. So if one registers a signal in a small MWPC placed behind a calorimeter then one can be confident that the particle is a muon.

4.8 To be detected, the event must have $150° < \theta < 30°$, i.e. $|\cos\theta| < 0.866$. Setting $x = \cos\theta$, the fraction of events in this range is

$$f = \int_{-0.866}^{+0.866} \frac{d\sigma}{dx}\,dx \Big/ \int_{-1.0}^{+1.0} \frac{d\sigma}{dx}\,dx = \left[x + x^3/3\right]_{-0.866}^{+0.866} \Big/ \left[x + x^3/3\right]_{-1.0}^{+1.0} = 0.812.$$

The total cross-section is given by

$$\sigma = \int \frac{d\sigma}{d\Omega}\, d\Omega = \int_0^{2\pi} d\phi \int_{-1}^{+1} d\cos\theta \, \frac{d\sigma}{d\Omega} = 2\pi \frac{\alpha^2 \hbar^2 c^2}{4E_{cm}^2} \int_{-1}^{+1} \left[1 + \cos^2\theta \right] d\cos\theta.$$

Using $E_{cm} = 10\,\text{GeV}$, gives $\sigma = 4\pi\alpha^2\hbar^2 c^2 / 3E_{cm}^2 = 0.866\,\text{nb}$. The rate of production of events is given by $L\sigma$ and since L is a constant, the total number of events produced will be $L\sigma t = 86\,600$.

The $\tau^\pm$ decay too quickly to leave a visible track in the drift chamber. The e^+ and the μ^- will leave tracks in the drift chamber and the e^+ will produce a shower in the electromagnetic calorimeter. If it has enough energy, the μ^- will pass through the calorimeters and leave a signal in the muon chamber. There will be no signal in the hadronic calorimeter.

4.9 Referring to the figure below, the distance between two positions of the particle Δt apart in time is $v\,\Delta t$. The wave fronts from these two positions have a difference in their distance travelled of $c\,\Delta t/n$.

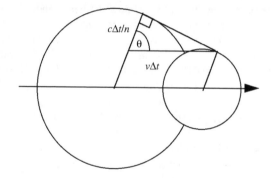

These constructively interfere at an angle θ, where

$$\cos\theta = \frac{c\,\Delta t/n}{v\,\Delta t} = \frac{1}{\beta n}.$$

The maximum value of θ corresponds to the minimum of $\cos\theta$ and hence the maximum of β. This occurs as $\beta \to 1$, when $\theta_{max} = \cos^{-1}(1/n)$. This value occurs in the ultra-relativistic or massless limit.

The quantity β may be expressed as $\beta = pc/E = pc[p^2c^2 + m^2c^4]^{-1/2}$. Hence,

$$\cos\theta = \frac{1}{n} \frac{\sqrt{p^2c^2 + m^2c^4}}{pc},$$

which rearranging, gives $x \equiv (mc^2)^2 = p^2c^2(n^2\cos^2\theta - 1)$. Differentiating this formula gives $dx/d\theta = -2p^2c^2n^2\cos\theta\sin\theta$ and the error on x is then given by

$\sigma_x = |dx/d\theta| \, \sigma_\theta$. For very relativistic particles, the derivative can be approximated by using θ_{max}, for which $\cos\theta_{max} = 1/n$, $\sin\theta_{max} = \sqrt{n^2 - 1}/n$. Hence

$$\sigma_x \approx 2p^2 c^2 n^2 \frac{1}{n} \frac{\sqrt{n^2 - 1}}{n} \sigma_\theta = 2p^2 c^2 \sqrt{n^2 - 1}\, \sigma_\theta.$$

4.10 The average distance between collisions of a neutrino and an iron nucleus is the mean free path $\lambda = 1/n\sigma_\nu$, where $n \approx \rho/m_p c^2$ is the number of nucleons per cm^3. Using the data given, $n \approx 4.7 \times 10^{24}\, cm^{-3}$ and $\sigma_\nu \approx 3 \times 10^{-36}\, cm^2$, so that $\lambda \approx 7.1 \times 10^{10}\, cm$. Thus if 1 in 10^9 neutrinos is to interact, the thickness of iron required is 71 cm.

4.11 Radiation energy losses are given by $-dE/dx = E/L_R$, where L_R is the radiation length. This implies that $E = E_0 \exp(-x/L_R)$, where E_0 is the initial energy. Using $E_0 = 2\,GeV$, $L_R = 36.1\, cm$, $x = 10\, cm$, gives $E = 1.51\,GeV$. Radiation losses at fixed E are proportional to m^{-2}, where m is the mass of the projectile. Thus for muons, they are negligible at this energy.

4.12 The total cross section is $\sigma_{tot} = \sigma_{el} + \sigma_{cap} + \sigma_f = 4 \times 10^2\, b$ and the attenuation is $\exp(-nx\sigma_{tot})$ where $nx = 10^{-1} N_A/A = 2.56 \times 10^{23}\, m^{-2}$. Thus $\exp(-nx\sigma_{tot}) = 0.9898$, i.e 1.02 per cent of the incident particles interact and of these the fraction that elastically scatter is given by the ratio of the cross-sections, i.e. $3 \times 10^{-2}/4 \times 10^2 = 0.75 \times 10^{-4}$. Thus the intensity of elastically-scattered neutrons is $0.75 \times 10^{-4} \times 0.0102 \times 10^6 = 0.765\, s^{-1}$ and finally the flux at 5 m is $0.765/(4 \times \pi \times 5^2) = 2.44 \times 10^{-3}\, m^{-2}\, s^{-1}$.

4.13 The total centre-of-mass energy is given by $E_{CM} \approx (2mc^2 E_L)^{\frac{1}{2}} = 0.23\,GeV$ and so the cross-section is $\sigma = 1.64 \times 10^{-34}\, m^2$. The interaction length is $\ell = 1/n\sigma$, where n is the number density of electrons in the target. This is given by $n = \rho N_A Z/A$, where N_A is Avogadro's number and for lead, $\rho = 1.14 \times 10^7\, kg\, m^{-3}$ is the density, $Z = 82$ and $A = 208$. Thus $n = 2.7 \times 10^{33}\, m^{-3}$ and $\ell = 2.3\, m$.

4.14 The target contains $n = 1.07 \times 10^{25}$ protons and so the total number of interactions per second is $N = n \times$ flux $\times \sigma_{tot} = (1.07 \times 10^{25}) \times (2 \times 10^7) \times (40 \times 10^{-31}) = 856\, s^{-1}$. There are thus 856 photons/s produced from the target.

4.15 For small v, the Bethe–Bloch formula may be written

$$S \equiv -\frac{dE}{dx} \propto \frac{1}{v^2} \ln\left(\frac{2m_e v^2}{I}\right) \quad \text{with} \quad \frac{dS}{dv} \propto \frac{2}{v^3}\left[1 - \ln\left(\frac{2m_e v^2}{I}\right)\right].$$

The latter has a maximum for $v^2 = eI/2m_e$. Thus for a proton in iron we can use $I = 10Z\,eV = 260\,eV$, so that $E_p = \frac{1}{2}m_p v^2 = m_p le/4m_e = 324\,keV$.

4.16 From Equation (4.24), $E(r) = V/r \ln(r_c/r_a)$ and at the surface of the anode this is $0.5/(20 \times 10^{-6}) \ln(500) = 4023\, kV\, m^{-1}$. Also, if $E_{threshold}(r) = 750\, kV\, m^{-1}$, then from Equation (4.24) $r = 0.107\, mm$ and so the distance to the anode is 0.087 mm.

This contains 22 mean free paths and so assuming each collision produces an ion pair, the multiplication factor is $2^{22} = 4.2 \times 10^6 = 10^{6.6}$.

Chapter 5

5.1 We have $m = \alpha + \beta + \gamma > n = \bar{\alpha} + \bar{\beta} + \bar{\gamma}$, where the inequality is because baryon number $B > 0$. Using the values of the colour charges I_3^C and Y^C from Table 5.1, the colour charges for the state are:

$$I_3^C = (\alpha - \bar{\alpha})/2 - (\beta - \bar{\beta})/2 \quad \text{and} \quad Y^C = (\alpha - \bar{\alpha})/3 + (\beta - \bar{\beta})/3 - 2(\gamma - \bar{\gamma})/3.$$

By colour confinement, both these colour charges must be zero for observable hadrons, which implies $\alpha - \bar{\alpha} = \beta - \bar{\beta} = \gamma - \bar{\gamma} \equiv p$ and hence $m - n = 3p$, where p is a non-negative integer. Thus the only combinations allowed by colour confinement are of the form

$$(3q)^p (q\bar{q})^n \quad (p, n \geq 0).$$

It follows that a state with the structure qq is not allowed, as no suitable values of p and n can be found.

5.2 (a)

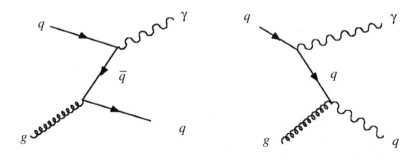

(b)

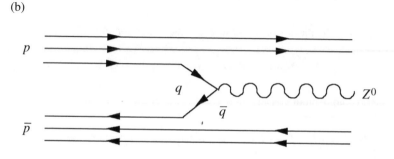

(c)

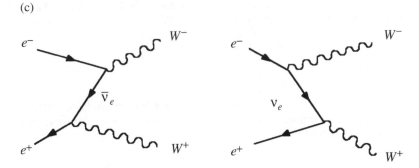

5.3 The Feynman diagram is:

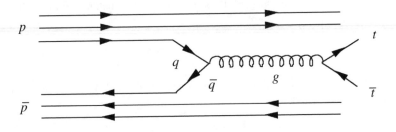

The four-momenta are:

$$P(p) = (E/c, \ \mathbf{p}) \quad \text{and} \quad P(\bar{p}) = (E/c, \ -\mathbf{p}),$$

with

$$P^2 = m^2c^2 = E^2/c^2 - \mathbf{p}^2 \quad \text{and} \quad m = m_p = m_{\bar{p}}.$$

Now $P(q) = (xE/c, \ x\mathbf{p})$ and $P(\bar{q}) = (xE/c, \ -x\mathbf{p})$ with $x = \frac{1}{6}$, so

$$E_{\text{CM}}^2 = x^2c^2[P(p) + P(\bar{p})]^2 = x^2[2m^2c^4 + 2E^2 + 2\mathbf{p}^2c^2].$$

Neglecting the masses of the proton and the antiproton at these energies, gives

$$E = 3E_{\text{CM}} \quad \text{and} \quad p = 3 \times 350 = 1050 \text{ GeV/c}.$$

5.4 Energy–momentum conservation gives,

$$W^2c^4 = [(E - E') + E_P]^2 - [(\mathbf{p} - \mathbf{p}') + \mathbf{P}]^2 c^2 = \text{invariant mass of } X.$$

Using, $Q^2 = (\boldsymbol{p} - \boldsymbol{p}')^2 - (E - E')^2/c^2$ and $M^2 c^4 = E_P^2 - P^2 c^2$, where M is the mass of the proton, gives

$$W^2 c^4 = -Q^2 c^2 + M^2 c^4 + 2E_P(E - E') - 2\boldsymbol{P} \cdot (\boldsymbol{p} - \boldsymbol{p}')c^2.$$

Also, $2M\nu \equiv W^2 c^2 + Q^2 - M^2 c^2$ and so, in the rest frame of the proton $(\boldsymbol{P} = \boldsymbol{0}, E_P = Mc^2)$, $\nu = E - E'$.

Since *some* energy must be transferred to the outgoing electron, it follows that $E \geq E'$, i.e. $\nu \geq 0$. Also, since the lightest state X is the proton, $W^2 \geq M^2$. Thus,

$$2M\nu = Q^2 + (W^2 - M^2)c^2 \geq Q^2.$$

From the definition of x, it follows that $x \leq 1$. Finally, $x > 0$ because both Q^2 and $2M\nu$ are positive.

5.5 In the quark model, $\Lambda = uds, p = uud, K^- = s\bar{u}, n = udd$ and $\pi^+ = u\bar{d}$. From the flavour independence of the strong interaction, we can set $\sigma(qq) = \sigma(ud) = \sigma(sd)$ etc. and $\sigma(q\bar{q}) = \sigma(u\bar{d}) = \sigma(s\bar{u})$ etc.. Then $\sigma(\Lambda p) = \sigma(pp) = 9\sigma(qq)$ and $\sigma(K^- n) = \sigma(\pi^+ p) = 3\sigma(qq) - 3\sigma(q\bar{q})$. The result follows directly.

5.6 By analogy with the QED formula, we have $\Gamma(3g) = 2(\pi^2 - 9)\alpha_s^6 m_c c^2/9\pi$, where $m_c \approx 1.5\,\text{GeV}/c^2$ is the constituent mass of the c-quark. Evaluating this gives $\alpha_s = 0.31$. In the case of the radiative decay, $\Gamma(gg\gamma) = 2(\pi^2 - 9)\alpha_s^4 \alpha^2 m_b c^2/9\pi$, where $m_b \approx 4.5\,\text{GeV}/c^2$ is the constituent mass of the b-quark. Evaluating this gives $\alpha_s = 0.32$. (These values are a little too large because in practice α is replaced by $\frac{4}{3}\alpha_s$.)

5.7 From Equation (5.38a)

$$F_2^{\ell p}(x) = x\left[\frac{1}{9}(d + \bar{d}) + \frac{4}{9}(u + \bar{u}) + \frac{1}{9}(s + \bar{s})\right]$$

and from Equations (5.38b) and (5.39)

$$F_2^{\ell n}(x) = x\left[\frac{4}{9}(d + \bar{d}) + \frac{1}{9}(u + \bar{u}) + \frac{1}{9}(s + \bar{s})\right],$$

so that

$$\int_0^1 [F_2^{ep}(x) - F_2^{en}(x)]\frac{dx}{x} = \frac{1}{3}\int_0^1 [u(x) + \bar{u}(x)]\,dx - \frac{1}{3}\int_0^1 [d(x) + \bar{d}(x)]dx.$$

However, summing over all contributions we must recover the quantum numbers of the proton, i.e.

$$\int_0^1 [u(x) - \bar{u}(x)]\,dx = 2; \quad \int_0^1 [d(x) - \bar{d}(x)]\,dx = 1.$$

Eliminating the integrals over u and d gives the Gottfried sum rule.

5.8 Substituting Equation (5.22) into Equation (5.23) and setting $N_C = 3$, gives

$$R = 3(1 + \alpha_s/\pi) \sum e_q^2,$$

where α_s is given by Equation (5.11) evaluated at $Q^2 = E_{CM}^2$ and the sum is over those quarks that can be produced in pairs at the energy considered. At 2.8 GeV the u, d and s quarks can contribute and at 15 GeV the u, d, s, c and b quarks can contribute. Evaluating R then gives $R \approx 2.17$ at $E_{CM} = 2.8$ GeV and $R \approx 3.89$ at $E_{CM} = 15$ GeV. When E_{CM} is above the threshold for $t\bar{t}$ production, R rises to $R = 5(1 + \alpha_s/\pi)$.

5.9 A proton has the valence quark content $p = uud$. Thus from isospin invariance the u quarks in the proton carry twice as much momentum as the d quarks, which implies $a = 2b$. In addition, we are told that

$$\int_0^1 x F_u(x)\,dx + \int_0^1 x F_d(x)\,dx = \frac{1}{2}.$$

Using the form of the quark distributions with $a = 2b$ gives $a = \frac{4}{3}$ and $b = \frac{2}{3}$.

5.10 The peak value of the cross-section is where $E = M_W c^2$, i.e.

$$\sigma_{max} = \frac{\pi(\hbar c)^2 (2/M_W c^2)^2 \Gamma_{u\bar{d}}}{3\Gamma} = \frac{4}{3}\frac{\pi(\hbar c)^2}{(M_W c^2)^2} \mathrm{br}(W^+ \to u\bar{d}) = 84 \,\mathrm{nb}.$$

The required integral is

$$\sigma_{p\bar{p}}(s) = \int_0^1 \int_0^1 \sigma_{u\bar{d}}(E)\, u(x_u)\, d(x_d)\, dx_u\, dx_d$$

where we have used C-invariance to relate the distribution functions for protons and antiprotons. In the narrow width approximation and using the quark distributions from Question 5.9,

$$\sigma_{p\bar{p}}(s) = C \int_0^1 \int_0^1 \frac{(1-x_u)^3}{x_u}\frac{(1-x_d)^3}{x_d} \delta\left(1 - \frac{x_u s}{(M_W c^2)^2}x_d\right) dx_u\, dx_d$$

where $C \equiv (8\pi\Gamma_W \sigma_{max})/(9 M_W c^2)$ and we have used $E^2 = x_u x_d s$. Thus,

$$\sigma_{p\bar{p}}(s) = C \int_k^1 \frac{(1-x_u)^3}{x_u}\left(1 - \frac{k}{x_u}\right)^3 dx_u,$$

where $k \equiv (M_W c^2)/s$ and the lower limit is because $k < x_u < 1$. The integral yields

$$\sigma_{p\bar{p}}(s) = \frac{8\pi}{9} \frac{\Gamma_W}{M_W c^2} \sigma_{max} \left\{ -(1 + 9k + 9k^2 + k^3)\ln(k) - \frac{11}{3} - 9k + 9k^2 + \frac{11}{3}k^3 \right\}.$$

Evaluating this for $\sqrt{s} = 1\,\text{TeV}$ gives $k = 0.0064$ and $\sigma_{p\bar{p}} = 9.3\,\text{nb}$, which is about a factor of two larger than experiment.

Chapter 6

6.1 A charged current weak interaction is one mediated by the exchange of charged $W^{\pm}$ boson. A possible example is $n \to p + e^- + \bar{\nu}_e$. A neutral current weak interaction is one mediated by a neutral Z^0 boson. An example is $\nu_\mu + p \to \nu_\mu + p$. Charged current weak interactions do not conserve the strangeness quantum number, whereas neutral current weak interactions do. For $\nu_\mu + e^- \to \nu_\mu + e^-$, the only Feynman diagram that conserves both L_e and L_μ is:

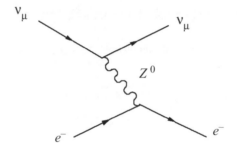

which is a weak neutral current. However, for $\nu_e + e^- \to \nu_e + e^-$, there are two diagrams:

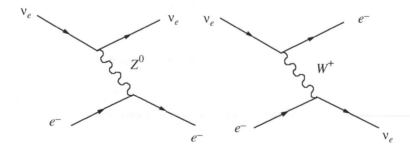

Thus the reaction has both neutral and charged current components and is not unambiguous evidence for weak neutral currents.

6.2 The lowest-order electromagnetic Feynman diagram is

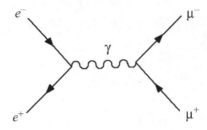

The total cross-section is given by

$$\sigma = \int_0^{2\pi} d\phi \int_{-1}^1 d\cos\theta \frac{d\sigma}{d\Omega} = \frac{2\pi\alpha^2\hbar^2c^2}{4E_{\mathrm{CM}}^2}\left[\cos\theta + \frac{1}{3}\cos^3\theta\right]_{-1}^1$$

$$= \frac{4\pi\alpha^2\hbar^2c^2}{3E_{\mathrm{CM}}^2} = 0.44 \text{ nb.}$$

The lowest-order weak interaction diagram is

With the addition of the weak interaction term,

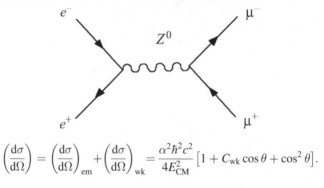

$$\left(\frac{d\sigma}{d\Omega}\right) = \left(\frac{d\sigma}{d\Omega}\right)_{\mathrm{em}} + \left(\frac{d\sigma}{d\Omega}\right)_{\mathrm{wk}} = \frac{\alpha^2\hbar^2c^2}{4E_{\mathrm{CM}}^2}\left[1 + C_{\mathrm{wk}}\cos\theta + \cos^2\theta\right].$$

Then, using

$$\sigma_F = C\int_0^1\left[1 + C_{\mathrm{wk}}\cos\theta + \cos^2\theta\right]d\cos\theta$$

and

$$\sigma_B = C\int_{-1}^0\left[1 + C_{\mathrm{wk}}\cos\theta + \cos^2\theta\right]d\cos\theta.$$

where $C \equiv 2\pi\alpha^2\hbar^2c^2/4E_{CM}^2$, gives

$$\sigma_F = C\left[\frac{4}{3} + \frac{C_{\mathrm{wk}}}{2}\right] \quad \text{and} \quad \sigma_B = C\left[\frac{4}{3} - \frac{C_{\mathrm{wk}}}{2}\right]$$

and so

$$A_{\mathrm{FB}} = \frac{C_{\mathrm{wk}}}{2(4/3)}, \quad \text{i.e. } 8A_{\mathrm{FB}} = 3C_{\mathrm{wk}}.$$

6.3 The Feynman diagram is

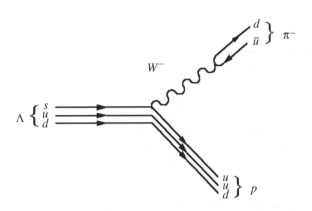

The amplitude has two factors of the weak coupling g_{W} and one W propagator carrying a momentum q, i.e.

$$\text{amplitude} \propto \frac{g_{\mathrm{W}}^2}{q^2 c^2 - M_W^2 c^4} \propto \frac{g_{\mathrm{W}}^2}{M_W^2},$$

because $qc \approx M_\Lambda c^2 \ll M_W c^2$. Now, $\Gamma(\Lambda \to p\pi^-) \propto (\text{amplitude})^2 \propto g_{\mathrm{W}}^4 / M_W^4$ and so doubling g_{W} and reducing M_W by a factor of four will increase the rate by a factor $[2^4]/[(1/4)^4] = 4096$.

6.4 The most probable energy is given by

$$\frac{\mathrm{d}}{\mathrm{d}E_e}\left(\frac{\mathrm{d}\omega}{\mathrm{d}E_e}\right) = 0, \quad \text{which gives} \quad \frac{2G_{\mathrm{F}}^2 \, m_\mu^2}{(2\pi)^3 (\hbar c)^6}\left(2E_e - \frac{4E_e^2}{m_\mu c^2}\right) = 0, \quad \text{i.e } E_e = m_\mu c^2/2.$$

When $E_e \approx m_\mu c^2/2$, the electron has its maximum energy and the two neutrinos must be recoiling in the opposite direction. Only left-handed particles (and right-handed antiparticles) are produced in weak interactions. Since the masses of all particles are neglected, states of definite handiness are also states of definite helicity, so the orientations of the momenta and spins are therefore as shown:

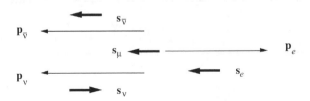

Integrating the spectrum gives

$$\Gamma = \frac{2G_F^2(m_\mu c^2)^2}{(2\pi)^3(\hbar c)^6} \int\limits_0^{m_\mu c^2/2} \left[E_e^2 - \frac{4E_e^3}{3m_\mu c^2} \right] dE_e = \frac{G_F^2(m_\mu c^2)^5}{192\pi^3(\hbar c)^6}.$$

Numerically, $\Gamma \approx 3.0 \times 10^{-19}$ GeV, which gives a lifetime $\tau = \hbar/\Gamma \approx 2.2 \times 10^{-6}$ s.

6.5 (a) In addition to the decay $b \to c + e^- + \bar{\nu}_e$, there are two other leptonic decays ($\ell = \mu^-, \tau^-$) and by lepton universality they will all have equal decay rates. There are also hadronic decays of the form $b \to c + X$ where $Q(X) = -1$. Examining the allowed $Wq\bar{q}$ vertices using lepton–quark symmetry shows that the only forms that X can have, if we ignore Cabibbo-suppressed modes, are $d\bar{u}$ and $s\bar{c}$. Each of these hadronic decays has a probability three times that of a leptonic decay because the quarks exist in three colour states. Thus, there are effectively six hadronic channels and three leptonic ones. So finally, $BR(b \to c + e^- + \bar{\nu}_e) = \frac{1}{9}$.

(b) The argument is similar to that of (a) above. Thus, in addition to the decay $\tau^- \to e^- + \bar{\nu}_e + \nu_\tau$, there is also the leptonic decay $\tau^- \to \mu^- + \bar{\nu}_\mu + \nu_\tau$ with equal probability and the hadronic decays $\tau^- \to \nu_\tau + X$. In principle, $X = d\bar{u}$ and $s\bar{c}$, but the latter is not allowed because $m_s + m_c > m_\tau$. So the only allowed hadronic decay is $\tau^- \to d + \bar{u} + \nu_\tau$ with a relative probability of three because of colour. So finally, $BR(\tau^- \to e^- + \bar{\nu}_e + \nu_\tau) = \frac{1}{5}$. (The measured rate is 0.18, but we have neglected kinematic corrections.)

6.6 For neutrinos, $g_R(\nu) = 0$; $g_L(\nu) = \frac{1}{2}$. So, $\Gamma_{\nu_e} = \Gamma_{\nu_\mu} = \Gamma_{\nu_\tau} = \Gamma_0/4$, where

$$\Gamma_0 = \frac{G_F M_Z^3 c^6}{3\pi\sqrt{2}(\hbar c)^3} = 668 \text{ MeV}.$$

Thus the partial width for decay to neutrino pairs is $\Gamma_\nu = 501$ MeV. For quarks, $g_R(u, c, t) = -\frac{1}{6}$ and $g_L(u, c, t) = \frac{1}{3}$. Thus, $\Gamma_u = \Gamma_c = \frac{10}{72}\Gamma_0$. Also, $g_R(d, s, b) = \frac{1}{12}$ and $g_L(b, s, d) = -\frac{5}{12}$. Thus, $\Gamma_d = \Gamma_s = \Gamma_b = \frac{13}{72}\Gamma_0$. Finally, $\Gamma_q = \sum_i \Gamma_i$, where $i = u, c, d, s, b$ – no top quark because $2M_t > M_Z$. So,

$$\Gamma_q = \left(\frac{3 \times 13}{72} + \frac{2 \times 10}{72} \right) \Gamma_0 = \frac{59}{72}\Gamma_0 = 547 \text{ MeV}.$$

Hadron production is assumed to be equivalent to the production of $q\bar{q}$ pairs followed by fragmentation with probability unity. Thus $\Gamma_{\text{hadron}} = 3\Gamma_q$, where the factor of three is because each quark exists in one of three colour states. Thus $\Gamma_{\text{hadron}} = 1641$ MeV.

If there are N_ν generations of neutrinos with $M_\nu < M_Z/2$, so that $Z^0 \to \nu\bar{\nu}$ is allowed, then $\Gamma_{\text{tot}} = \Gamma_{\text{had}} + \Gamma_{\text{lep}} + N_\nu\Gamma_{\nu\bar{\nu}}$ where $\Gamma_{\nu\bar{\nu}}$ is the width to a specific $\nu\bar{\nu}$ pair. Thus

$$N_\nu = \frac{\Gamma_{\text{tot}} - \Gamma_{\text{had}} - \Gamma_{\text{lep}}}{\Gamma_{\nu\bar{\nu}}} = \frac{(2490 \pm 7) - (1738 \pm 12) - (250 \pm 2)}{167}$$

$$= 3.01 \pm 0.05,$$

which rules out values of N_ν greater than 3.

6.7   The quark compositions are: $D^0 = c\bar{u}$; $K^- = s\bar{u}$; $\pi^+ = u\bar{d}$. Since preferentially $c \to s$, we have

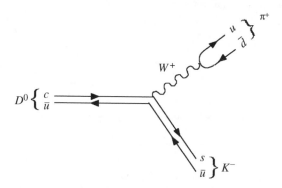

i.e. a lowest-order charge current weak interaction. However, for $D^+ \to K^0 + \pi^+$, we have $D^+ = c\bar{d}$; $K^0 = d\bar{s}$; $\pi^+ = u\bar{d}$. Thus we could arrange $c \to d$ via W emission and the W^+ could then decay to $u\bar{d}$, i.e. π^+. However, this would leave the $\bar{d}$ quark in the D^+ to decay to an $\bar{s}$ quark in the K^0 which is not possible as they both have the same charge.

6.8   The relevant Feynman diagrams are:

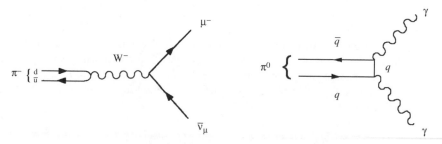

In the case of the charged pion, there are two vertices of strength $\sqrt{\alpha_W}$, and there will be a propagator

$$\frac{1}{Q^2 + M_W^2c^2} \approx \frac{1}{M_W^2c^2},$$

because the momentum transfer (squared) Q^2 carried by the W is very small. Thus the decay rate will be proportional to

$$\left(\frac{\sqrt{\alpha_W}\sqrt{\alpha_W}}{M_W^2}\right)^2 = \frac{\alpha_W^2}{M_W^4}.$$

In the case of the neutral pion, there are two vertices of strength $\sqrt{\alpha_{em}}$, but no propagator. Thus the decay rate will be proportional to α_{em}^2 and since $\alpha_{em} \approx \alpha_W$, the decay rate for the charged pion will be much smaller than that for the neutral decay, i.e. the lifetime of the π^0 will be much shorter.

6.9 The two Feynman diagrams are:

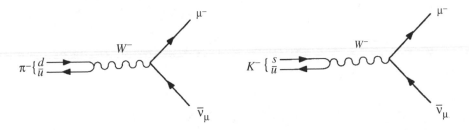

Using lepton–quark symmetry and the Cabibbo hypothesis, the two hadron vertices are given by $g_{udW} = g_W \cos\theta_C$ and $g_{usW} = g_W \sin\theta_C$. So, if we ignore kinematic differences and spin effects, we would expect the ratio of decay rates is given by

$$R = \frac{\text{Rate}\,(K^- \rightarrow \mu^- + \bar{\nu}_\mu)}{\text{Rate}\,(\pi^- \rightarrow \mu^- + \bar{\nu}_\mu)} \propto \frac{g_{usW}^2}{g_{udW}^2} = \tan^2\theta_C \approx 0.05$$

The measured ratio is actually about 1.3, which shows the importance of the neglected effects. For example, the Q-value for the kaon decay is almost 20 times that for pion decay.

6.10 To a first approximation the difference in the two decay rates is due to two effects. First, $\Sigma^- \rightarrow n + e^- + \bar{\nu}_e$ has $|\Delta S| = 1$ and hence is proportional to $\sin^2\theta_C$, where θ_C is the Cabibbo angle, whereas $\Sigma^- \rightarrow \Lambda + e^- + \bar{\nu}_e$ has $|\Delta S| = 0$ and is proportional to $\cos^2\theta_C$. Secondly, the Q-values are different for the two reactions. Thus, using Sargent's Rule,

$$R \approx \frac{\sin^2\theta_C}{\cos^2\theta_C}\left(\frac{Q_{\Sigma n}}{Q_{\Sigma\Lambda}}\right)^5 \approx 0.053\left(\frac{257}{81}\right)^5 = 17.0.$$

(The experimental value is 17.8.) Whereas, $\Sigma^- \rightarrow n + e^- + \bar{\nu}_e$ is a first-order weak interaction, no Feynman diagram with a single W-boson exchanged can be drawn for $\Sigma^+ \rightarrow n + e^+ + \nu_e$ (try it), i.e. it is higher-order and hence very heavily suppressed – in practice not seen.

6.11 The required number of events produced must be 20 000, taking account of the detection efficiency. If the cross-section is $60\,\text{fb} = 6 \times 10^{-38}\,\text{cm}^2$, then the integrated luminosity required is $2 \times 10^4/6 \times 10^{-38} = (1/3) \times 10^{42}\,\text{cm}^{-2}$ and hence the instantaneous luminosity must be $3.3 \times 10^{34}\,\text{cm}^{-2}\,\text{s}^{-1}$.

The branching ratio for $Z^0 \to b\bar{b}$ is found from the partial widths to be 15 per cent. Thus, if b quarks are detected, the much greater branching ratio for $H \to b\bar{b}$ will help distinguish this decay from the background of $Z^0 \to b\bar{b}$.

6.12 By 'adding' an $I = \frac{1}{2}$ particle to the initial state we can assume isospin invariance holds. Consider $\Xi^- + S^0 \to \Lambda + \pi^-$. The final state is $|I = 1, I_3 = -1\rangle$ and so is the initial state because $I_3(S^0) = -\frac{1}{2}$. Thus the transition is pure $I = 1$ and the rate is $|M_1|^2$. For $\Xi^0 + S^0 \to \Lambda + \pi^0$, the final state is again pure $I = 1$ but with $I_3 = 0$. However, the initial state is an equal mixture of $I = 0$ and $I = 1$, i.e.

$$\left|\Xi^- S^0\right\rangle = \frac{1}{\sqrt{2}}|I = 1, I_3 = 0\rangle \pm \frac{1}{\sqrt{2}}|I = 0, I_3 = 0\rangle$$

and so the rate is $\frac{1}{2}|M_1|^2$. Thus $R = 2$. (The measured value is about 1.8.)

6.13 Integrating the differential cross-sections over y (from 0 to 1) gives for a spin-$\frac{1}{2}$ target with a specific quark distribution

$$\frac{\sigma^{NC}(\nu)}{\sigma^{CC}(\nu)} = \left[\int_0^1 [g_L^2 + g_R^2(1 - y)^2]\,dy\right]\left[\int_0^1 dy\right]^{-1} = g_L^2 + \frac{1}{3}g_R^2$$

and

$$\frac{\sigma^{NC}(\bar{\nu})}{\sigma^{CC}(\bar{\nu})} = \left[\int_0^1 [g_L^2(1 - y)^2 + g_R^2]\,dy\right]\left[\int_0^1 (1 - y)^2\,dy\right]^{-1} = g_L^2 + 3g_R^2.$$

For an isoscalar target, we must add the contributions for u and d quarks in equal amounts, i.e.

$$\frac{\sigma^{NC}(\nu)}{\sigma^{CC}(\nu)}\,(\text{isoscalar}) = g_L^2(u) + \frac{1}{3}g_R^2(u) + g_L^2(d) + \frac{1}{3}g_R^2(d)$$

and

$$\frac{\sigma^{NC}(\bar{\nu})}{\sigma^{CC}(\bar{\nu})}\,(\text{isoscalar}) = g_L^2(u) + 3g_R^2(u) + g_L^2(d) + 3g_R^2(d).$$

Substituting for the couplings finally gives for an isoscalar target

$$\frac{\sigma^{NC}(\nu)}{\sigma^{CC}(\nu)} = \frac{1}{2} - \sin^2\theta_W + \frac{20}{27}\sin^4\theta_W, \quad \frac{\sigma^{NC}(\bar{\nu})}{\sigma^{CC}(\bar{\nu})} = \frac{1}{2} - \sin^2\theta_W + \frac{20}{9}\sin^4\theta_W.$$

Chapter 7

7.1   For the $^7_3$Li nucleus, $Z = 3$ and $N = 4$. Hence the configuration is

$$\text{protons:}\quad (1s_{1/2})^2(1p_{3/2})^1; \quad \text{neutrons:}\quad (1s_{1/2})^2(1p_{3/2})^2.$$

By the pairing hypothesis, the two neutrons in the $1p_{3/2}$ sub-shell will have a total orbital angular momentum and spin $\mathbf{L} = \mathbf{S} = \mathbf{0}$ and hence $\mathbf{J} = \mathbf{0}$. Therefore they will not contribute to the overall nuclear spin, parity or magnetic moment. These will be determined by the quantum numbers of the unpaired proton in the $1p_{3/2}$ sub-shell. This has $J = \frac{3}{2}$ and $\ell = 1$, hence for the spin-parity we have $J^P = \frac{3}{2}^-$. The magnetic moment is given by

$$\mu = j\,g_{\text{proton}} = j + 2.3 \ \left(\text{since } j = \ell + \frac{1}{2}\right) = 1.5 + 2.3$$

$$= 3.8 \ \text{nuclear magnetons.}$$

If only protons are excited, the two most likely excited states are:

$$\text{protons:}\quad (1s_{1/2})^2(1p_{1/2})^1; \quad \text{neutrons:}\quad (1s_{1/2})^2(1p_{3/2})^2,$$

which corresponds to exciting a proton from the $p_{3/2}$ sub-shell to the $p_{1/2}$ sub-shell, and

$$\text{protons:}\quad (1s_{1/2})^{-1}(1p_{3/2})^2; \quad \text{neutrons:}\quad (1s_{1/2})^2(1p_{3/2})^2,$$

which corresponds to exciting a proton from the $s_{1/2}$ sub-shell to the $p_{3/2}$ sub-shell.

7.2   A state with quantum number $j(= \ell \pm \frac{1}{2})$ can contain a maximum number $N_j = 2(2j + 1)$ nucleons. Therefore, if $N_j = 16$ it follows that $j = \frac{7}{2}$ and $\ell = 3$ or 4. However, we know that the parity is odd and since $P = (-1)^\ell$, it follows that $\ell = 3$.

7.3   The configuration of the ground state is

$$\text{protons:}\quad (1s_{1/2})^2(1p_{3/2})^4(1p_{1/2})^2(1d_{5/2});$$
$$\text{neutrons:}\quad (1s_{1/2})^2(1p_{3/2})^4(1p_{1/2})^2.$$

To get $j^P = \frac{1}{2}^-$, one could promote a $p_{1/2}$ proton to the $d_{5/2}$ shell, giving

$$\text{protons:}\quad (1s_{1/2})^2(1p_{3/2})^4(1p_{1/2})^{-1}(1d_{5/2})^2.$$

Then by the pairing hypothesis, the two $d_{5/2}$ protons could combine to give $j^P = 0^+$, so that the total spin-parity would be determined by the unpaired $p_{1/2}$ neutron, i.e. $j^P = \frac{1}{2}^-$. Alternatively, one of the $p_{3/2}$ protons could be promoted to the $d_{5/2}$ shell, giving

$$\text{protons:}\quad (1s_{1/2})^2(1p_{3/2})^{-1}(1p_{1/2})^2(1d_{5/2})^2$$

and the two $d_{5/2}$ protons could combine to give $j^P = 2^+$, so that when this combines with the single unpaired $j^P = \frac{3}{2}^-$ proton the overall spin-parity is $j^P = \frac{1}{2}^-$. There are many other possibilities.

7.4 For $^{93}_{41}$Nb, $Z = 41$ and $N = 52$. From the filling diagram Figure 7.4, the configuration is predicted to be:

$$\text{proton: } \ldots (2p_{3/2})^4 (1f_{5/2})^6 (2p_{1/2})^2 (1g_{9/2})^1; \quad \text{neutron: } \ldots (2d_{5/2})^2.$$

So $\ell = 4$, $j = \frac{9}{2} \Rightarrow j^P = \frac{9}{2}^+$ (which agrees with experiment). The magnetic dipole moment follows from the expression for j_{proton} in Equations (7.31) with $j = \ell + \frac{1}{2}$, i.e. $\mu = (j + 2.3)\mu_N = 6.8\mu_N$. (The measured value is $6.17\mu_N$.)

For $^{33}_{16}$S, $Z = 16$ and $N = 17$. From the filling diagram Figure 7.4, the configuration is predicted to be:

$$\text{proton } \cdots (1d_{5/2})^6 (2s_{1/2})^2; \quad \text{neutron: } \cdots (1d_{5/2})^7 (2s_{1/2})^2 (1d_{3/2})^1.$$

So $\ell = 2$, $j = \frac{3}{2} \Rightarrow j^P = \frac{3}{2}^+$ (which agrees with experiment). The magnetic dipole moment follows from the expression for j_{neutron} in Equations (7.31) with $j = \ell - \frac{1}{2}$, i.e. $\mu = (1.9j)/(j+1)\mu_N = 1.14\mu_N$. (The measured value is $0.64\mu_N$.)

7.5 From Equation (7.32),

$$eQ = \int \rho(2z^2 - x^2 - y^2)\mathrm{d}\tau$$

with $\rho = Ze/(\frac{4}{3}\pi b^2 a)$ and the integral is through the volume of the spheroid $(x^2 + y^2)/b^2 + z^2/a^2 \leq 1$. The integral can be transformed to one over the volume of a sphere by the transformations $x = bx'$, $y = by'$ and $z = az'$. Then

$$Q = \frac{3Z}{4\pi} \int\int\int \mathrm{d}x'\mathrm{d}y'\mathrm{d}z' \, (2a^2 z'^2 - b^2 x'^2 - b^2 y'^2).$$

But

$$\int\int\int x'^2 \mathrm{d}x' \, \mathrm{d}y'\mathrm{d}z \text{ (i.e. } z') = \frac{1}{3} \int_0^1 r'^2 4\pi r'^2 \mathrm{d}r' = \frac{4\pi}{15},$$

and similarly for the other integrals. Thus, by direct substitution, $Q = \frac{2}{5}Z(a^2 - b^2)$.

7.6 From Question 7.5 we have $Q = \frac{2}{5}Ze(a^2 - b^2)$ and using $Z = 67$ this gives $a^2 - b^2 = 13.1\,\text{fm}^2$. Also, from Equation (2.32) we have $A = \frac{4}{3}\pi ab^2 \rho$, where $\rho = 0.17\,\text{fm}^{-3}$ is the nuclear density. Thus, $ab^2 = 231.7\,\text{fm}^3$. The solution of these two equations gives $a \approx 6.85\,\text{fm}$ and $b \approx 5.82\,\text{fm}$.

7.7 From Equation (7.53), $t_{1/2} = \ln 2/\lambda = CR \ln 2 \exp(G)$, where C is a constant formed from the frequency and the probability of forming α-particles in the nucleus.

Thus $t_{1/2}(\text{Th}) = t_{1/2}(\text{Cf}) \exp[G(\text{Th}) - G(\text{Cf})]$. The Gamow factors may be calculated from the data given. Some intermediate quantities are: $r_C = 45.96$ fm (Th); 37.72 fm (Cf); $R = 9.268$ fm (Th); 9.439 fm (Cf) (using $R = 1.21\,(A^{1/3} + 4^{1/3})$ and recalling that (Z, A) refer to the daughter nucleus). These give $G = 66.5$ (Th); 54.9 (Cf) and $t_{1/2}(\text{Th}) = e^{11.6}\,t_{1/2}(\text{Cf}) = 4.0$ years. (The measured value is 1.9 years).

7.8 The J^P values of the Σ^0 and the Λ are both $\frac{1}{2}^+$ (see Chapter 3), so the photon has $L = 1$ and as there is no change of parity the decay proceeds via an M1 transition. The Δ^0 has $J^P = \frac{3}{2}^+$ and again there is no parity change. Therefore both M1 and E2 multipoles could be involved, with M1 dominant (see Section 7.8.2). If we assume that the reduced transition probabilities are equal in the two cases, then from Equations (7.80), in an obvious notation,

$$\tau(\Sigma^0) = \left[\frac{E_\gamma(\Delta^0)}{E_\gamma(\Sigma^0)} \right]^3 \tau(\Delta^0),$$

i.e. $\tau(\Sigma^0) = (292/77)^3 \times (0.6 \times 10^{-23})/0.0056 = 5.8 \times 10^{-20}$ s (the measured value is $(7.4 \pm 0.7) \times 10^{-20}$ s).

7.9 In the centre-of-mass system, the threshold for $^{34}\text{S} + p \rightarrow n + {}^{34}\text{Cl}$ is $6.45 \times (34/35) = 6.27$ MeV. Correcting for the neutron–proton mass difference gives the Cl–S mass difference as 5.49 MeV and since in the positron decay $^{34}\text{Cl} \rightarrow {}^{34}\text{S} + e^+ + \nu_e$, $Q = M(A, Z) - M(A, Z - 1) - 2m_e$, the maximum positron energy is 4.47 MeV.

7.10 From Equation (7.71) the electron energy spectrum may be written $I(E) = AE^{1/2} \times (E_0 - E)^2$, where E is the electron energy, E_0 is the end-point, A is a constant and we have neglected the Fermi screening correction and set the neutrino mass to be zero. We need to calculate the fraction

$$F \equiv \left[\int\limits_{E_0 - \Delta}^{E_0} I(E)\,\mathrm{d}E \right] \left[\int\limits_{0}^{E_0} I(E)\,\mathrm{d}E \right]^{-1}$$

where Δ is a small quantity. Using $\int x^{1/2}(a - x)^2 \mathrm{d}x = [\frac{1}{2}a^2x^2 - \frac{2}{3}ax^3 + \frac{1}{4}x^4]^{1/2}$, gives, using $E_0 = 18.6 \times 10^3$ eV and $\Delta = 10$ eV, $F = 3.1 \times 10^{-10}$.

7.11 The mean energy $\bar{E}$ is defined by

$$\bar{E} \equiv \left[\int\limits_{0}^{E_0} E\,\mathrm{d}\omega(E) \right] \left[\int\limits_{0}^{E_0} \mathrm{d}\omega(E) \right]^{-1}.$$

The integrals are:

$$\int E^{3/2}(E_0 - E)^2 \mathrm{d}E = \frac{2}{315} E^{5/2} \left[63E_0^2 - 90E_0E + 35E^2 \right]$$

and

$$\int E^{1/2}(E_0 - E)^2 dE = \frac{2}{105}E^{3/2}\left[35E_0^2 - 42E_0E + 15E^2\right].$$

Substituting the limits gives $\bar{E} = \frac{1}{3}E_0$, as required.

7.12 The possible transitions are as follows:

Initial	Final	L	ΔP	Multipoles
$\frac{3}{2}^-$	$\frac{5}{2}^-$	1, 2, 3, 4	No	M1, E2, M3, ...
$\frac{3}{2}^-$	$\frac{1}{2}^-$	1, 2	No	M1, E2
$\frac{5}{2}^-$	$\frac{1}{2}^-$	2, 3	No	E2, M3

From Figure 7.13, the dominant multipole for a fixed transition energy will be M1 for the $\frac{3}{2}^- \to \frac{5}{2}^-$ and $\frac{3}{2}^- \to \frac{1}{2}^-$ transitions and E2 for the $\frac{5}{2}^- \to \frac{1}{2}^-$ transition. Thus we need to calculate the rate for an M1 transition with $E_\gamma = 178$ keV. This can be done using Equations (7.80) and gives $\tau_{1/2} \approx 3.9 \times 10^{-12}$ s. The measured value is 3.5×10^{-10} s, which confirms that the Weisskopf approximation is not very accurate.

7.13 Set $L = 3$ in Equation (7.78a), substitute the result into Equation (7.77) and use $\Gamma_\gamma = \hbar T$ to give $\Gamma_\gamma(E3) = (2.3 \times 10^{-14})E_\gamma^7 A^2$ eV, where E_γ is expressed in MeV.

Chapter 8

8.1 To balance the number of protons and neutrons, the fission reaction must be

$$n + {}^{235}_{92}U \to {}^{92}_{37}Rb + {}^{140}_{55}Cs + 4n,$$

i.e. four neutrons are produced. The energy released is the differences in binding energies of the various nuclei, because the mass terms in the SEMF cancel out. We have, in an obvious notation,

$$\Delta(A) = 3; \quad \Delta(A^{2/3}) = -9.26; \quad \Delta\left[\frac{(Z-N)^2}{4A}\right] = 0.28; \quad \Delta\left[\frac{Z^2}{A^{1/3}}\right] = 485.0.$$

The contribution from the pairing term is negligible (about 1 MeV). Using the numerical values for the coefficients in the SEMF, the energy released per fission $E_F = 157.9$ MeV.

The power of the nuclear reactor is $P = nE_F = 100$ MW $= 6.25 \times 10^{20}$ MeV s^{-1}, where n is the number of fissions per second. Since one neutron escapes per fission and contributes to the flux, the flux F is equal to the number of fissions per unit area per second, i.e.

$$F = \frac{n}{4\pi r^2} = \frac{P}{4\pi r^2 E_F} = \frac{6.25 \times 10^{20} \, \text{MeVs}^{-1}}{(157.9 \, \text{MeV}) \times (12.57 \, \text{m}^2)} = 3.15 \times 10^{17} \, \text{s}^{-1} \, \text{m}^{-2}.$$

The interaction rate R is given by $R = \sigma \times F \times$ (number of target particles). The latter is given by $n_T = n \times N_A$, where N_A is Avogadro's number and n is found from the ideal gas law to be $n = PV/RT$, where R is the ideal gas constant. Using $T = 298 \, \text{K}$, $P = 1 \times 10^5 \, \text{Pa}$ and $R = 8.31 \, \text{Pa m}^3 \, \text{mol}^{-1} \, \text{K}^{-1}$, gives $n = 52.5 \, \text{mol}$ and hence $n_T = 3.2 \times 10^{25}$. Using the cross-section $\sigma = 10^{-31} \, \text{m}^2$, the rate is $1.0 \times 10^{12} \, \text{s}^{-1}$.

8.2 The neutron speed in the CM system is $v - mv/(M+m) = Mv/(M+m)$ and if the scattering angle in the CM system is θ, then after the collision the neutron will have a speed $v(m + M\cos\theta)/(M+m)$ in the original direction and $Mv\sin\theta/(M+m)$ perpendicular to this direction. Thus the kinetic energy is

$$E(\cos\theta) = \frac{mv^2(M^2 + 2mM\cos\theta + m^2)}{2(M+m)^2}$$

and the average value is

$$E_{\text{final}} = \bar{E} \equiv \left[\int_{-1}^{1} E(\cos\theta) \, \mathrm{d}\cos\theta \right] \left[\int_{-1}^{1} \mathrm{d}\cos\theta \right]^{-1} = RE_{\text{initial}},$$

where the reduction factor is $R = (M^2 + m^2)/(M+m)^2$. For neutron scattering from graphite, $R \approx 0.86$ and after N collisions the energy will be reduced to $E_{\text{final}} = R^N E_{\text{initial}}$. The average initial energy of fission neutrons from ^{235}U is 2 MeV and to thermalize them their energy would have to be reduced to about 0.025 eV. Thus $N \approx \ln(E_{\text{final}}/E_{\text{initial}})/\ln(0.86) = 116$.

8.3 From Equation (1.44a), for the fission of ^{235}U, $W_f = JN(235)\sigma_f$ and the total power output is $P = W_f E_f$, where E_f is the energy released per fission. For the capture by ^{238}U, $W_c = JN(238)\sigma_c$. Eliminating the flux J, gives

$$W_c = \frac{N(238)\sigma_c}{N(235)\sigma_f} \left(\frac{P}{E_f} \right).$$

Using the data supplied, gives $W_c = 1.08 \times 10^{19} \, \text{atoms s}^{-1} \approx 135 \, \text{kg year}^{-1}$.

8.4 Consider fissions occurring sequentially separated by a small time interval δt. The instantaneous power is the sum of the power released from all the fissions up to that time. If E is the energy released in each fission, then over the lifetime of the reactor, i.e. up to time T, the power is given by $P_0 = nE/T$, where n is the total number of fissions and $\delta t = E/P_0$.

The power after some time t after the reactor has been shut down is

$$P(t) = 3(T+t)^{-1.2} + 3(T+t-\delta t)^{-1.2} + 3(T+t-2\delta t)^{-1.2} \cdots + 3t^{-1.2}.$$

In this formula, the first term is the power released from the first fission and the last term is the power released from the last fission before the reactor was shut down. To sum this series, we convert it to an integral:

$$P(t) = 3 \sum_{n=0}^{n=P_0 T/E} (T+t-nE/P_0)^{-1.2} \approx 3 \int_0^{TP_0/E_F} (T+t-nE/P_0)^{-1.2} \, dn.$$

Setting $u = (T+t-nE/P_0)$, gives

$$P(t) = -3\frac{P_0}{E} \int_{T+t}^{t} u^{-1.2} \, du = 0.075 P_0 \left[t^{-0.2} - (T+t)^{-0.2} \right].$$

Using $T = 1$ year and $t = 0.5$ year, gives a power output of approximately 1.1 MW after 6 months.

8.5 The PPI chain overall is: $4(^1\text{H}) \rightarrow {}^4\text{He} + 2e^+ + 2\nu_e + 2\gamma + 24.68$ MeV. Two corrections have to be made to this. Firstly, the positrons will annihilate with electrons in the plasma releasing a further $2m_e = 1.02$ MeV per positron. Secondly, each neutrino carries off 0.26 MeV of energy into space that will not be detected. So, making these corrections, the total output per hydrogen atom is $\frac{1}{4}(24.68 + 2.04 - 0.52) = 6.55$ MeV. The total energy produced to date is 5.60×10^{43} J $= 3.50 \times 10^{56}$ MeV. Thus, the total number of hydrogen atoms consumed is 5.34×10^{55} and so the fraction of the Sun's hydrogen used is $5.34 \times 10^{55}/9 \times 10^{56} = 5.9$ per cent and as this corresponds to 4.6 billion years, the Sun has another 73 billion years to burn before its supply of hydrogen is exhausted.

8.6 A solar constant of 8.4 J cm^{-2} s^{-1} is equivalent to 5.25×10^{13} MeV cm^{-2} s^{-1} of energy deposited. If this is due to the PPI reaction $4(^1\text{H}) \rightarrow {}^4\text{He} + 2e^+ + 2\nu_e + 2\gamma$, then this rate of energy deposition corresponds to a flux of $(5.25 \times 10^{13}/2 \times 6.55) \approx 4 \times 10^{12}$ neutrinos cm^{-2} s^{-1}.

8.7 For the Lawson criterion to be just satisfied, from Equation (8.46),

$$L = \frac{n_d \langle \sigma_{dt} v \rangle t_c Q}{6kT} = 1.$$

We have $kT = 10$ keV and from Figure 8.7 we can estimate $\langle \sigma_{dt} v \rangle \approx 10^{-22}$ m^3 s^{-1}. Also, from Equation (8.45), $Q = 17.6$ MeV. So, finally, $n_d = 6.8 \times 10^{18}$ m^{-3}.

8.8 The mass of a d–t pair is $5.03\,\mathrm{u} = 4.69 \times 10^9\,\mathrm{eV}/c^2 = 8.36 \times 10^{-24}\,\mathrm{g}$. The number of d–t pairs in a 1 mg pellet is therefore 1.2×10^{20}. From Equation (8.45), each d–t pair releases 17.6 MeV of energy. Thus, allowing for the efficiency of conversion, each pellet releases $5.3 \times 10^{26}\,\mathrm{eV}$. The output power is $750\,\mathrm{MW} = 4.7 \times 10^{27}\,\mathrm{eV/s}$. Thus the number of pellets required is $8.9 \approx 9\,\mathrm{s}^{-1}$.

8.9 Assume a typical body mass of 70 kg, half of which is protons. This corresponds to 2.1×10^{28} protons and after 1 year the number that will have decayed is $2.1 \times 10^{28}[1 - \exp(-1/\tau)]$, where τ is the lifetime of the proton in years. Each proton will eventually deposit almost all of its rest energy, i.e. approximately 0.938 GeV, in the body. Thus in 1 year the total energy in Joules deposited per kg of body mass would be $4.5 \times 10^{16}[1 - \exp(-1/\tau)]$ and this amount will be lethal if greater than 5 Gy. Expanding the exponential gives the result that the existence of humans implies $\tau > 0.9 \times 10^{16}$ years.

8.10 The approximate rate of whole-body radiation absorbed is given by Equation (8.48a). Substituting the data given, we have

$$\frac{dD}{dt}(\mu\mathrm{Sv\,h^{-1}}) = \frac{A(\mathrm{MBq}) \times E_\gamma(\mathrm{MeV})}{6r^2(\mathrm{m}^2)} = \frac{(40 \times 10^{-3}) \times (1.173 + 1.333)}{6}$$
$$= 1.67 \times 10^{-2}\,\mu\mathrm{Sv\,h^{-1}}$$

and so in 18 h, the total absorbed dose is $0.30\,\mu\mathrm{Sv}$.

8.11 If the initial intensity is I_0, then from Equation (4.18), the intensities after passing through bone, I_b, and tissue, I_t, are

$$I_b \approx I_0 \exp[-(\mu_b b + 2\mu_t t)] \quad \text{and} \quad I_t \approx I_0 \exp[-\mu_t(b + 2t)].$$

Thus $R = \exp[-b(\mu_b - \mu_t)] = 0.7$ and hence $b = -\ln(0.7)/(\mu_b - \mu_t) = 2.5\,\mathrm{cm}$.

8.12 From Figure 4.8, the rate of ionization energy losses is only slowly varying for momenta above about 1 GeV/c and given that living matter is mainly water and hydrocarbons a reasonable estimate is $3\,\mathrm{MeV\,g^{-1}\,cm^2}$. Thus the energy deposited in 1 year is $2.37 \times 10^9\,\mathrm{MeV\,kg^{-1}}$, which is $3.8 \times 10^{-4}\,\mathrm{Gy}$.

8.13 In general, the nuclear magnetic resonance frequency is $f = |\mu|B/jh$. The numerical input we use is:

$$j = 7/2, \quad B = 1\,\mathrm{T}, \quad \mu = 3.46\,\mu_N, \quad \mu_N = 3.15 \times 10^{-14}\,\mathrm{MeV\,T^{-1}}$$

and $$h = 4.13 \times 10^{-21}\,\mathrm{MeV\,s},$$

giving $f = 7.5\,\mathrm{MHz}$.

Appendix B

B.1 (a) From the definitions of s, t and u, we have

$$(s+t+u)c^2 = (p_A^2 + 2p_Ap_B + p_B^2) + (p_A^2 - 2p_Ap_C + p_C^2) + (p_A^2 - 2p_Ap_D + p_D^2)$$

which, using $p_A^2 = m_A^2c^2$ etc., becomes

$$(s+t+u)c^2 = 3m_A^2c^2 + m_B^2c^2 + m_C^2c^2 + m_D^2c^2 + 2p_A(p_B - p_C - p_D).$$

However, from four-momentum conservation, $p_A + p_B = p_C + p_D$, so that

$$(s+t+u)c^2 = 3m_A^2c^2 + m_B^2c^2 + m_C^2c^2 + m_D^2c^2 - 2p_A^2$$

and hence

$$(s+t+u) = \sum_{j=A,B,C,D} m_j{}^2.$$

(b) From the definition of t,

$$c^2t = p_A^2 + p_C^2 - 2p_Ap_C = m_A^2c^2 + m_C^2c^2 - 2\left(\frac{E_AE_C}{c^2} - \mathbf{p}_A \cdot \mathbf{p}_C\right).$$

For elastic scattering, $A \equiv C$. Thus $E_A = E_C$ and $|\mathbf{p}_A| = |\mathbf{p}_C| = p$, so that $\mathbf{p}_A \cdot \mathbf{p}_C = p^2 \cos\theta$. Then $c^2t = 2m_A^2c^2 - 2(E_A^2/c^2 - p^2\cos\theta)$ and using $E_A^2 = p^2 c^2 + m_A^2c^4$, gives $t = -2p^2(1 - \cos\theta)/c^2$.

B.2 Energy conservation gives $E_\pi = E_\mu + E_\nu$, where

$$E_\pi = \gamma m_\pi c^2, \quad E_\mu = c(m_\mu^2c^2 + p_\mu^2)^{1/2}, \quad E_\nu = p_\nu c$$

and hence

$$\left(\gamma m_\pi c^2 - p_\nu c\right)^2 = c^2\left(m_\mu^2c^2 + p_\mu^2\right). \tag{1}$$

However, three-momentum conservation gives

$$p_\mu \cos\theta = p_\pi = \gamma m_\pi v, \quad p_\mu \sin\theta = p_v = E_\nu/c. \tag{2}$$

Eliminating p_μ and p_ν between (1) and (2) and simplifying, gives

$$\tan\theta = \frac{(m_\pi^2 - m_\mu^2)}{2\beta\gamma^2 m_\pi^2}.$$

B.3 Conservation of four-momentum is $p_\mu = p_\pi - p_\nu$, from which $p_\mu^2 = p_\pi^2 + p_\nu^2 - 2p_\pi p_\nu$. Now $p_j^2 = m_j^2 c^2$ for $j = \pi, \mu$ and ν, and

$$p_\pi p_\nu = \frac{E_\pi E_\nu}{c^2} - \mathbf{p}_\pi \cdot \mathbf{p}_\nu = m_\pi E_\nu = m_\pi |\mathbf{p}_\nu| c,$$

because $\mathbf{p}_\pi = \mathbf{0}$ and $E_\pi = m_\pi c^2$ in the rest frame of the pion. However, $|\mathbf{p}_\nu| = |\mathbf{p}_\mu| \equiv p$ because the muon and neutrino emerge back-to-back. Thus, $p = (m_\pi^2 - m_\mu^2)\, c/2m_\pi$; but $p = \gamma m_\mu v$, from which $v = pc\left[p^2 + m_\mu^2 c^2\right]^{-\frac{1}{2}}$. Finally, substituting for p gives

$$v = \left(\frac{m_\pi^2 - m_\mu^2}{m_\pi^2 + m_\mu^2}\right) c.$$

B.4 By momentum conservation, the momentum components of X^0 are: $p_x = -0.743$ (GeV/c), $p_y = -0.068$ (GeV/c), $p_z = 2.595$ (GeV/c) and hence $p_X^2 = 7.291$. Also, $p_A^2 = 4.686$ (GeV/c)^2 and $p_B^2 = 0.304$ (GeV/c)^2.

Under hypothesis (a):
$E_A = (m_\pi^2 c^4 + p_A^2 c^2)^{1/2} = 2.169 \,\text{GeV}$ and $E_B = (m_K^2 c^4 + p_B^2 c^2)^{1/2} = 0.740 \,\text{GeV}$.
Thus $E_X = 2.909 \,\text{GeV}$ and $M_X = (E_X^2 - p_X^2 c^2)^{1/2} c^{-2} = 1.082 \,\text{GeV/c}^2$.
Under hypothesis (b):
$E_A = (m_p^2 c^4 + p_A^2 c^2)^{1/2} = 2.359 \,\text{GeV}$ and $E_B = (m_\pi^2 c^4 + p_B^2 c^2)^{1/2} = 0.569 \,\text{GeV}$.
Thus $E_X = 2.928 \,\text{GeV}$ and $M_X = (E_X^2 - p_X^2 c^2)^{1/2} c^{-2} = 1.132 \,\text{GeV/c}^2$.
Since $M_D = 1.86 \,\text{GeV/c}^2$ and $M_\Lambda = 1.12 \,\text{GeV/c}^2$, the decay is $\Lambda \to p + \pi^-$.

B.5 If the four-momenta of the initial and final electrons are $p = (E/c, \mathbf{q})$ and $p' = (E'/c, \mathbf{q}')$, respectively, the squared four-momentum transfer is defined by

$$Q^2 \equiv -(p' - p)^2 = -2m^2 c^2 + 2EE'/c^2 - 2\mathbf{q} \cdot \mathbf{q}'.$$

However, $E = E'$ and $|\mathbf{q}| = |\mathbf{q}'| \equiv q$, so neglecting the electron mass, $Q^2 \approx 2q^2 \times (1 - \cos\theta)$. The laboratory momentum may be found from Equation (B.36):

$$q^2 = \frac{c^2}{4m_p^2}\left[s - (m_p - m_e)^2\right]\left[s - (m_p + m_e)^2\right] \approx \frac{c^2(s - m_p^2)^2}{4m_p^2},$$

where the invariant mass squared s is defined by $s \equiv (p + P)^2/c^2$ and P is the four-momentum of the initial proton, i.e. $P = (m_p c, \mathbf{0})$. Thus,

$$s = m_e^2 + m_p^2 + 2m_p E/c^2 \approx m_p^2 + 2m_p E/c^2.$$

Substituting into the expression for Q^2 gives $Q^2 \approx 2E^2(1 - \cos\theta)/c^2$.

B.6 The total four-momentum of the initial state is $p_{\text{tot}} = \left[(E + m_p c^2)/c, \mathbf{p}_L\right]$. Hence the invariant mass W is given by $(Wc^2)^2 = (E_L + m_p c^2)^2 - p_L^2 c^2$, where $p_L \equiv |\mathbf{p}_L|$. The invariant mass squared in the final state evaluated in the centre-of-mass frame has a minimum value $(4m_pc)^2$ when all four particles are stationary. Thus, E_{min} is given by

$$(E_{min} + m_pc^2)^2 - p_L^2c^2 = (4m_pc^2)^2$$

which expanding and using $E_{min}^2 - p_L^2c^2 = m_p^2c^4$, gives $E_{min} = 7m_pc^2 = 6.6$ GeV.

For a bound proton, the initial four-momentum of the projectile is $(E_L'/c, \mathbf{p}_L')$ and that of the target is $(E/c, -\mathbf{p})$, where $\mathbf{p}$ is the internal momentum of the nucleons, which we have taken to be in the opposite direction to the beam because this gives the maximum invariant mass for a given E_L'. The invariant mass W' is now given by

$$(W'c^2)^2 = (E_L' + E)^2 - (p_L' - p)^2c^2 = 2m_p^2c^4 + 2EE_L' + 2pp_L'c^2.$$

Since the thresholds E_{min} and E_{min}' correspond to the same invariant mass $4m_p$, we have $2m_pc^2E_{min} = 2EE_{min}' + 2pp_{min}'c^2$. Finally, since the internal momentum of the nucleons is ~ 250 MeV/c (see Chapter 7), $E \approx m_pc^2$, while for the relativistic incident protons $p_{min} \approx E_{min}/c$, so using these gives

$$E_{min}' \approx \left(1 - p/m_pc\right)E_{min} = 4.8 \text{ GeV}.$$

B.7 The initial total energy is $E_i = E_A = m_Ac^2$ and the final total energy is $E_f = E_B + E_C$, where $E_B = (m_B^2c^4 + p_B^2c^2)^{\frac{1}{2}}$, and $E_C = (m_C^2c^4 + p_C^2c^2)^{\frac{1}{2}}$, with $p_B = |\mathbf{p}_B|$ and $p_C = |\mathbf{p}_C|$. However, by momentum conservation, $\mathbf{p}_B = -\mathbf{p}_C \equiv \mathbf{p}$ and so

$$\left[m_Ac^2 - (m_B^2c^4 + p^2c^2)^{\frac{1}{2}}\right]^2 = (m_C^2c^4 + p^2c^2),$$

which on expanding gives $E_B = (m_A^2 + m_B^2 - m_C^2)c^2/2m_A$.

B.8 If the four-momenta of the photons are $p_i = (E_i/c, \mathbf{p}_i)(i = 1, 2)$, then the invariant mass of M is given by $M^2c^4 = (E_1 + E_2)^2 - (\mathbf{p}_1 + \mathbf{p}_2)c^2 = 2E_1E_2(1 - \cos\theta)$, since $\mathbf{p}_1 \cdot \mathbf{p}_2 = E_1E_2(1 - \cos\theta)/c^2$ for zero-mass photons. Thus, $\cos\theta = 1 - M^2c^4/2E_1E_2$.

B.9 A particle with velocity v will take time $t = L/v$ to pass between the two counters. Relativistically, $p = mv\gamma$ with $\gamma = (1 - v^2/c^2)^{-\frac{1}{2}}$. Solving, gives $v = c(1 + m^2c^2/p^2)^{-\frac{1}{2}}$ and hence the difference in times-of-flight (assuming $m_1 > m_2$) is

$$\Delta t = \frac{L}{c}\left[\left(1 + \frac{m_1^2c^2}{p^2}\right)^{\frac{1}{2}} - \left(1 + \frac{m_2^2c^2}{p^2}\right)^{\frac{1}{2}}\right].$$

Using $m_1c^2 = m_pc^2 = 0.983$ GeV, $m_2c^2 = m_\pi c^2 = 0.140$ GeV and $pc = 2$ GeV gives $\Delta t = [1.114 - 1.002](L/c)$ and $L_{min} = 0.54$ m.

B.10 In an obvious notation, the kinematics in the lab frame are:

$$\gamma(E_\gamma, \mathbf{p}_\gamma) + e^-(mc^2, 0) \rightarrow \gamma(E_\gamma', \mathbf{p}_\gamma') + e^-(E, \mathbf{p}).$$

Energy conservation gives $E_\gamma + mc^2 = E'_\gamma + E$ and momentum conservation gives $\mathbf{p}_\gamma = \mathbf{p}'_\gamma + \mathbf{p}$. From the latter we have $E^2 - m^2c^4 = c^2(\mathbf{p}^2_\gamma + \mathbf{p}'^2_\gamma - 2\mathbf{p}_\gamma \cdot \mathbf{p}'_\gamma)$. But $p_\gamma c = E_\gamma$, $p'_\gamma c = E'_\gamma$ and the scattering angle is θ, so we have $E^2 - m^2c^4 = E^2_\gamma + E'^2_\gamma - 2E_\gamma E'_\gamma \cos\theta$. Eliminating E between this equation and the equation for energy conservation gives $E'_\gamma = E_\gamma[1 + E_\gamma(1 - \cos\theta)/mc^2]^{-1}$. Finally, using $E_\gamma = E'_\gamma/2$ and $\theta = 60^0$, gives $E_\gamma = 2mc^2 = 1.02\,\text{MeV}$.

Appendix C

C.1 The assumptions are: ignore the recoil of the target nucleus because its mass is much greater than the total energy of the projectile α-particle; use non-relativistic kinematics because the kinetic energy of the α-particle is very much less that its rest mass; assume the Rutherford formula (i.e. the Born approximation) is valid for small-angle scattering. The relevant formula is then Equation (C.13) and it may be evaluated using $z = 2$, $Z = 83$, $E_{kin} = 20\,\text{MeV}$ and $\theta = 20°$. The result is $d\sigma/d\Omega = 98.3\,\text{b/sr}$.

C.2 From Figure C.2, the distance of closest approach d is when $x = 0$. For $x < 0$, the sum of the kinetic and potential energies is $E_{tot} = \frac{1}{2}mv^2$ and the angular momentum is mvb. At $x = 0$, the total mechanical energy is $\frac{1}{2}mu^2 + Zze^2/4\pi\varepsilon_0 d$ and the angular momentum is mud, where u is the instantaneous velocity. From angular momentum conservation, $u = vb/d$ and using this in the conservation of total mechanical energy gives $d^2 - Kd - b^2 = 0$ where, using Equation (C.9), $K \equiv 2b/\cot(\theta/2)$. The solution for $d \geq 0$ is $d = b[1 + \text{cosec}(\theta/2)]/\cot(\theta/2)$.

C.3 The result for small-angle scattering follows directly from Equation (C.9) in the limit $\theta \to 0$. Evaluating b, we have, using the data given,

$$b = \frac{zZe^2}{2\pi\varepsilon_0 mv^2\theta} = 2zZ\left(\frac{e^2}{4\pi\varepsilon_0\hbar c}\right)\frac{\hbar c}{mc^2}\frac{1}{(v/c)^2\theta} = 1.55 \times 10^{-13}\,\text{m}.$$

The cross-section for scattering through an angle greater than $5°$ is thus $\sigma = \pi b^2 = 7.55 \times 10^{-26}\,\text{m}^2$ and the probability that the proton scatters through an angle greater than $5°$ is $P = 1 - \exp[-n\sigma t]$, where n is the number density of the target. Using $n = (6.022 \times 10^{26}/194) \times 21450 = 6.658 \times 10^{28}\,\text{m}^{-3}$, gives $P = 4.91 \times 10^{-2}$. Since P is very small but the number of scattering centres is very large, the scattering is governed by the Poisson distribution and the probability for a single scatter is $P_1(m) = me^{-m} = 4.91 \times 10^{-2}$, giving $m \approx 0.052$. Finally, the probability for two scattering is $P_2 = m^2\exp(-m)/2! \approx 1.3 \times 10^{-3}$.

References

Aj90   F. Ajzenberg-Selove (1990) Energy levels of light nuclei., *Nucl. Phys.*, **A506** 1–158.

Am95   P. Amaudruz *et al.* (1995) A re-evaluation of the nuclear structure function ratios for D, He, ^{6}Li, C and Ca, *Nucl. Phys.*, **B441**, 3–11.

Ar95   M. Arneodo *et al.* (1995) The structure function ratios F_2^C / F_2^D and F_2^{Ca}/F_2^D at small x. *Nucl. Phys.*, **B441**, 12–30.

Ar97   M. Arneodo *et al.* (1997) Measurements of the proton and deuteron structure functions F_2^p and F_2^d and of the ratio σ_L/σ_T. *Nucl. Phys.*, **B483**, 3–43.

As04   Y. Ashie. *et al.* (2004) Evidence for an oscillatory signature in atmospheric neutrino oscillations. *Phys. Rev. Letters*, **93**, 101801/1–5.

At82   W. B. Atwood (1982) Lectures on Lepton Nucleon Scattering and Quantum Chromodynamics, *Progress in Physics Vol. 4.*

Ba77   R. C. Barrett and D. F. Jackson (1977) *Nuclear Sizes and Structures*, Clarendon Press, Oxford, UK.

Be67   J. B. Bellicard *et al.* (1967) Scattering of 750-MeV electrons by calcium isotopes. *Phys. Rev. Letters*, **19**, 527–529.

Bl52   J. M. Blatt and V. F. Weisskopf (1952), *Theoretical Nuclear Physics*, John Wiley and Sons, New York, USA.

Bo69   A. Bohr and B. R. Mottelson (1969) *Nuclear Structure Vol. 1*, W. A. Benjamin Inc., New York, USA.

Ch97   *Chart of the Nuclides* (1997) General Electric Company, Schenectady, New York, USA.

Co01   W. N. Cottingham and D. A. Greenwood (2001) *An Introduction to Nuclear Physics*, 2nd edn. Cambridge University Press, Cambridge, UK.

De99   P. P. Dendy and B. Heaton (1999) *Physics for Diagnostic Radiology*, 2nd edn., Institute of Physics Publishing, UK.

Ei04   S. Eidelman *et al.* (2004) Review of Particle Physics. *Physics Letters* **B592**, 1–1109.

En66   H. A. Enge (1966) *Introduction to Nuclear Physics*, Addison-Wesley Publishing Company.

Fe86   R. Fernlow (1986) *Introduction to Experimental Particle Physics*, Cambridge University Press, Cambridge, UK.

Fo61   D. B. Fossan *et al.* (1961) Neutron total cross sections of Be, B^{10}, B, C, and O. *Phys. Rev.*, **123**, 209–218.

Fr83   B. Frois (1983) *Proc. Int. Conf. Nucl. Phys.*, Florence eds. P. Blasi and R. A. Ricci, (Tipografia Compositori, Bolgona) Vol. 2, p. 221.

Ga76   D. I. Garber and R. R. Kinsey (1976) *Neutron Cross-sections Vol 2*, Brookhaven National Laboratory Report BNL-325, Upton, NY, USA.

Gj74   S. Gjesdal *et al.* (1974) A measurement of the $K_L - K_S$ mass difference from the charge asymmetry in semi-leptonic kaon decays. *Phys. Letters*, **B52**, 113–119.

Go86   K. Gottfried and V. F. Weisskopf (1986) *Concepts of Particle Physics, vol. 2*, Oxford University Press, Oxford, UK.

Gr96   C. Grupen (1996) *Particle Detectors*. Cambridge University Press, Cambridge, UK.

Gr87   D. Griffiths (1987) *Introduction to Elementary Particles*. John Wiley and Sons, New York, USA.

Ha84   F. Halzen and A. D. Martin (1984) *Quarks and Leptons*. John Wiley and Sons, New York, USA.

He97   W. R. Hendee (1997) Physics and applications of medical imaging. *Rev. Mod. Phys.*, **71**, S444–S450.

Hi68   F. Hintenberger *et al.* (1968) Inelastic scattering of 52 MeV deuterons. *Nucl. Phys.*, **A115**, 570–592.

Ho97   P. E. Hodgson, E. Gadioli and E. Gadioli Erba (1997) *Introductory Nuclear Physics*. Oxford University Press, Oxford, UK.

Ho97a   R. K. Hobbie (1997) *Intermediate Physics for Medicine and Biology*, 3rd edn.. Springer and AIP Press, Germany.

Ja75   J. D. Jackson (1975) *Classical Electrodynamics*, 2nd edn.. John Wiley & Sons, New York, USA.

Je90   N. A. Jelley (1990) *Fundamentals of Nuclear Physics*. Cambridge University Press, Cambridge, UK.

Ke82   D. Keefe (1982) Inertial confinement fusion. *Ann. Rev. Nucl. Part. Sci.*, **32**, 391–441.

Kl86   K. Kleinknecht (1986) *Detectors for Particle Radiation*. Cambridge University Press, Cambridge, UK.

Kr88   K. S. Krane (1988) *Introductory Nuclear Physics*. John Wiley and Sons, New York, USA.

Li01   J. Lilley (2001) *Nuclear Physics. Principles and Applications*, John Wiley and Sons. Ltd., Chichester, UK.

Ma92   F. Mandel (1992) *Quantum Mechanics*, 2nd edn.. John Wiley and Sons Ltd., Chichester, UK.

Ma97   B. R. Martin and G. Shaw (1997) *Particle Physics*, 2nd edn.. John Wiley and Sons Ltd., Chichester, UK.

McR03   D. W. McRobbie, *et al.* (2003) *MRI From Picture to Proton*. Cambridge University Press, Cambridge, UK.

Mo94   L. Montanet *et al.* (1994) Review of particle properties. *Phys. Rev.*, **D50**, 1173–1814.

Me61   E. Merzbacher (1961) *Quantum Mechanics*. John Wiley and Sons, New York, USA.

NRC99   Report of the Board on Physics and Astronomy of the National Research Council, USA (1999) *Nuclear Physics: The Core of Matter, The Fuel of Stars*. National Academies Press, Washington, D.C., USA.

Pe00   D. H. Perkins (2000) *Introduction to High Energy Physics*, 4th edn. Cambridge University Press, Cambridge, UK.

Pe03   D. H. Perkins (2003) *Particle Astrophysics*. Oxford University Press, Oxford, UK.

Ph94   A. C. Phillips (1994) *The Physics of Stars*. John Wiley and Sons. Ltd., Chichester, UK.

Po99   B. Povh *et al.* (1999) *Particles and Nuclei. An Introduction to the Physical Concepts*. Springer, Germany.

Sa67   G. R. Satchler (1967) Optical model for 30 MeV proton scattering. *Nucl. Phys.*, **A92**, 273–305.

Sc68   L. I. Schiff (1968) *Quantum Mechanics*, 3rd edn.. McGraw-Hill, New York, USA.

Sc01   W. Scheider (2001) News Update to *A Serious But Not Ponderous Book About Nuclear Energy*, Cavendish Press, Ann Arbour, USA.

Se80   E. Segrè (1980) *From X-Rays to Quarks*. W. H. Freeman and Company, USA.

Se97   W. G. Seligman *et al.* (1997) Improved determination of α_s from neutrino-nucleon scattering. *Phys. Rev. Letters*, **79**, 1213–1216.

Si75   I. Sick *et al.* (1975) Shell structure of the ^{58}Ni charge density. *Phys. Rev. Letters*, **35**, 910–913.

Tr75   G. L. Trigg (1975) *Landmark Experiments in 20th Century Physics*. Crane, Russak and Co. Ltd, New York, USA.

Zu00   A. Zurlo *et al.* (2000) The role of proton therapy in the treatment of large irradiation volumes: a comparative planning study of pancreatic and biliary tumors. *Int. J. Radiat. Oncol. Biol. Phys*, **48**, 277–288.

Bibliography

Below are brief notes on a few books on nuclear and particle physics at the appropriate level which I have found particularly useful. Other, more specialized, texts are listed in the References section which follows.

1 Nuclear Physics

Two very readable concise texts at about the level of the present book although covering more topics are: W. N. Cottingham and D. A. Greenwood, *An Introduction to Nuclear Physics* 2nd edn., Cambridge University Press, 2001, and N. A. Jelley, *Fundamentals of Nuclear Physics*, Cambridge University Press, 1990. Both deal with theoretical aspects only – there is nothing about experimental methods. Both provide some problems for each chapter with either answers or brief hints on solutions. Another good book at this level is: J. Lilley, *Nuclear Physics – Principles and Applications*, John Wiley and Sons, 2001. This is in two parts. The first covers the principles of nuclear physics, including experimental techniques, and the second discusses an unusually wide range of applications, including industrial and biomedical uses. An extensive range of problems is provided, with detailed notes on their solutions.

Two good examples of comprehensive texts covering both theory and experiment are: K. S. Krane, *Introductory Nuclear Physics*, John Wiley and Sons, 1988, and P. E. Hodgson, E. Gadioli and E. Gadioli Erba, *Introductory Nuclear Physics*, Oxford University Press, 1997. Both provide problems, but without solutions.

Finally there is the unique set of (hand written!) lecture notes by Fermi: E. Fermi, *Nuclear Physics*, University of Chicago Press, 1950. Although old, these are still well worth reading.

2 Particle Physics

There are several books covering particle physics at the appropriate level, For obvious reasons, the one closest to the present book is: B. R. Martin and G. Shaw,

Nuclear and Particle Physics   B. R. Martin
© 2006 John Wiley & Sons, Ltd

Particle Physics, 2nd edn., John Wiley and Sons, 1997, and some of the material on particle physics in the present book has been developed from this previous book. It covers both theory and experimental methods. Problems with full solutions are provided for each chapter.

Some of the other texts available are now rather dated, but one that is not is: D. H. Perkins, *Introduction to High Energy Physics*, 4th edn., Cambridge University Press, 2000. This book is well-established and has changed substantially over the years. It goes further than the present book in its use of relativistic calculations. The latest edition (the fourth) has far less discussion of experimental methods than earlier editions, but an expanded chapter on astroparticle physics. It is therefore worth looking at the third edition also. Problems are provided, some with answers.

Another older book, but still relevant, is: D. Griffiths, *Introduction to Elementary Particle Physics*, John Wiley and Sons, 1987. Griffiths' book is written in an unconventional conversational style, with interesting footnotes (and extensive notes at the end of most chapters), giving further details and background. It is exclusively theoretical – there is nothing on experimental techniques. It goes well beyond the present text, as at least half of the book involves the detailed evaluation of Feynman diagrams. A wealth of interesting problems is provided at the end of each chapter, but no solutions are given.

3   Nuclear and Particle Physics

There are not many books that treat nuclear and particle physics together and some of those are out-of-date. Five that are appropriate are:

R. A. Dunlap, *The Physics of Nuclei and Particles*, Thomson Learning – Brooks/ Cole, 2004;

A. Das and T. Ferbel, *Introduction to Nuclear and Particle Physics*, John Wiley and Sons, 1994;

W. S. C. Williams, *Nuclear and Particle Physics*, Oxford University Press, 1991;

W. E. Burcham and M. Jobes, *Nuclear and Particle Physics*, Longman Scientific and Technical, 1995;

B. Povh, K. Rih, C. Scholz and F. Zetsche, *Particles and Nucle*, 2nd edn., Springer, 1995.

The first two books are concise readable introductions, although in Dunlap's case the particle physics part is very short – just 50 pages. This book is exclusively about theory, whereas the book of Das and Ferbel also discusses experimental methods. Both books provide problems, but neither supplies solutions. The book by Williams is fairly comprehensive, although now a little old. The style is rather discursive. There is a wealth of illustrations and many problems are given, with answers to some of them supplied. (A full solutions manual is available as a separate volume.) The book by Burcham and Jobes is also comprehensive and goes further than the present text. There are many problems, all with solutions. Both of the latter two books treat nuclear and particle physics as almost independent subjects. The book by Povh *et al.* is closest in its coverage to the present book and at a similar level, although experimental methods are only discussed in a brief appendix. Some problems with solutions are provided for all chapters.

Index